Statistics for Economists

By the same author

AN INTRODUCTION TO MULTIVARIATE STATISTICS FOR THE SOCIAL AND BEHAVIOURAL SCIENCES
(with Spencer Bennett)

TRADE UNIONS AND THE ECONOMY
(with Brian Burkitt)

Statistics for Economists

DAVID BOWERS

First published 1982
Reprinted 1987

Published by
MACMILLAN EDUCATION LTD
Houndmills, Basingstoke, Hampshire RG21 2XS
and London
Companies and representatives
throughout the world

Printed in Hong Kong

ISBN 0-333-30110-2 (hardcover)
ISBN 0-333-30111-0 (paperback)

To Eve, Sarah and Imogen

Christ you know it ain't easy
You know how hard it can be

JOHN LENNON

Contents

Preface

While the main thrust of this book is in the area of economics and socio-economics, it should also be useful to students in the related areas of the social sciences, e.g. politics, geography, social planning, sociology, etc., and many of the worked examples have been deliberately chosen from these other fields.

The book aims to provide all the statistics which would be required by students following courses in these subjects. The book starts at the very beginning, with a chapter on the organisation of data, followed by chapters on the description of data. After a chapter on probability and probability distributions, there follows three chapters on statistical inference (sampling distributions, estimation and hypothesis testing). The book concludes with an examination of simple and multiple regression analysis.

The mathematical ideas have been kept as simple as possible (for example, I have eschewed the use of matrix algebra when dealing with multiple regression) and the book is therefore suitable for students with no mathematics beyond 'O' level.

The book is designed for the following groups of people:

(a) Students following degree (or equivalent) courses in economics or related social sciences who require a formal course in introductory statistics.
(b) Research or postgraduate students either taking up statistics for the first time, or wishing to refresh their memory in certain areas.
(c) People working in the professions who need to understand (or re-understand) some statistics.

The author wishes to acknowledge the help of his editors, the assistance of an anonymous reviewer and the skill of all those who contributed to the typing of the manuscript.

Burton-in-Kendal
September 1981

DAVID BOWERS

Acknowledgements

The author and publishers with to thank the following who have kindly given permission for the use of copyright material:

Biometrika Trustees, Imperial College of Science and Technology, for abridged tables from *Biometrika*, vol. 33 (1943), *Biometrika*, vol. 38 (1951) and *Biometrika Tables for Statisticians*, vol. 1, 3rd edition (1966).

Department of Statistical Science, University College, London, for Tables of Random Sampling Numbers, by M. G. Kendall and B. Babington Smith.

Macmillan Publishing Co. Inc. for a table from *Elements of Econometrics* by Jan Kmenta (1971), and an abridged table from *Statistical Methods for Research Workers*, 14th edition, by R.A. Fisher, Copyright © 1970 University of Adelaide.

Every effort has been made to trace all the copyright holders, but if any have been inadvertently overlooked, the publishers will be pleased to make the necessary arrangements at the first opportunity.

1 Introduction

In a recent newspaper article discussing the effects of changes in the duty on cigarettes the following paragraph appeared:

> Buckmaster and Moore [stockbrokers] predict that the 29 per cent rise in cigarette prices this year will lead to a 15 per cent fall in the consumption of cigarettes and other tobacco products this year. This compares with a 10 per cent drop implied by Treasury figures.
>
> (*The Guardian*, 6 July 1981)

How did the Treasury and the firm of stockbrokers arrive at their figures? Why are the figures different? Who is more likely to be correct? How could we go about deciding between the two figures? Clearly both groups have acknowledged the theoretical relationship between the price of a product and the demand for it. And both have to some extent probably examined recent data on cigarette prices and sales. However, they have obtained different results. This relationship between theory and data, the interface between the two, forms the subject-matter of this book.

Theoreticians may make statements such as 'unemployment causes crime', or 'money supply influences inflation', or 'unions cause strikes', and all of these statements may be based on a rational and logical deployment of the appropriate theoretical arguments. However, they must remain theories until they can be tested by using relevant data, and even then the results may be inconclusive or contradictory.

We can think therefore of economics (or indeed of any other social or natural science) as being broadly divisible into two areas, one theoretical, the other empirical. The subject-matter of this book, a discussion of a variety of statistical techniques, offers a way of bridging these two areas. The material to be examined will enable us to test the veracity of economic theories by appealing to some relevant body of data. Notice that we proceed from theory to data via the interface of some appropriate statistical method. In general, this will be the normal method of proceeding. In some cases, however, the data themselves may suggest the existence of some previously unsuspected relationship between variables and lead to a re-examination of the accepted theory or even to the development of some new theory. In Figure 1.1 we illustrate this process schematically.

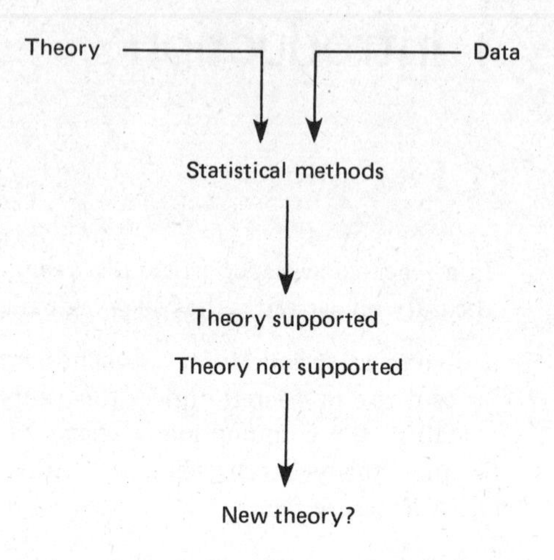

Figure 1.1 The role of statistics in linking theory and data

The aims and objectives of this book

The object of this book is to discuss and examine a series of statistical concepts and techniques, in some sort of rational order, which will enable the reader to

1. Understand and critically appraise the statistical content of the work of other economists and social scientists.
2. Investigate the relationships between variables by applying the most appropriate statistical method.

Although the book is aimed principally at economists (in the sense that most of the examples discussed are normally considered to be in the domain of the economist), the techniques which are examined are powerful and wide-ranging in application and could be just as usefully applied by practitioners in other branches of the social and behavioural sciences.

This is not a book about mathematical economics (which could be thought of as a rephrasing of conventional economic theory in mathematical terms) or about economic statistics (which could be thought of as a study of economic data *per se*, i.e. the collection, ordering and interpretation of data), though it might from time to time touch upon both of these areas.

In terms of prerequisites the book assumes no more than a familiarity with elementary arithmetic, algebra and co-ordinate geometry, such as is covered by an 'O'-level mathematics syllabus.

The book is aimed at first- or second-year undergraduate students who wish for an understanding of some very necessary statistical techniques either as an end in itself or as a basis for a more advanced course in, for example, econometrics, at post-graduate students who may suddenly find an unexpected need for expertise in some elementary statistics, and at those working in the professions who have a similar requirement.

Plan of the book

The book is divided into four parts. Part I is concerned with the arrangement and presentation of data on some single variable, and with ways of measuring the average value taken by that variable and the degree to which the individual observations are spread out around this average. The material covered in Part I is known collectively as 'descriptive statistics'.

In Part II we introduce the concept of probability. This enables us to make judgements about the probability of certain events occurring. This material is essential for most of what is to follow in the remaining chapters.

Part III introduces the notion of statistical inference. In Part I of the book, for example, if we were measuring the average level of absenteeism among British executives, we assumed that we had data on the absenteeism record level of *all* British executives. In practice, of course, this will not usually be the case; normally we will have to work with a *sample* of executives. Statistical inference is really a study of the implications of studying samples rather than whole populations.

In Part IV we examine ways in which the application of a very powerful class of statistical techniques, known as regression analysis, can be used to investigate the relationship *between* variables. Thus in Parts I or III of the book we might be interested in the average level of absenteeism, in Part IV we would be interested in investigating the form and significance of the relationship between absenteeism and salary level or age or both.

The book does not aim to give a rigorous treatment of statistical theory, and for this reason many results are not proved in detail, or at all. Such proofs if required can be found in the plethora of statistical books available. The maths is kept at as simple a level as possible (for example, topics such as matrix algebra are not included), and should present no problems to the reasonably intelligent general reader.

PART I

DESCRIPTIVE STATISTICS

2 Frequency Distributions

OBJECTIVES

After reading this chapter and working through the examples students should understand the meaning of the following terms:

raw data
range
class
frequency
frequency table
frequency distribution
class limits
class boundaries
class width
class mid-point
relative frequency
cumulative frequency
joint frequency

Students should be able to:

arrange raw data into a suitable frequency table
determine class parameters
construct relative frequency distributions
construct cumulative frequency distributions
construct joint frequency distributions
interpret any of the above

2.1 Introduction

Very often economists and other social scientists are faced with an impressively large but disorganised body of *raw* data. By raw data we mean observations on some variable which appear as they were perhaps originally collected. Data which have not been arranged in any sort of order, either size or chronology or importance, which have not been subject to any form of investigation or analysis, those data remain amorphous and uniformative.

In this section we want to investigate some fairly simple ways of rearranging such data so that any information contained within them which is of interest to us becomes much easier to extract, i.e. so that the principal features of the data become readily apparent.

When we examine a set of raw data, which someone (you, me or anyone) has collected, we usually have some idea of what it is of interest that we wish to determine. For example, we may want to know the lowest and highest values, we may want some idea as to whether the data cluster around any particular value, how they are spread out, etc. The techniques described in this section are a first step in making these objectives easier to achieve. We should be cautious about approaching the analysis of data with too limited a viewpoint. There may well be (hidden in the raw data) features and characteristics which are of considerable interest but which are not anticipated at the outset. If we take too narrow an approach, concentrating perhaps only on one or two characteristics, we may well miss these additional features.

2.2 Frequency tables and frequency distributions

Consider the raw data shown in Table 2.1, which represents the number of strikes or stoppages in UK industry for each month of the four-year period 1973–6. There are thus forty-eight observations in all. (The data in Table 2.1 are seasonally corrected. This means that any effect on the figures due to time of year has been removed. For those interested the *Department of Employment Gazette* gives the unadjusted figures.)

We will suppose that there is some aspect of industrial relations during this period which is of interest to us and we decide to start by looking at the actual number of strikes in each month. So what does an inspection of Table 2.1 tell us? A glance reveals immediately that in a good many of the months the number of stoppages seems to be either in the 200s or the 300s. Looking a little more carefully we see that the lowest number of strikes

Table 2.1 Number of industrial stoppages in the United Kingdom per month, 1973–6 (seasonally corrected)

	1973	1974	1975	1976
January	207	104	189	166
February	243	116	235	154
March	293	251	220	203
April	234	300	261	157
May	249	292	229	156
June	262	323	257	175
July	178	188	235	162
August	261	236	149	172
September	239	289	157	179
October	327	401	170	190
November	309	309	115	199
December	71	113	65	103

Source: *Department of Employment Gazette.*

was 65 in December 1975 and the highest number was in October 1974. That is, the monthly number of strikes varies from a low of 65 to a high of 401. The difference between the highest and lowest values in a set of data is known as the *range*; thus in this case the range is 401 – 65 = 336.

The minimum value and the maximum value, together with the resulting range, represent therefore the first summary measures of our data (see Table 2.2). There is not much more that a casual

Table 2.2 Minimum, maximum and range of monthly strike values from Table 2.1

Minimum number of strikes per month =	65
Minimum number of strikes per month =	401
Range =	336

inspection of Table 2.1 can tell us. However, we might be able to improve matters somewhat by arranging the values in ascending or descending order; let us do this and see if it helps. If we rearrange the values in ascending order we have a choice of either retaining or casting aside the chronological information (i.e. the year and month of each value).

Let us first retain the chronology associated with the values by noting the year and month of each. We can use a notation such as 1974.09 to mean the ninth month of 1974. Doing this gives us Table 2.3. Producing Table 2.3 is a somewhat tedious procedure and quite liable to error, though fortunately any mistakes in the order become very obvious. However, as we shall see shortly, it does prove to be a useful and rewarding exercise.

Table 2.3 Monthly stoppages arranged in ascending order (with chronology)

65 (1975.12)	162 (1976.07)	207 (1973.01)	261 (1973.08)
71 (1973.12)	166 (1976.01)	220 (1975.03)	261 (1975.04)
103 (1976.12)	170 (1975.10)	229 (1975.05)	262 (1973.06)
104 (1974.01)	172 (1976.08)	234 (1973.04)	289 (1974.09)
113 (1974.12)	175 (1976.06)	235 (1975.02)	292 (1974.05)
115 (1975.11)	178 (1973.07)	235 (1975.07)	293 (1973.03)
116 (1974.02)	179 (1976.09)	236 (1974.08)	300 (1974.04)
149 (1975.08)	188 (1974.07)	239 (1973.09)	309 (1973.11)
154 (1976.02)	189 (1975.01)	243 (1973.02)	309 (1974.11)
156 (1976.05)	190 (1976.10)	249 (1973.05)	323 (1974.06)
157 (1975.09)	199 (1976.11)	251 (1974.03)	327 (1973.10)
157 (1976.04)	203 (1976.03)	257 (1975.06)	401 (1974.10)

Table 2.3 tells us two things (neither of which is immediately obvious from Table 2.1). First, the lower values tend to occur in months 12 and 1, i.e. December and January. This may be due to the smaller number of working days available for industrial stoppages to occur (because of extended Christmas and New Year holidays). The second thing we observe is that the single highest value, 401 stoppages, is *considerably* higher than values immediately below it. It seems to be untypically high and we may well have good reason to check this value to make sure we have not copied it down incorrectly from the published tables. If 401 is untypically high, it means that the range of 336 shown in Table 2.2 is exaggerated and should be treated with caution. This is a useful piece of additional information (leaving the value of 401 out of our calculations would give us a range of $327 - 65 = 262$, which is about 20 per cent less). We may finally note that Table 2.3 confirms the immediate impression we gained from Table 2.1, that the majority of months have stoppages in the 100s and 200s (it is easy to see that these represent forty out of the forty-eight values).

Table 2.3, while providing us with marginally more information than Table 2.1, still contains a perhaps confusingly large number of values, takes up a lot of room on the page and a lot of time to copy out. What we want is a way of further rearranging the data so that we can present them more concisely without losing too much of the detail.

Table 2.4 represents a possible first step in the right direction. Here we have arranged the values in classes of 10, i.e. all those in the 70s, all those in the 80s, and so on (at the same time we have dropped the chronological information, assuming that this is not of any great interest to us).

So, for example, there are three months in which the number of industrial stoppages is in the 110s, and the values, taken from

Table 2.4 Initial grouping of data in Table 2.3

Number of stoppages	Number of months	Number of stoppages	Number of months	Number of stoppages	Number of months	Number of stoppages	Number of months	Number of stoppages	Number of months
in the 0s	0	in the 100s	2	in the 200s	2	in the 300s	3	in the 400s	1
in the 10s	0	in the 110s	3	in the 210s	0	in the 310s	0	in the 410s	0
in the 20s	0	in the 120s	0	in the 220s	2	in the 320s	2	in the 420s	0
in the 30s	0	in the 130s	0	in the 230s	5	in the 330s	0	in the 430s	0
in the 40s	0	in the 140s	1	in the 240s	2	in the 340s	0	in the 440s	0
in the 50s	0	in the 150s	4	in the 250s	2	in the 350s	0	in the 450s	0
in the 60s	1	in the 160s	2	in the 260s	3	in the 360s	0	in the 460s	0
in the 70s	1	in the 170s	5	in the 270s	0	in the 370s	0	in the 470s	0
in the 80s	0	in the 180s	2	in the 280s	1	in the 380s	0	in the 480s	0
in the 90s	0	in the 190s	2	in the 290s	2	in the 390s	0	in the 490s	0

Table 2.3, are 113, 115 and 116. Of course, we could have produced the classification structure of Table 2.4 directly from the raw-data case. However, it is easier to do it from Table 2.3 and since we have it available we might as well use it.

You would be right to think that Table 2.4 does not seem much of an improvement on Table 2.3; in fact it occupies even more of the page. Moreover we have actually *lost* data. We know there are three values in the 110s and five in the 230s, and so on, but Table 2.4 does not tell us what they are. However, as we shall see, Table 2.4 does have considerable potential. One of the things wrong with Table 2.4 as it stands is that there are too many classes (there are fifty in all). Second, a lot of these classes have zero entries. Thus we could improve upon this presentation by reducing the number of classes and getting rid, wherever possible, of redundant entries.

In Table 2.4 the number of stoppages is organised into classes of 10. Let us see what happens if we reorganise it into classes of 50; the result is shown in Table 2.5.

Table 2.5 Final classification of data in Table 2.3

Number of stoppages	Number of months
0–49	0
50–99	2
100–49	6
150–99	15
200–49	11
250–99	8
300–49	5
350–99	0
400–49	1
	48

Thus the first class of Table 2.5, the class 0–49 stoppages, is the total of the first five classes of Table 2.4; as we can see, there are no months in which the number of stoppages lies between 0 and 49, and so the first entry in the 'Number of months' column of Table 2.5 is zero. The second class in Table 2.5, group 50–99, is the aggregate of the next five classes of Table 2.4, i.e. the 50s, 60s, 70s, 80s and 90s. We see there is a total of two months (one in the 60s and one in the 70s) in this class, and so the second entry in the 'Number of months' column of Table 2.5 is 2; and so on. Finally, we have added a 'total' value for the months column as a check that we have not left any observations out.

Comparison of Table 2.5 with Table 2.4 shows that the number of classes is reduced to a more manageable nine, while there are only two classes with zero entries. In terms of gaining information by a rapid inspection Table 2.5 is a considerable improvement on Table 2.4.

There are several points we can immediately make about Table 2.5. First of all we could have left out the first class, 0–49, without any loss of information since there are no months when the number of industrial stoppages was less than fifty. However, we might have it in mind to use this table to record the number of stoppages in future years. If this is a possibility we should leave this first class in because it is quite conceivable that some month in the future might experience a number of stoppages in the 0–49 range. The second point is that the table emphasises the isolation of the single entry in the 400–49 class and confirms that it is perhaps rather untypical. If this single entry were not present, we could have stopped Table 2.5 with the 300–49 class. The class 350–99 contains no entries and is redundant but we have to include it in the table (for the moment anyway).

The third point about Table 2.5 is that we can gain an immediate impression of the most commonly occurring number of stoppages per month. This would appear to be between 150 and 199, since there are fifteen months in which the number of stoppages lies in this range, almost half as many again as the next highest, eleven months, in which the number of stoppages lies between 200 and 249. However, we should be cautious about placing too much reliance on this figure since it depends very much upon the design of the frequency table. To see this, consider Table 2.6, which has the same class intervals but a different arrangement of class limits. As we can see the most commonly occurring number of stoppages per month is fourteen, during which there were between 220 and 269 stoppages.

Table 2.6

Number of stoppages	Number of months
20– 69	1
70–119	6
120– 69	7
170–219	11
220– 69	14
270–319	6
320– 69	2
370–419	1
	48

The final point about Table 2.5 is that by increasing the size of the classes we have again lost information. In Table 2.5 we can see that there were two months in which the number of stoppages lay between 50 and 99, but we do not know what these values were; they could both have been 51 or both 99, and so on. With Table 2.4 we knew that in one of these months the number of stoppages was in the 60s, in the other it was in the 70s. And of course Table 2.4 showed a loss of information compared with Table 2.3. Thus every time we condense the data by grouping the data into classes we lose information. The smaller the number of classes, the more information is lost.

Thus producing tables such as Table 2.5 is a compromise between an increase in immediate comprehension of the broad pattern of variation in the data on the one hand and a loss of information on the other.

There are no rules which tell us exactly how broad or how narrow our grouping of the data should be, i.e. how many classes we should have. That is something we have to decide for ourselves from situation to situation. As a rough guide 5–15 classes will often prove to be a suitable compromise, telling us enough without concealing too much.

Table 2.5 is known as a *frequency table* and is said to describe the *frequency distribution* of the number of industrial stoppages.

The *frequency* of any particular class is the number of months in which the number of stoppages falls in that class. For example, for the class in which the number of stoppages lies in the range 100–49 the frequency is six because between 1973 and 1976 there were six months in which the number of stoppages per month lay in this range. Similarly there were eight months in which the number of stoppages lay in the range 250–99, and so the frequency of the class 250–99 is eight. The frequency *distribution* simply means the way in which these frequencies (i.e. the number of months) are shared out or distributed between the various classes. As a result the columns of a frequency table such as Table 2.5 are often labelled as shown in Table 2.7. Notice that the last class in Table 2.7 is now shown as '350 and over'. This enables us to reduce the number of classes by one and eliminates the redundant class 350–99 by amalgamation. Such a class is known as an *open-ended* class since we set no limit on its upper value. The trouble is that we again lose a little information since the frequency table of Table 2.7 tells us simply that there was one month when the number of stoppages was in excess of 349. Without access to the original data we have no way of knowing what this value is, it could be 350, it could be 600, or indeed any value. However, open-ended classes are often a useful way of abbreviating what might otherwise be unwieldy frequency tables,

Table 2.7 Frequency table for data of Table 2.1

Class (number of stoppages)	Frequency (number of months)
0–49	0
50–99	2
100–49	6
150–99	15
200–49	11
250–99	8
300–49	5
350 and over	1
	48

and we can use an open-ended class at the beginning as well as at the end of a frequency table. In this respect we could, if we wished (although in this example there is not a lot to be gained), replace the first two classes of the frequency table of Table 2.7 with class '99 and under', which would have a frequency of 2.

We have spent some time describing the construction of the frequency table shown in Table 2.7. As we stated earlier, we would normally construct the frequency table directly from the raw data rather than by proceeding through the various intermediate stages. Let us consider now a second example and describe a suitable procedure more formally. We will also take the opportunity of defining a few terms which we need to establish before we go much further.

Table 2.8 contains raw data on the weekly expenditure on food (in £) by each of fifty households. We want to construct a suitable frequency table to enable us to gain some idea of the principal characteristics of this data.

The first step is to find the minimum and maximum values of expenditure on food; this will give us some idea of the number and size of classes needed. Inspection of Table 2.8 reveals a minimum value of £18.60 per week and a maximum of £95.90. The range is equal to the difference between these two values, i.e.

Minimum value = £18.60 per week
Maximum value = £95.90 per week
Range = £95.90 – £18.60 = £77.30 per week

Table 2.8 The weekly expenditure on food (£) by each of fifty households

79.22	18.60	30.45	59.90	55.00	50.96	48.92	60.00	40.50	53.75
44.66	36.84	83.27	61.23	62.45	52.00	49.08	57.23	34.71	52.81
78.94	71.24	42.91	54.99	69.62	41.00	55.12	43.00	76.60	58.91
65.55	63.89	50.15	74.56	87.31	20.21	40.81	65.04	95.90	73.12
24.65	68.89	24.65	68.09	39.00	48.92	28.93	31.26	38.01	46.81

Thus if we want about ten separate classes, then the width of each class needs to be of the order of 77.30 divided by 10, i.e. about £8 per week. However, working in classes each £8 wide may be a little confusing and we prefer if possible to round this up to £10 (units of five or ten or multiples thereof are the most common class widths used by statisticians).

Notice that these data are different from those on industrial stoppages. Here we have some decimal places whereas previously all the values were integers (i.e. whole numbers). This means that in designing the classes we have decimal points in each class so that each data point can be adequately and correctly located in its appropriate class.

For example, if the first two classes were defined as being

10.0–19.9
20.0–29.9

we would not know into which class to allocate an expenditure of, say, £19.95: £19.95 is slightly too large for the first class and too small for the second. We must therefore have as many decimal points in the classes as in our data.

Finally, having decided to have classes of expenditure increasing in £10 steps we have to decide on the first class. Since the lowest value is £18.60, our first two classes could be

18.60–28.59
28.60–38.59

and so on, but such a class structure is certainly not something we can appraise and understand quickly at a glance, as the numbers are too awkward. It would be better to start the first class at £10.00 and proceed thus:

10.00–19.99
20.00–29.99

It might be even better to start at a zero level of expenditure if we anticipate using the same frequency table for other data in the future when conceivably we might encounter a value of expenditure of less than £10 per week. It is often a good idea to start the first class of a frequency table at zero since it provides an easily understood benchmark for the data (always provided that we can do this without stretching or distorting the table too much – as with many aspects of frequency table construction, experience will eventually indicate the best way of dealing with these points).

We construct the frequency table by using a tally column to help us count each observation as we deal with it. The method is to work through the raw data, observation by observation, putting a dash or a tick in the tally column of the appropriate class and crossing out each observation in the raw data as we deal with it. It is common practice to count the tallies in fives; four vertical strokes followed by a cross-stroke; this makes adding up of the tally-score in each class easier. The total number of tally strokes in each class is of course the frequency of that class. Table 2.9 illustrates this procedure for the food expenditure data of Table 2.8.

Table 2.9 Tally method used to produce frequency table of food expenditure data in Table 2.8

Class (expenditure in £ per week)	Tally	Frequency (number of households)
0.00– 9.99		0
10.00–19.99	1	1
20.00–29.99	1111	4
30.00–39.99	~~1111~~ 1	6
40.00–49.99	~~1111~~ ~~1111~~	10
50.00–59.99	~~1111~~ ~~1111~~ 1	11
60.00–69.99	~~1111~~ 1111	9
70.00–79.99	~~1111~~ 1	6
80.00–89.99	11	2
90.00–99.99	1	1
		50

The frequency distribution to be observed in Table 2.9 shows that the commonest level of weekly household expenditure on food (i.e. the most frequently occurring) was between £50.00 and £59.99 (eleven households), while the next most common was between £40.00 and £49.99 (ten households). It is also possible at a glance to gain an appreciation of how the frequency (i.e. the number of households) varies across the classes. Finally the frequency table tells us something of the range of the data. Compared with the raw data of Table 2.8, the organisation of the data into a frequency table such as that in Table 2.9 represents a significant improvement in understanding the principal summary features of the data. We now want to define some terms associated with frequency tables.

2.3 Some definitions

Class limits

Each class in a frequency table is defined by a class *lower limit* and a class *upper limit*. In the food expenditure example of Table 2.9, the lower limit of the first class is £0.00, the upper limit £9.99. For the second class the lower limit is £10.00, the upper limit £19.99, and so on.

Class boundaries

The lower and upper limits of each class as defined above are the *apparent* class limits. Each class also has *true* upper and lower limits known as class *boundaries*. Between the upper limit of the first class of Table 2.9, £9.99, and the lower limit of the second class, £10.00, there is a small gap (equal to £10.00 – £9.99 = £0.01). There is a similar gap at both ends of each class. The class boundary (or true limit) is defined as the point halfway between the upper limit of one class and the lower limit of the next class. Thus the lower boundary of the second class is £9.995 and the upper boundary is £19.995, and so on. The lower boundary of the first class is usually placed as much below its lower limit as the upper boundary is above the upper limit. Thus the lower boundary of the first class is –£0.005 (although such a negative value can have little or no meaning in this example). Similarly the upper boundary of the largest class is 99.995. The difference between the largest and smallest class boundaries in a frequency table is the *range* of the frequency distribution.

Class width or interval

This is the difference between the upper and lower boundary values of each class. For the first class of Table 2.9, the class width is therefore given by £9.995 – (–£0.005) = £10.000. For the second class the class width is £19.995 – £9.995 = £10.000. In the frequency table of Table 2.9 all the classes have the same width. This will not always be the case. We normally use w_i to denote the width of the ith class.

Class mid-point

Known also as the class-mark, this is the middle value of each class and is located halfway between the upper limit (or boundary) and the lower limit (or boundary) of each class. For the first class of Table 2.9 the class mid-point is equal to

$$\tfrac{1}{2}[£9.995 - (-£0.005)] = \tfrac{1}{2} \times £10.000 = £5.000$$

For the second class it is £15.00, and so on. In general we can use the formula:

$$\text{class mid-point} = \text{lower limit} + \tfrac{1}{2} \times (\text{upper limit} - \text{lower limit})$$

or

$$\text{class mid-point} = \text{lower boundary} + \tfrac{1}{2} \times (\text{upper boundary} - \text{lower boundary})$$

whichever is most convenient. We normally use the symbol x_i to denote the mid-point of the ith class.

To reinforce an understanding of these definitions let us apply them to the frequency table shown in Table 2.10, which relates to the number of working days lost and the number of workers involved through stoppages in the United Kingdom in 1976.

Table 2.10 Number of working days lost and number of workers involved, as a result of industrial stoppages, United Kingdom, 1976

Class (number of workers involved)	Frequency (number of working days lost, thousands)
Under 100	293
100 and under 250	399
250 and under 500	570
500 and under 1000	563
1000 and under 2500	773
2500 and under 5000	426
500 and over	485

Source: *Annual Abstract of Statistics,* 1977, table 6.15.

We may note two things immediately about this frequency table: it has an open-ended class at each end and the widths of each class are not the same. This table tells us, for example, that 399 000 working days were lost due to stoppages involving between 100 and 249 workers, that 485 000 working days were lost due to stoppages involving 5000 or more workers, and so on.

If we are to determine what the class parameters are (i.e. the widths, limits, boundaries and mid-points), we will have to make some assumptions about the two open-ended classes. For the 'under 100' class this is not difficult since the lowest number of workers who can be involved in a stoppage is one (although in practice this is an improbably small figure, it is the only lower limit we can be sure of). The '5000 and over' class presents more difficulty. Without access to the raw data and perhaps with no other knowledge concerning the UK labour scene, what intelligent guess can we make as to a possible upper limit? In this case we could make this class twice as wide as the previous class and hope that such a wide class interval captures most of the unknown data points. Thus we assume an upper limit of 'under 10 000' for this class. Having made these two assumptions we can proceed to determine the class parameters. These are shown in Table 2.11, along with the original frequency distribution. (We should note that in practical situations we may be able to do better than make an intelligent guess as to the limit of an open-ended class. Very often we will have secondary information, for example knowledge of the average value of a class, which will enable us to arrive at a more rationally based value.)

Hopefully the entries in Table 2.11 are self-explanatory. However, one or two comments might illuminate any remaining difficulties. Let us see how our definitions apply to the second

Table 2.11 Parameters of the frequency table of Table 2.9

Class (number of workers involved)	Frequency f_i (number of working days lost, thousands)	Class limits		Class boundaries		Class width w_i	Class mid-point x_i
		Lower	Upper	Lower	Upper		
Under 100	293	1*	99	0.5	99.5	99	50.0
100 and under 250	399	100	249	99.5	249.5	150	174.5
250 and under 500	570	250	499	249.5	499.5	250	374.5
500 and under 1000	563	500	999	499.5	999.5	500	749.5
1000 and under 2500	773	1000	2499	999.5	2499.5	1500	1749.5
2500 and under 5000	426	2500	4999	2499.5	4999.5	2500	2749.5
5000 and over	485	5000	9999*	4999.5	9999.5	5000	7499.5

* Assumed values.

class, '100 and under 250':

The lower limit is obviously 100.
The upper limit is (perhaps slightly less obviously) 249.
The class mid-point is halfway between the lower limit (or boundary), i.e. (using limits) = $100 + \frac{1}{2}(249 - 100) = 174.5$.
The class width is equal to the upper boundary – lower boundary = $249.5 - 99.5 = 150.0$.

2.4 Relative frequency

Sometimes we may wish to compare two frequency tables to see how their frequency distributions differ. To do this efficiently, however, the total frequency of each table should be equal. To illustrate what we mean by this, consider the frequency distributions shown in Table 2.12. These refer to the age of UK merchant ships in 1960 and in 1976.

Table 2.12 Age distribution of UK merchant vessels

Class (age in years)	Frequency f_i (number of vessels)	
	1960	1976
Under 5	742	490
5 and under 10	596	441
10 and under 15	658	272
15 and under 20	512	244
20 and under 25	208	78
25 years and over	186	48
	2902	1573

Source: *Annual Abstract of Statistics*, 1971 and 1977.

Let us suppose that we wish to compare the distribution of ships in each age group to see, for the UK merchant fleet, whether this has changed between the two years. We could try to make such a comparison by examining Table 2.12, but this is made difficult by the considerable difference in the total number of vessels between the two years. We can eliminate this difficulty by expressing the frequency in each class for each year as a percentage (or as a proportion, whichever we wish) of the total number of that year. Such frequencies are known as percentage or relative frequencies. If we do this for the data in Table 2.12, expressing the frequencies in percentages, we obtain Table 2.13.

Table 2.13 Relative percentage frequency table for UK merchant shipping data

Class (age in years)	Relative frequency (number of vessels) (%)	
	1960	1976
Under 5	25.6	31.2
5 and under 10	20.5	28.0
10 and under 15	22.7	17.3
15 and under 20	17.6	15.5
20 and under 25	7.2	5.0
25 years and over	6.4	3.0
	100 %	100 %

Notice as a check that the two totals now sum to 100 per cent (allowing for rounding errors.)[1] We can thus make a direct class-by-class comparison between the two years. The immediate impression is that the UK merchant shipping fleet had a greater percentage of newer ships in 1976 than in 1960. In the latter year the number of ships under five years old was 25.6 per cent of the total, but by 1976 this had risen to 31.2 per cent. In the same way the percentage of very old ships (i.e. 25 years old or more) had by 1976 fallen to less than half of the 1960 figure (6.4 per cent down to 3.0 per cent). By adding together the first two classes we see that in 1960, 46.1 per cent (25.6 + 20.5) of vessels were under 10 years old, and by 1976 this had risen to 59.2 per cent (31.2 + 28.0).

This adding together, or cumulating, of the classes in a frequency table will often prove to be a useful way of giving us extra information that is perhaps not immediately obvious from a frequency table and will enable us to answer questions relating to cumulative frequency values which may not otherwise be easy to deal with.

2.5 Cumulative frequency tables

Take a look at the first three columns of the frequency table in Table 2.14. This describes the frequency distribution of male recipients of sickness and invalidity benefit by age in the United Kingdom in 1975. The third column of Table 2.14 is found by

[1] Note that if we had expressed the relative frequencies as proportions, e.g. the first value in 1960 would be 0.256, then the totals would be equal to 1.0 and not 100 per cent.

Table 2.14 Age distribution of male supplementary benefit recipients, UK 1975, cumulative frequency

Class (age of recipient, years)	Frequency, f_i (number of recipients thousands)	Cumulative frequency (thousands)	Relative cumulative frequency (%)
Under 20	17	17	2.3
20–29	73	90	12.2
30–39	91	181	24.6
40–49	122	303	41.2
50–59	204	507	68.9
60–64	204	711	96.6
65 and over	25	736	100.0
	736		

Source: *Annual Abstract of Statistics*, 1977.

successively adding the frequencies in the second column, starting at the top of the table, e.g. 17 + 73 = 90, 90 + 91 = 181, etc. Cumulative frequency tables are useful for answering such questions as 'How many men in the total of 736 receiving benefit were under 40?' (the answer to this question is 181 000, which can be read directly from the third column); or 'How many men receiving benefit were aged under 65?' (the answer, read directly from the third column, is 711 000).

It is often more convenient to work in percentage terms and we can do this by calculating a column of percentage relative cumulative frequency; this is shown as the fourth column of Table 2.14 and is found by expressing each cumulative frequency value in the third column as a percentage of the total of 736. In answer to a question such as 'What percentage of men receiving benefit were aged under 50?' we would answer 41.2 per cent, reading this value directly from the fourth column.

Suppose we had asked the question 'Of the men receiving benefit how many were aged *more* than 49?' Note that this is a slightly different form of question; we are now concerned with frequency totals which are *more than* some value rather than less than, as before. The answer to this question can be found by totalling the last three frequency values in the second column, i.e. 204 + 204 + 25 = 433. But just as with the 'less than' questions it is often more convenient to work out a cumulative frequency column. To answer questions of the more than or greater than type, we form the appropriate cumulative frequency by adding the values in the frequency column starting at the bottom of the table. Table 2.15 shows this for the frequency distribution of Table 2.14.

Table 2.15 'More than' cumulative frequency for supplementary benefit receivers

Class (age of recipient, years)	Frequency f_i (number of recipients, thousands)	'More than' cumulative frequency (thousands)	'More than' relative cumulative frequency (%)
Under 20	17	736	100.0
20–29	73	719	97.7
30–39	91	646	87.8
40–49	122	555	75.4
50–59	204	433	58.8
60–64	204	229	31.1
65 and over	25	25	3.4
	736		

In answer to the question 'How many of the men receiving benefit were aged more than 49?' We can read the answer directly from the third column: it is 433 000 (or 58.8 per cent if we wish to answer in percentage terms).

Often in descriptive statistics only the 'less than' type of cumulative frequency of Table 2.14 is needed, in which case we have no need to distinguish between the two types and can refer to it simply as 'cumulative frequency'. However, if we wish to investigate frequencies of both types we can distinguish between them by labelling them as 'less than' and 'more than'.

As a final point on cumulative frequencies notice that we can only answer questions which refer to values less than or more than exact class limit values. In other words we cannot answer questions such as 'What percentage of recipients were aged less than 47 or more than 33?' since these values are not the same as any of the class limits but fall somewhere outside a class. However, in a later section we will examine a way in which such questions can be answered.

We would like to conclude this section by working through an example to consolidate the material we have covered so far.

The raw data in Table 2.16 relate to the number of employees

Table 2.16 Numbers employed (rounded to nearest thousand) in the forty-three industries of the UK services sector in 1976

104	188	61	219	223	197	21	79	72
80	431	226	29	284	599	1284	121	263
321	105	87	31	91	1886	112	1288	32
109	102	99	94	266	165	246	106	67
96	54	26	439	5	660	966		

Source: *Annual Abstract of Statistics.*

(measured in thousands) in the forty-three industries of the services sector of the UK economy in 1976 (we have excluded five miscellaneous or non-classified industries). Let us apply what we have learnt so far to this data and see what we can discover.

Inspection of a table such as Table 2.16 does not, as we have seen already, reveal a great deal immediately (and often we may have many more than forty-three observations to deal with).

The first step in constructing a frequency table is to find the smallest and largest values and the range:

Smallest value = 5
Largest value = 1886
Range = 1886 – 5 = 1881

This suggests that if we have ten classes, each class should be 200 wide (1881 divided by 10) and since the lowest value is five this means the first class should be 0–199, the second 200–399, and so on. Designing the table in this way will give us the frequency distribution shown in Table 2.17.

Table 2.17 Initial design of frequency table for employment data of Table 2.15

Class (numbers employed) thousands)	Tally	Frequency f_i (number of industries)
0– 199	~~1111~~ ~~1111~~ ~~1111~~ ~~1111~~ ~~1111~~ 11	27
200– 399	~~1111~~ 111	8
400– 599	111	3
600– 799	1	1
800– 999	1	1
1000–1199		0
1200–1399	11	2
1400–1599		0
1600–1799		0
1800–1999	1	1
		43

Unfortunately, as we can see, this table is unsatisfactory on several grounds. First of all, nearly two-thirds of the observations fall in the first class (which means an unacceptably high loss of information), and second there are several classes with zero entries. So we reject this design. If we split the first class into several smaller classes and amalgamate some of the higher value classes, we can achieve a more satisfactory distribution. Such a frequency table is shown in Table 2.18. We should perhaps add that it is always desirable to have classes all the *same* size in a frequency

Table 2.18 Final design for frequency table for employment data of Table 2.16

Class (numbers employed, thousands)	Tally	Frequency f_i (number of industries)
0– 49	~~1111~~ 1	6
50– 99	~~1111~~ ~~1111~~ 1	11
100–199	~~1111~~ ~~1111~~	10
200–399	~~1111~~ 111	8
400–599	111	3
600–799	1	1
800–999	1	1
1000 and over	111	3
		43

table if this can be achieved without producing a table that is otherwise unsatisfactory. Equal class sizes make for an easier and quicker understanding of the frequency distribution.

What does an inspection of Table 2.18 tell us about the size distribution of industries in the UK services sector? Remember that we have produced this frequency distribution because we believe it will be more immediately meaningful than the untreated raw data of Table 2.16. We can see that the largest number of firms have between 50 000 and 99 000 employees, with the second largest group having between 100 000 and 199 000. So if we had to make a guess at the average size of firm (in terms of labour-force size), we could do worse than to guess at 100,000 employees (we will see later exactly what we mean by an average and how we can define and measure it mathematically). We can further see that there are few firms employing more than 600 000 workers and if we had labelled the top class as 1000–1999 (instead of 1000 and over) we could have said that there were none employing more than 1 999 000 (in retrospect it might have been better to so label this class but the design of frequency tables is always a matter of debate and compromise).

Finally, let us construct the relative and the cumulative relative frequency columns; these are shown in Table 2.19. Notice that this time we have not calculated the simple cumulative frequencies – instead we have added up the percentage values in the relative frequency column to provide cumulative relative frequencies (and, furthermore, we have not calculated a 'more than' cumulative frequency, being content with one of the 'less than' type).

From the relative frequency column we can immediately see that the largest number of industries, 25.6 per cent, employ between 50 000 and 99 000 employees, while (for example) only

Table 2.19 Relative and cumulative relative frequency for employment data

Class (numbers employed, thousands)	Frequency f_i (number of firms)	Relative frequency (%)	Cumulative relative frequency (%)
0– 49	6	14.0	14.0
50– 99	11	25.6	39.6
100–199	10	23.2	62.8
200–399	8	18.6	81.4
400–599	3	7.0	88.4
600–799	1	2.3	90.7
800–999	1	2.3	93.0
1000 and over	3	7.0	100.0
	43	100.0	

7 per cent of industries employ 1 000 000 or more. From the cumulative relative frequency column we can see (for example) that 39.6 per cent of firms employ 99 000 workers or less, 90.7 per cent employ 799 000 or less, and so on.

2.6 Joint frequency distributions

Finally, we want to consider briefly the problems involved in studying the behaviour of two (usually) interrelated variables, for example household income and expenditure, unemployment and strikes, inflation and money supply, investment and profits, etc. To illustrate the procedure consider the ficticious data in Table 2.20, which relate to the weekly income and expenditure of fifty households (measured in £, and to the nearest £).

We could arrange this raw data into conventional frequency table form as in Table 2.21, but an alternative and potentially more rewarding form, known as the *joint* frequency distribution, is shown in Table 2.21.

For example in the first row and first column position of Table 2.22 is entered the number of households (i.e. 5) which have *both* income *and* expenditure in the 0–19 £ per week class.

In the second row of the first column is entered the number of households (i.e. 1) who have incomes in the 20–39 class but expenditures in the 0–19 class, and so on. Notice as a check that the column and row totals of the joint frequency distribution are the same as the entries in the two columns of Table 2.21, i.e. the last column is just the frequency distribution of income, the last row just the frequency distribution of expenditure.

Table 2.20 Raw ficticious data for income and expenditure of fifty households (£ per week)

Household number	Household income	Household expenditure
1	20	18
2	72	65
3	110	98
4	26	25
5	38	32
6	97	84
7	56	51
8	23	23
9	71	65
10	16	15
11	67	60
12	24	20
13	99	89
14	50	45
15	64	61
16	18	14
17	80	70
18	49	44
19	114	112
20	35	34
21	82	80
22	43	40
23	30	28
24	104	91
25	132	112
26	35	31
27	89	82
28	59	53
29	19	16
30	70	67
31	38	37
32	69	65
33	24	22
34	76	71
35	51	50
36	85	81
37	44	39
38	33	29
39	78	68
40	52	46
41	51	49
42	41	38
43	60	55
44	29	28
45	118	107
46	18	17
47	113	110
48	121	109
49	149	136
50	16	15

Table 2.21 Frequency distributions for income and expenditure data of the fifty households shown in Table 2.20

Class (£ per week)	Frequency (no. of households)	
	Income	Expenditure
0– 19	5	6
20– 39	12	13
40– 59	10	9
60– 79	9	9
80– 99	6	7
100–119	5	5
120 and over	3	1

2.7 Summary

In this chapter, we have seen that we can usually improve the immediate information content of raw data by rearranging and grouping them into frequency tables which reveal much more clearly the distribution of frequencies from class to class. We have seen that frequency tables are something of a compromise; and although they give us a much quicker appreciation of the broad pattern of the data, they do suppress information. The most appropriate design of frequency table varies considerably from one set of data to another, and is in any case a matter of subjective judgement.

As well as discussing frequency tables and frequency distributions we also examined simple extensions of these, i.e. relative (or

Table 2.22 Joint frequency distribution for household income and expenditure data

Income (£ per week)	Expenditure (£ per week)							
	0–19	20–39	40–59	60–79	80–99	100–119	120 and over	TOTALS
0– 19	5							5
20– 39	1	11						12
40– 59		2	8					10
60– 79			1	8				9
80– 99				1	5			6
100–119					2	3		5
120 and over						2	1	3
	6	13	9	9	7	5	1	

percentage) frequencies and cumulative frequencies (both *less than* and *more than* types). Finally, we encountered some formal definitions relating to the parameters of a frequency table, e.g. class limits, boundaries, widths, and so on, which we shall find useful later on.

In the next chapter we shall examine methods by which data can be interpreted graphically. As we shall see, this can further improve our ability to make sense out of data.

EXERCISES

2.1 The data in Table 2.23 give details of the average hourly earnings (in pence) of manual workers in fifty-five selected industries in the United Kingdom in 1974 by sex (the industries selected are those employing the most women). Rearrange the data into two frequency tables using suitable class intervals. Comment on what these tables reveal which is not immediately obvious from the raw data.

Table 2.23

Industry	Men	Women
1. Bread and flour confectionery	89.28	57.71
2. Biscuits	100.79	69.82
3. Bacon, meat and fish	99.05	72.36
4. Milk and milk products	102.39	76.27
5. Cocoa, chocolate, etc.	101.44	71.57
6. Fruit and vegetables	102.83	73.73
7. Food industries (nes)	108.83	75.69
8. Other drinks industries	108.62	85.33
9. Tobacco	132.77	108.14
10. General chemicals	121.48	77.86
11. Pharmaceuticals	108.33	73.63
12. Other chemical industries	108.29	77.73
13. Other machinery	106.95	78.47
14. Other mechanical engineering (nes)	110.75	80.62
15. Watches and clocks	106.18	73.34
16. Scientific instruments	100.42	70.87
17. Electrical machinery	108.06	74.85
18. Insulated wires and cables	115.09	85.55
19. Telephone apparatus	105.35	81.61
20. Electronic components	101.45	70.96
21. Television and hi-fi	95.23	71.68
22. Electronic capital goods	104.52	74.04
23. Domestic appliances	104.03	78.21
24. Other electrical goods	109.60	77.16
25. Motor vehicles	127.01	93.13

Industry	Men	Women
26. Aerospace	116.24	81.72
27. Metal industries (nes)	106.00	70.73
28. Cotton spinning, etc.	93.45	70.72
29. Weaving	102.38	72.78
30. Woollens	91.20	67.56
31. Hosiery and knitwear	103.35	66.91
32. Carpets	106.67	79.09
33. Made-up textiles	86.66	59.78
34. Textile finishing	93.69	68.55
35. Weatherproof outerwear	92.56	69.10
36. Male tailored outerwear	93.15	70.00
37. Female tailored outerwear	96.76	64.96
38. Men's wear	87.22	63.03
39. Women's wear	89.44	62.15
40. Dress industries (nes)	88.85	64.33
41. Footwear	108.86	78.58
42. Pottery	100.76	71.18
43. Glass	120.05	87.68
44. Furniture	111.53	82.43
45. Packaging products	114.92	75.91
46. Stationery	107.66	77.19
47. Other printing, etc.	117.73	76.97
48. Rubber	116.98	74.86
49. Toys, sports equipment, etc.	96.03	65.31
50. Plastic products (nes)	108.09	71.39
51. Road passenger transport	97.98	83.87
52. Other transport and communication	115.65	80.92
53. Laundries	81.70	52.80
54. National government	84.24	74.13
55. Local government	87.79	73.34

nes = not elsewhere specified.

Source: *Department of Employment Gazette*, 1975.

2.2 Determine the boundary values, the class marks and the class intervals for the frequency tables constructed in exercise 2.1.

2.3 Construct relative frequency tables for the data in exercise 2.1. What percentage of industries in each case were in

(a) the lowest paid earnings group in your frequency table?

(b) the highest paid earnings group?

2.4 Construct 'less than' and 'more than' cumulative and cumulative relative frequency tables for the data in exercise 2.1. What sort of questions do these tables enable us to answer?

2.5 Construct a joint frequency table for the data in exercise 2.1 of male and female earnings. What does this table tell us about the relationship between the two measures?

2.6 The data in Table 2.24 relate to the distribution of personal wealth of individuals in the United Kingdom in 1977. Construct cumulative and cumulative relative frequency tables of the number of individuals and their total wealth. Answer the following questions:

(a) About 41 per cent of all individuals own less than what percentage of the total wealth?

(b) Less than 1 per cent of all individuals own more than what percentage of the total wealth?

(c) About 50 per cent of the total wealth is owned by what percentage of all individuals?

Table 2.24

Over (£)	Not over (£)	Number of individuals (thousands)	Total wealth (£ billion)
	1 000	2013	1.0
1 000	3 000	2860	6.0
3 000	5 000	2223	9.1
5 000	10 000	3638	27.0
10 000	15 000	2816	35.2
15 000	20 000	1451	25.1
20 000	25 000	761	17.4
25 000	50 000	1090	37.8
50 000	100 000	351	24.5
100 000	200 000	93	13.6
200 000		28	13.1

Source: *Annual Abstract of Statistics*, 1980, table 15.3.

3 Frequency Graphs

OBJECTIVES

After reading this chapter and working through the examples students should understand the meaning of the following terms:

histogram
frequency density
relative frequency histograms
frequency polygon
cumulative frequency polygons
ogives
frequency curve
normal curve
skewed distribution

Students should be able to:

draw a histogram from a frequency distribution
calculate frequency density
draw relative frequency histograms
draw cumulative frequency polygons or ogives
use the ogives to answer questions relating to intermediate values in a frequency distribution
draw frequency curves

3.1 Introduction

There is little doubt that most people gain information more quickly from a visual device, such as a graph or bar chart, than they would from the underlying numeric data – certainly this is the case in the short term.

For passing on the broader features of data, there is little doubt that a pictorial or visual presentation is very efficient, and many authors strongly advocate graphical presentation of data not only as a preliminary step but as an intrinsically powerful analytic device (e.g. see Tukey, 1977), which if used properly may reveal much of the underlying and sometimes hidden features of a body of data.

In Chapter 2 we spent some time considering in detail the construction of frequency tables and associated concepts. Here we shall examine several ways in which frequency distributions can be displayed graphically, concentrating much of our effort on describing the design and construction of what are known as *histograms.*

Construction of a histogram is usually straightforward but sometimes difficulties do arise, for example due to unequal class sizes, and we want to see how we can deal with such problems. It is also possible to graph cumulative frequency distributions and we shall examine ways of doing this. Finally, we shall consider the very important concept of a *frequency curve* and the meaning of normal and skewed distributions.

3.2 The histogram

Let us return to the example on household expenditure on food, which we considered earlier in Chapter 2. The frequency distribution of Table 2.9 is reproduced here for convenience (and relabelled as Table 3.1).

Drawing a histogram for the frequency table of Table 3.1 is a comparatively straightforward exercise. Essentially we plot frequency on the vertical axis and class boundary on the horizontal axis. We then draw a vertical column on each class base equal in height to the frequency of that class. The result is shown as Figure 3.1.

It is worth mentioning that it is common practice to mark the horizontal axis with class limits. While this is not entirely accurate, it may offer the advantage of enhanced clarity to the diagram since class limit values will usually be shorter than boundaries. Class mid-points are also used although less frequently. As long as

Table 3.1 Household food expenditure (£ per week) for fifty households

Class (expenditure in £ per week)	Frequency f_i (number of households)
0.00– 9.99	0
10.00–19.99	1
20.00–29.99	4
30.00–39.99	6
40.00–49.99	10
50.00–59.99	11
60.00–69.99	9
70.00–79.99	6
80.00–89.99	2
90.00–99.99	1
	50

the result is in no way misleading and as long as the columns of adjacent classes *touch*, then this method is suitable.

The point in drawing the histogram is to produce something which is immediately informative in the visual domain. Comparison of Figure 3.1 with Table 3.1 will allow the reader to decide whether this is true in this particular example. Certainly we obtain a clearer impression of the relative size of the frequencies and at least as good an idea as to the over-all pattern and the concentration of frequencies over certain expenditure values.

Sometimes the frequency of each class is marked at the top of each column to save the eye the task of constantly flitting sideways to read the approximate value from the vertical axis. Whether

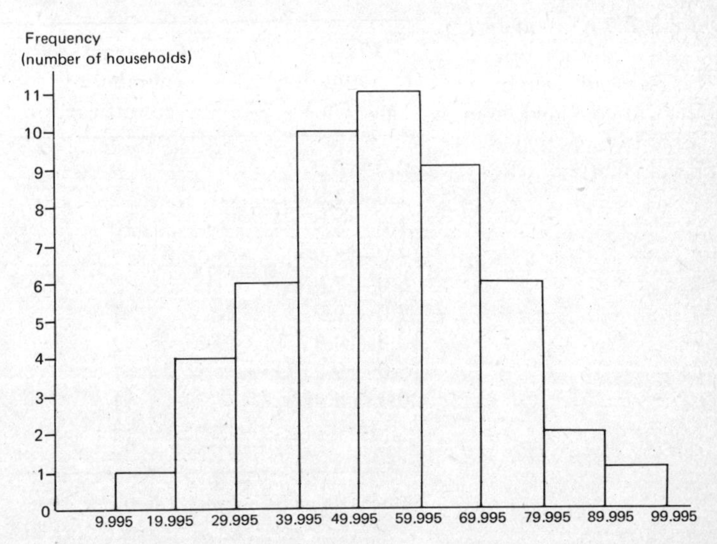

Figure 3.1 Histogram of household food expenditure

this is done or not is a matter of individual choice. Often it is not necessary (and may impede quick appraisal of the situation by cluttering up the diagram) – besides which, the frequency table and its corresponding histogram are often presented together.

Area of the histogram

The area of a histogram can be made to equal 1.0. To do this consider the table and histogram shown in Table 3.1 and Figure 3.1. The area of the first column is equal to its base (i.e. the class width) multiplied by its height (i.e. frequency). Thus in Figure 3.1 the area of the first column = 10 x 0 = 0; similarly, the area of the second column = 10 x 1 = 10; and so on for the remaining bars. The total area of the histogram is found by adding together the areas of each of the ten columns. This gives a total area equal to 0 + 10 + 40 + . . . + 10 = 500. If we now express the area of each column as a proportion of this total of 500, then the area of the first column as a proportion of 500 equals 0/500 = 0, the area of the second column equals 10/500 = 0.02. Thus, as proportions, the total of the areas = 0 + 0.02 + 0.08 + . . . + 0.02 = 1.0. That is, the total area of the histogram is equal to 1.0. This fact is of some importance, as we shall see later.

Frequency density

As we have just seen, constructing a histogram from a frequency table is usually quite straightforward. This is certainly true when all the classes in the frequency table have equal widths. This, however, will not always be the case. Take, for example, the frequency table shown in Table 3.2. This table shows the fre-

Table 3.2 Frequency table of the amount of overseas aid given by the United Kingdom in 1976 to forty-two developing countries

Class (amount of aid, £m.)	Frequency f_i (number of countries)
0.0– 0.9	9
1.0– 1.9	9
2.0– 3.9	10
4.0– 5.9	7
6.0– 7.9	2
8.0– 9.9	1
10.0–14.9	2
15.0–19.9	1
20.0 and over	1
	42

Source: *Annual Abstract of Statistics*, 1976.

quency distribution of UK overseas aid to forty-two developing countries in 1976, in £ millions, rounded to the nearest £0.1 million (excluding India and several smaller unnamed countries).

Inspection of Table 3.2 reveals, for example, that more countries (ten in all) received aid in the £2–3.99 million band than any other amount, though almost as many countries (nine each) received between zero and £0.9 million and between £1 and £1.9 million. Only four countries (a quick cumulative count of the last three classes) received aid in excess of £10.0 million. Unless we make some assumption about the upper limit of the '20 and over' class we cannot make any decision as to the range of the amount of aid.

Now we come to consider an appropriate histogram for this frequency table. Although we stated previously that the height of each column is proportional to the frequency, this is only true when all the class widths are the same. In general it is the *area* of each bar of the histogram which should be proportional to its respective class frequency. This being the case, and since the area of a column is equal to the product of its base times its height, to draw the histogram as we described in the previous section would be to produce a misleading diagram. We must take into account the width of the base of each column (i.e. the class width) in the histogram and adjust the height (frequency) of the column accordingly. Thus if two classes have the same frequency but one is *twice* as wide as the other, the first one should have the height of its column, i.e. its frequency, halved (or the other column its height doubled).

Before we discuss a formal method for doing this we can illustrate how misleading a histogram which has not been constructed in this way can be. Figure 3.2 shows the histogram for the frequency table of Table 3.2 drawn without taking the unequal class widths into account. To draw a histogram at all we must make some assumption as to the upper limit of the open-ended class. Usually we have enough information from the data or from the nature of the problem to make an informal guess as to a reasonable upper limit, though sometimes we may have to guess rather more wildly than we would wish. In this example we have assumed a top limit for this '20 and over' class of 24.9.

To see how misleading this histogram is look at the first (0.0–0.9) column and the seventh (10.0–14.9) column (and remember that the area of each column is supposed to be proportional to its frequency). The area of the latter column is $2 \times 5 = 10$, while the area of the first column is only $9 \times 1 = 9$, yet the frequency of the first column is 4.5 times the frequency of the seventh (9 compared with 2).

Figure 3.2 Incorrectly drawn histogram of frequency table (Table 3.2): different class widths ignored

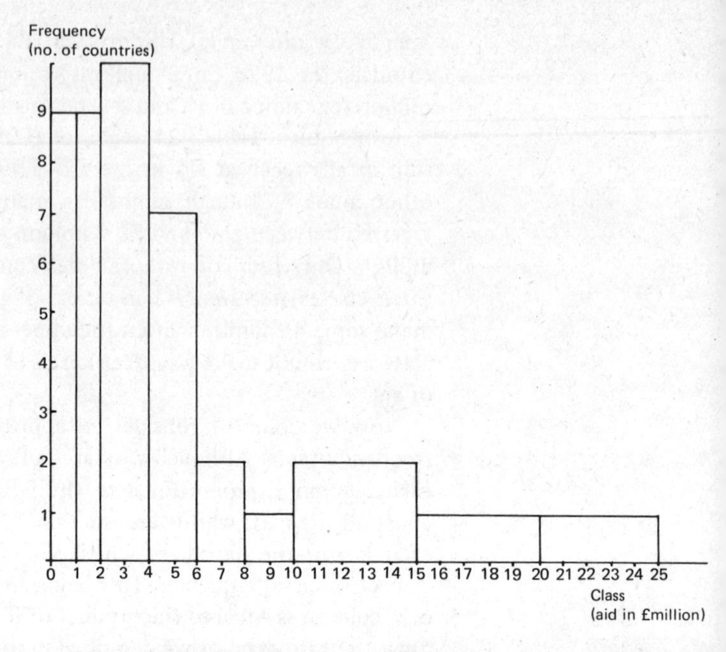

Thus we need to adjust the height of each column to take into account the differing class widths. We do this by calculating what is known as the *frequency density* of each class. This is simply the frequency divided by the width of each class. That is, the frequency density of the ith class equals

$$\frac{\text{Frequency of } i\text{th class}}{\text{Width of } i\text{th class}} = \frac{f_i}{w_i}$$

If we apply this procedure to the frequency table of Table 3.2, we obtain Table 3.3.

Table 3.3 Frequency densities for overseas aid problem

Class (amount of aid, £m.)	Frequency f_i (number of countries)	Class width w_i	Frequency density f_i/w_i
0.0– 0.9	9	1.0	9.0
1.0– 1.9	9	1.0	9.0
2.0– 3.9	10	2.0	5.0
4.0– 5.9	7	2.0	3.5
6.0– 7.9	2	2.0	1.0
8.0– 9.9	1	2.0	0.5
10.0–14.9	2	5.0	0.4
15.0–19.9	1	5.0	0.2
20.0–24.9*	1	5.0	0.2
	42		

* Assumed upper limit.

We can now draw the appropriate histogram but this time we plot frequency *density* on the vertical axis rather than just frequency. The histogram is shown in Figure 3.3.

Comparison of Figure 3.2 with Figure 3.3 shows how misleading the former histogram is because of the undue importance given to the frequencies of all the classes above the first two.

The difficulty with histograms drawn from frequency densities is that, although the relative importance of the bars is correctly shown, it is not easily possible from the diagram alone to determine what the *actual* frequency for any particular class value is. Of course, if we have access to the frequency table as well, this is not a problem.

Figure 3.3 Histogram of overseas aid using frequency densities

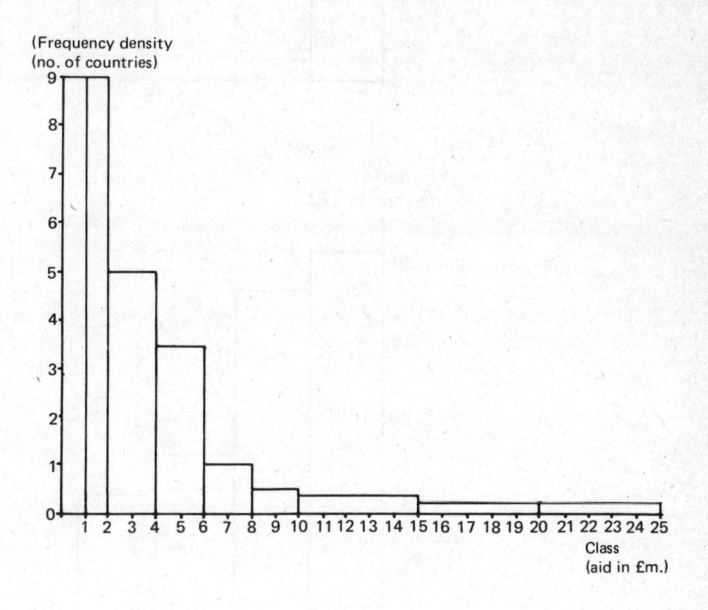

Relative frequency histograms

Histograms are useful devices for giving a quick visual appreciation of the broad distribution of frequencies between classes. We can also use histograms to compare the general patterns of two or more frequency distributions. When the total frequencies of the distributions are the same there is no problem – we simply draw the histograms as described above. However, when the total frequencies are not the same then we have to calculate *relative* frequencies and draw the histograms from these rather than from the original frequency values. Let us return to the example on the distribution by age of the UK shipping fleet in the years 1960 and 1976, shown in Table 2.12. The relative frequencies are shown in Table 2.13. If we draw a histogram for each year, we obtain the diagrams shown in Figure 3.4. Notice that we have assumed an

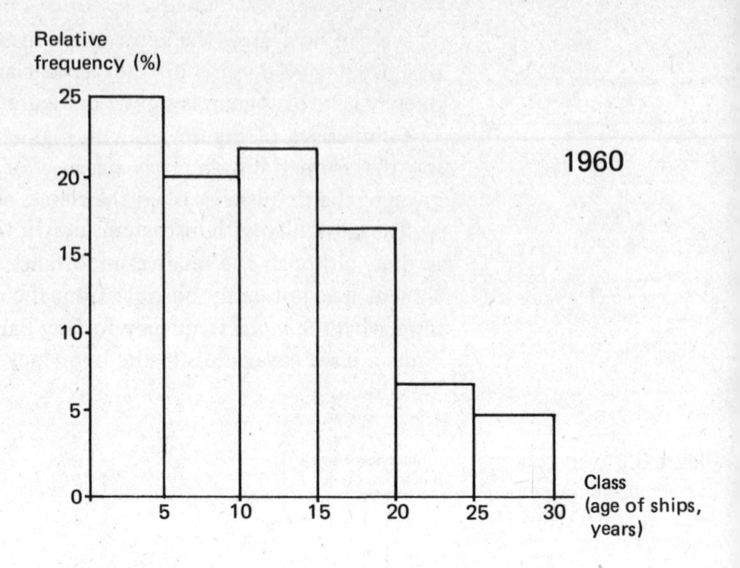

Figure 3.4 Relative frequency histograms of the age distribution of UK merchant shipping in 1960 and 1976 based on Table 2.13

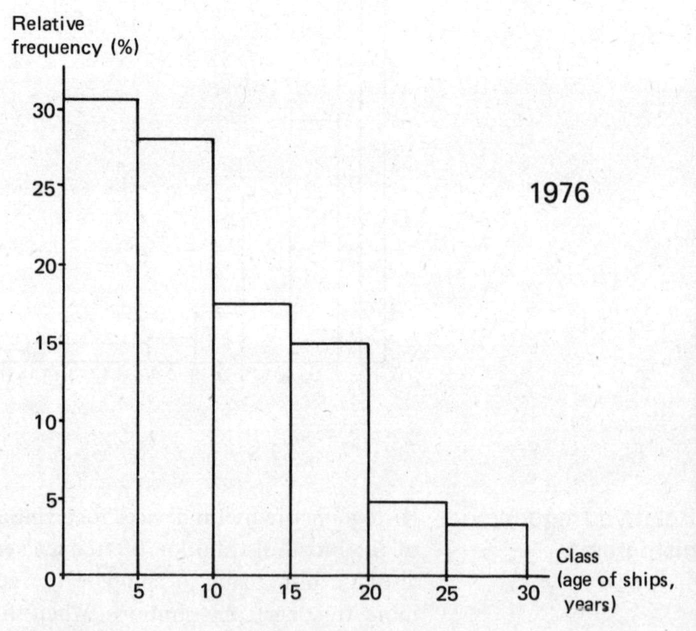

upper limit of 30 years for the last class in Table 2.12, thus giving all classes the same width. This means that drawing the histogram is straightforward. If the classes were not all of the same width (and with the total number of observations in each year being different), we would have to compute the *relative frequency density* in order to draw the histograms accurately.

Drawing one histogram directly beneath the other facilitates the comparison, and in this example it is quite easy to see the broad changes which took place in the age distribution of the UK merchant fleet in the period between 1960 and 1976. Essentially we are claiming that the visual presentation of the relative frequencies in the form of histograms, as in Figure 3.4, is more *immediately* informative of the over-all *pattern* of change than is the relative frequency table of Table 2.13. Of course we can detect the sharp decline in the number of vessels ten or more years old in 1976 by examining the actual data in Table 2.13, but the change is more readily observed in the 1976 histogram of Figure 3.4.

Of course, as we have seen, every improvement in the immediacy of understanding of frequency distribution is accompanied by some loss in detailed information. However, in most circumstances we are usually prepared to accept this.

3.3 The frequency polygon

There is another device for displaying the information contained in a frequency table, known as a *frequency polygon*. Quite simply it is a graph with frequency on the vertical axis and class mid-point on the horizontal. It can be drawn from a histogram by joining up all the points marking the mid-point at the top of each column in the histogram. Alternatively, if a histogram is not required or if we want to draw it separately, we can simply plot frequencies against class mid-points. It is important to remember to close the frequency polygon at each end. This involves assuming two imaginary classes one at each end of the frequency table having zero frequency (unless either of the classes already has a zero frequency). To see what we mean consider the frequency table shown as Table 3.4. This describes the number of widows

Table 3.4 Number and age of women receiving widows' benefit, United Kingdom, November 1976

Class (age in years)	Frequency, f_i (number of widows, thousands)	Class mid-points x_i
Under 30	3	24.5
30–39	16	34.5
40–49	77	44.5
50–59	296	54.5
60 and over	94	69.5
	486	

Source: *Annual Abstract of Statistics*, 1976.

receiving benefit in the United Kingdom in November 1976 (measured in thousands) analysed by age. The third *column* contains the class mid-points. We have assumed a lower limit of 20 for the first class and an upper limit of 69 for the highest class (we don't feel there will be a significant number of widows younger than 20 or older than 69 and choosing these values keeps the class widths uniform).

If we wish to display these data by means of a frequency polygon, we can either draw the histogram first and then join the middle points at the top of each column or draw the frequency polygon directly as described above. Figure 3.5 shows the frequency polygon resulting from the application of each of these methods. Notice that in this example the horizontal axis is marked by class limits rather than boundary values.

We should mention that the area of (or under) a frequency polygon is the same as the area under the corresponding histogram, i.e. it can be made to equal 1, provided all the class widths are the same (and will still be approximately true when this is not the

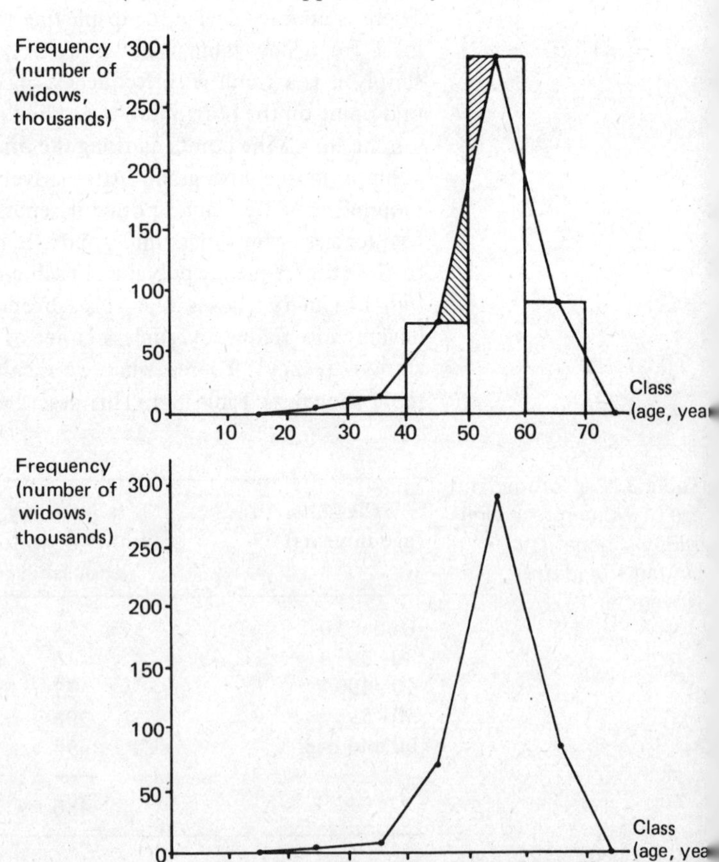

Figure 3.5 Frequency polygons for number and age distribution of widows' benefit

case). This is because the frequency polygon cuts off some area of each bar but adds extra and equal area at the same time. This is illustrated by the two shaded areas in the histogram-derived polygon of Figure 3.5. We loose the upper shaded area but gain the lower and since they are both equal the net area remains the same.

We should note that if we wish to plot a frequency polygon from a distribution containing unequal class intervals, we cannot simply plot frequency density against boundary values since this leads to an area under the resulting polygon which is not equal to 1.0. One way around this is to split up the larger class interval(s) into multiples of the smaller classes (splitting up the frequencies in the same way). We can then plot the polygon using the mid-points of these new classes. This procedure will preserve the area properties.

Cumulative frequency polygons (or ogives)

In this section we shall discuss a method by which cumulative frequencies can be displayed visually. To illustrate the procedure let us consider once again the frequency distribution of male recipients of sickness benefit shown in Table 2.14. This table also shows the 'less than' cumulative frequency and the 'less than' relative cumulative frequency. Table 2.15 shows the equivalent 'more-than' values. We combine these into one table (see Table 3.5). Remember that we need both types of cumulative frequency to answer two different types of question, for example of the type 'How many men aged *more* than a certain age were in receipt of benefit?' or of the type 'How many men aged *less* than a certain age . . . ?'[1]

Before we draw the cumulative frequency graphs (or *ogives*, as they are also known) we have to assume values for the limits of the two open-ended classes. We will assume a lower limit of 16 (the lowest age at which people can work) for the youngest class and an upper limit of 67 (as a guess) for the oldest class.

We can plot separate cumulative frequency curves for each of the last four columns of Table 3.5, but the basic rules are as follows:

(a) The vertical axis is marked with cumulative or relative cumulative frequency, the horizontal axis with class.
(b) The '*less than*' cumulative (or relative cumulative) frequency

[1] We could actually derive the answer to the 'more than' question by subtracting from the total frequency the answer to a 'less than' question. However, it is often more convenient to have both types of cumulative frequency.

value for each class is plotted against the *upper* boundary of that class.

(c) The '*more than*' cumulative (or relative cumulative) frequency value for each class is plotted against the *lower* boundary of that class.

(d) The maximum value of a cumulative frequency graph is equal to the sum of the frequencies (i.e. 736 in this example). The maximum value of a relative cumulative frequency graph is equal to 100 per cent.

Let us start by plotting the 'less than' cumulative frequency curve, i.e. column three of Table 3.5. The result is shown in Figure 3.6.

Having drawn the axes as shown in Figure 3.6 we take the first value in the 'less than' cumulative frequency column of Table 3.5. This value is 17 and is plotted against the upper boundary value, i.e. 19.5 of this first 16–20 class, i.e. 17 on the vertical axis is plotted against 19.5 on the horizontal axis (this is point A). The next 'less than' cumulative frequency value of 90 is plotted against 29.5 (the upper boundary of the next class) – this is point B – and so on. The last point is C (736 on the vertical axis against 67.5 on the horizontal). These points are all joined up with a series of *straight* lines. The bottom point on the 'less than' ogive is the origin. This is because we have assumed a lower limit of 16 and there is a zero frequency (i.e. *no* benefit recipients) below this, so zero on the vertical axis is plotted against 15.5 on the horizontal. Thus the curve from the origin to point C is the 'less than' cumulative frequency curve.

Table 3.5 Cumulative frequencies for age distribution of male sickness benefit recipients (1975)

Class (age of recipient, years)	Frequency f_i (number of recipients, thousands)	'Less than' cumulative frequency (thousands)	'Less than' relative cumulative frequency (%)	'More than' cumulative frequency (thousands)	'More than relative cumulative frequency (%)
16*–19	17	17	2.3	736	100.0
20 –29	73	90	12.2	719	97.7
30 –39	91	181	24.6	646	87.8
40 –49	122	303	41.2	555	75.4
50 –59	204	507	68.9	433	58.8
60 –64	204	711	96.6	229	31.1
65 –67†	25	736	100.0	25	3.4
	736				

* Assumed lower limit.
† Assumed upper limit.
Source: *Annual Abstract of Statistics*, 1977.

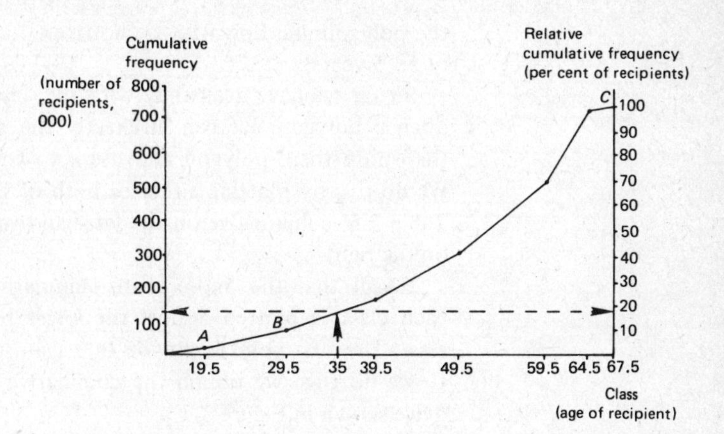

Figure 3.6 The 'less than' cumulative frequency polygon for male benefit recipients

We can also make it the 'less than' relative cumulative frequency curve by marking the right-hand vertical axis in percentage terms. Alternatively, we could have started with the relative cumulative frequency curve and transformed it into a cumulative frequency curve by adding a vertical axis in frequency rather than percentage terms.

Let us now consider some examples of how we might use this curve.

We have seen that we can use the 'less than' frequency polygon to answer questions such as 'How many men aged 39 or less received benefit?' The answer, which we can obtain directly from the frequency table (column three of Table 3.5), is 181 000. Indeed, we can use the table to answer directly any such question provided it relates to one of the class-limit values. However, suppose we wanted to know how many men aged 35 or less received benefit. We could derive the answer arithmetically by interpolation from the frequency table (we have to assume an even distribution of frequencies within classes), but it is much easier and quicker to read the answer off the 'less than' ogive in Figure 3.6. If we fix 35 on the horizontal axis, move up to the polygon and across to the left vertical axis we find a value of 130 000 or (moving to the right-hand vertical axis) a value of 18 per cent. The construction is shown by dashed lines in Figure 3.6. Naturally the accuracy of the answers depends upon the accuracy of the diagram. Thus we can answer questions relating to any age of recipient; we are not confined to class-limit values. We can also use the cumulative frequency polygon to answer questions such as '30 per cent of recipients were aged less than what?' The answer is found by finding 30 per cent on the right-hand vertical axis, reading across to the ogive and down to the horizontal axis, where the answer is 44.5 years. Or we could ask '400 000 recipients were aged less than what?' By reading from the left-hand vertical axis across to the polygon and down to the horizontal axis we find the answer of 53.5 years.

So far we have dealt only with the 'less than' cumulative frequency polygon. We can in exactly the same manner construct the 'more than' polygon and use it to answer similar questions. We do this by plotting either or both of the last two columns of Table 3.5, column five on the left-hand vertical axis, column six on the right.

Recall that the 'more than' cumulative frequency value for each class is plotted against the *lower* boundary value of that class. The first point is thus 73 000 against 15.5 years, and so on. If we do this, we obtain the cumulative frequency polygon or ogive shown in Figure 3.7.

Figure 3.7 The 'more than' cumulative frequency polygon for male benefit recipients

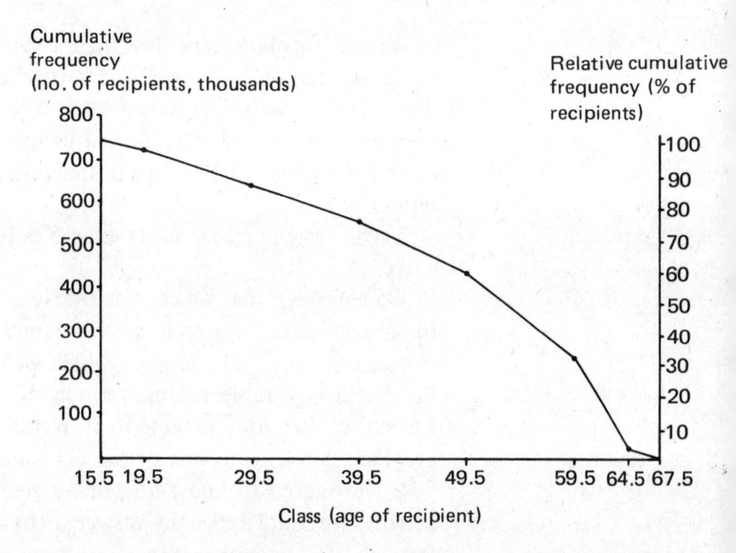

We can use this more than ogive to answer similar questions to those we asked with less than ogive, for example, 'How many recipients were aged more than 55 years?', or '90 per cent of recipients were aged more than what?' Answers of 315 000 and 27 years respectively can be found quickly using the graph. Finally, one question of some interest, to which we shall return later, concerns the halfway age value in the number of recipients. By this we mean either half the recipients were aged less than what, or half the recipients were aged more than what? We can answer by reading either from the 50 per cent value on the right-hand vertical axis or 368 (half of 736) on the left-hand vertical axis. (Strictly speaking, this should be the value halfway between 368 and 369 when the total number of observation, i.e. n, is even. We will consider this point further in the next chapter.) From the ogive of Figure 3.6 we read an answer of 52 years, from Figure

3.7 the same answer. In other words, if both cumulative frequency polygons were drawn on the same graph, they would intersect at the halfway point on the vertical axes.

Frequency curves, the normal curve, skewness

If we were to draw a smooth curve through the frequency polygon of Figure 3.5, we would have a *frequency curve*. The result of doing this is shown in Figure 3.8.

Frequency curves can take on many different shapes depending on the structure of the underlying frequency distribution. However, in economics (and in the social and behavioural sciences) the curves will often have the general form shown in Figure 3.8, i.e. a single peak with values rising and falling more or less smoothly on either side. There is a special type of frequency curve known as the *normal curve* which is of great importance (as we shall see later). The normal curve is symmetrical in shape around a vertical line drawn through the peak and has a smooth 'bell-shaped' appearance. A typical normal curve is shown in Figure 3.9.

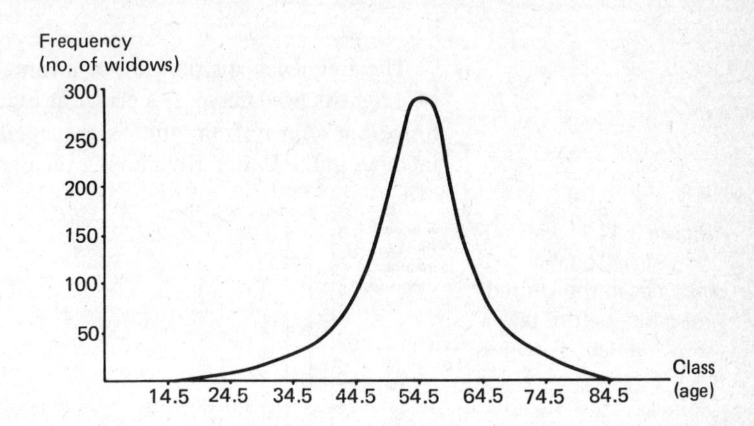

Figure 3.8 Frequency curve drawn from frequency polygon of Figure 3.5

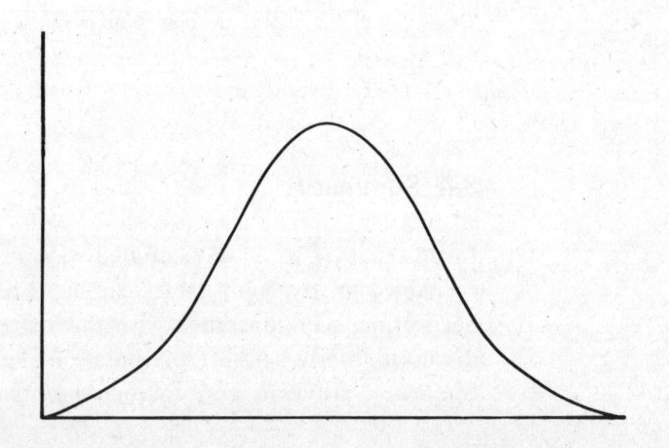

Figure 3.9 A normal frequency curve

Many frequency distributions in economics and related disciplines can be represented by normal or near-normal curves (or can be transformed into normal curves by taking logs).

Of course, many curves are not normal, or symmetrical, and we speak of curves such as those shown in Figure 3.10 as being *skewed*. A left-skewed frequency curve has a long tail of low values to the left; a right-skewed curve a long tail of high values to the right.

Figure 3.10 Skewed frequency curves

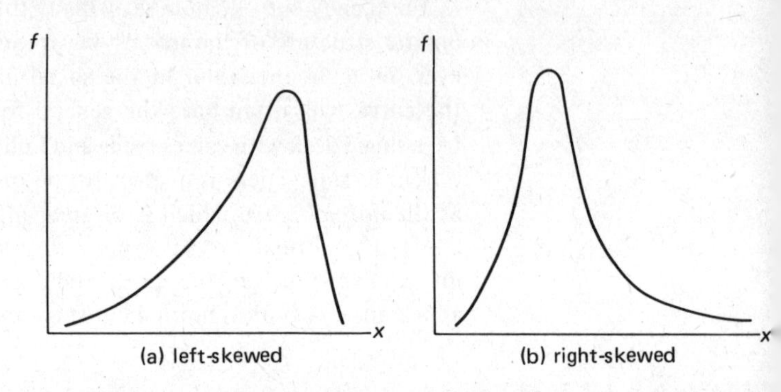

The frequency distribution of income is a notorious example of a right-skewed curve, as a glance at Figure 3.11 shows. This is a histogram drawn from the frequency distribution of personal incomes in the United Kingdom in the period 1973–4.

Figure 3.11 Distribution of personal incomes in the United Kingdom before tax showing right-skewness

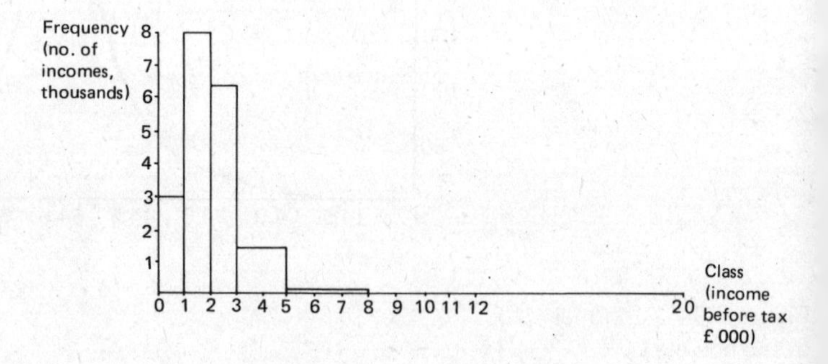

3.4 Summary

In this chapter we have examined ways in which the information contained in frequency tables can be displayed visually. We have been principally concerned with the histogram, but we have also discussed briefly ogives (representation of cumulative frequencies), frequency polygons and frequency curves. We noted that the normal frequency curve (or approximations to it) occurs quite often in economics.

We can hope to use the histogram and its extensions to give us a rapid impression of how and where data are located, i.e. a value around which many of the observations will seem to be concentrated, and the way in which the data are spread out or dispersed across the range of values.

As useful as they may be, these graphical devices have limitations. They are excellent in terms of giving a broad picture of the pattern of observations and giving it quickly, usually without too much analytical work. However, interpretation of a histogram is subjective and lacks precision, and for this reason we want to develop objective and precise analytical measures which will enable us to derive exact numerical summary measures describing the principal characteristics of data and frequency distributions.

EXERCISES

3.1 Draw histograms for the male and for the female earnings data given in exercise 2.1 (p. 30) and compare the two diagrams.

3.2 The data in Table 3.6 relate to the age distribution of the UK population in 1901 and 1971. Construct the appropriate relative frequency histograms and comment upon the differences thus revealed (the age group '85 and over' has been omitted).

Table 3.6

Age	Population (thousands)	
	1901	1971
0– 4	4381	4505
5– 9	4106	4670
10–14	3934	4213
15–19	3826	3832
20–24	3674	4237
25–29	3308	3610
30–34	2833	3529
35–39	2494	3169
40–44	2165	3331
45–49	1837	3544
50–54	1566	3273
55–59	1236	3360
60–64	1067	3206
65–69	743	2707
70–74	535	2005
75–79	313	1331
80–84	157	790

Source: *Annual Abstract of Statistics*, 1980, table 2.3.

3.3 The data in Table 3.7 relate to the age distribution of males in the United Kingdom in 1901 and 1971. Calculate the corresponding frequency density values and draw the appropriate histograms for each of the two years. Calculate the relative frequency density values and redraw the histograms, comparing changes in the distribution of males in the population between the two years.

Table 3.7

Age	Males (thousands)	
	1901	1971
Under 5	2190	2312
5–14	4024	4561
15–29	5191	5915
30–44	3597	4909
45–64	2705	6452
65–74	565	1976
75 and over	219	828

Source: *Annual Abstract of Statistics*, 1980, table 2.3.

3.4 The data in Table 3.8 refer to the percentage of persons in Great Britain in 1971 (by age and sex) who have had some form of higher education. Construct the appropriate histograms, frequency polygons and frequency curves. What do these diagrams reveal about the sex and age differences of those having had higher education?

Table 3.8

Age	Males (%)	Females (%)
15–19	0.2	0.2
20–24	7.5	6.7
25–29	13.6	9.8
30–39	14.2	10.0
40–49	10.7	6.7
50–59	6.1	4.5
60+	5.1	3.8

Source: *General Household Survey*, 1973, table 7.15.

3.5 Draw the 'less than' and 'more than' frequency polygons (ogives) for the data in exercise 3.3 and comment upon what they reveal.

3.6 Discuss the shape of the histograms, frequency polygons and frequency curves you have drawn for exercises 3.1 to 3.3 in terms of their symmetry and skewness.

4 Measures of Location

OBJECTIVES

After reading this chapter and working through the examples students should understand the meaning of the following terms:

arithmetic mean — **quartile**
summation — **mode**
median — **geometric mean**

Students should be able to:

- **calculate the mean of raw data**
- **calculate the mean of grouped data**
- **calculate the median of raw data**
- **calculate the mode of raw data**
- **choose the most appropriate measure of location**
- **calculate rates of growth using the geometric mean**

4.1 Introduction

In the previous two chapters we have examined ways in which raw data can be rearranged (in the form of frequency tables) or depicted graphically (for example, by histograms) to make it more immediately informative. Such treatment enables us to perceive over-all broad patterns in the data more quickly than would otherwise be the case, though applying the methods described above does cause us to lose the underlying detail available in the original observations. Usually we will be able to make a reasonable guess as to the value around which most observations are concentrated or located and we might refer to this as the 'average' value. Second we will be able to gain some idea of the way the data are spread around this average value and we might refer to this as the 'dispersion' of the data.

However, as we pointed out at the end of the previous chapter, our guesses or estimates can only be approximate; what would be very useful would be some accurate summary measures which we could use to describe a particular set of data, first in terms of some average value, and second in terms of the spread of the observations around this average. As well as being useful in themselves, such measures would enable us to compare with and differentiate from other perhaps broadly similar sets of data.

For example, consider the data contained in the frequency distribution of Table 4.1, which relates to the age of heads of households in Great Britain differentiated by colour (as perceived by the investigators) and reported in the *General Household Survey* published by the Central Statistical Office in 1973.

If we examine this distribution, we can see that white heads of household appear to be centred upon the age groups 30–44 and

Table 4.1 Age of heads of household by colour

Class (age of head of household)	Frequency (per cent of all heads of household)	
	White	Coloured
Less than 25	4.2	5.5
25–29	6.9	13.0
30–44	25.3	54.5
45–59	29.2	21.0
60–64	10.2	1.0
65–69	8.8	1.5
70–79	11.7	2.5
80 and over	3.7	1.0

Source: *General Household Survey*, 1973.

45–59, these two groups making up 54.5 per cent of all household heads. If we were asked to estimate or guess the average age of white heads of household, i.e. the central value around which the data appear to be clustered, a reasonable estimate would be fifty years.

For coloured heads of household the corresponding value is obviously lower and seems to be about forty years. But these are only approximate estimates based on a subjective (albeit intelligent or experienced) appraisal of the data. More useful would be a rigorous approach that would provide an accurate figure for the central or average value. That is the purpose of this chapter. As we shall soon see, there are three commonly used measures of *average* and we shall deal with each in turn, concluding with a brief discussion as to the circumstances under which one measure might be preferred over the other two.

As to some suitable measure of the spread or dispersion of the observations around this average value, we shall postpone discussion of this to Chapter 5. Even so, inspection of Table 4.1 indicates that the age distribution of household heads is more narrowly concentrated around the central value in the case of coloured heads of household than with whites. In the former case more than three-quarters of household heads are to be found in age groups 30–44 and 45–49 compared with only just over a half in white households.

4.2 The arithmetic mean

The first and most important of the three measures of location to be discussed is the *arithmetic mean* (or more usually just 'mean'), which corresponds to the idea of average as usually conceived by the person in the street.

Finding the arithmetic mean is analogous to finding that point which would balance a rule if all the observations were laid in order on top of the rule. For example, suppose we had five women whose average weekly earnings (in £) were as follows:

20 40 40 60 70

If we allow each of these women's earnings to be represented by a block of equal weight and if we place these blocks on an otherwise weightless rule, as shown in Figure 4.1, then the mean would be the value where a fulcrum would have to be placed to balance the rule, as shown.

More rigorously, the mean is that value which if multiplied by the number of observations equals the total value found by adding all the individual values together.

Figure 4.1 The arithmetic mean seen as a balancing fulcrum

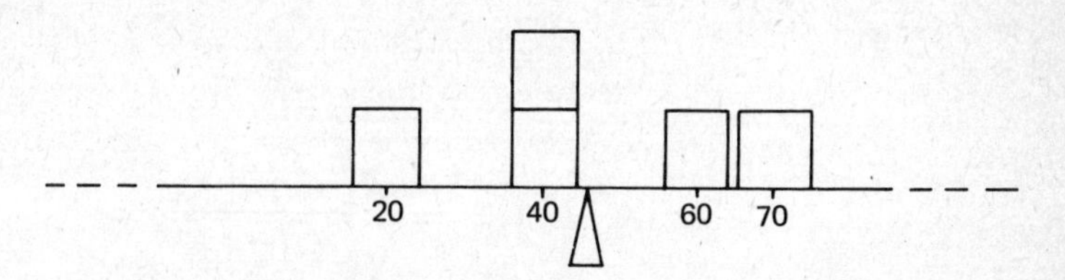

The mean of raw data

Calculation of the mean for a set of raw observations is very straightforward and almost certainly familiar to the reader. We simply add together all the observations and divide by the number of observations. In the above example of the earnings of five women the mean is thus calculated as

$$\frac{20 + 40 + 40 + 60 + 70}{5} = \frac{230}{5} = 46$$

One way of interpreting this result is to say that if all five women lumped their earnings together and then divided the total out equally between them they would each receive £46 per week.

If we now multiply the mean value of 46 by the number of data points, i.e. 5, we have a total equal to that obtained by adding all the observations together, which complies with the above definition, i.e.

$$5 \times 46 = 230$$

Calculation of the mean of raw data is thus a simple task, but before we proceed to consideration of data in frequency table form (both grouped and ungrouped) it would be convenient to introduce some algebraic notation which we shall find increasingly useful in what is to follow.

Summation notation

In the above example of women's earnings let us use X to denote weekly earnings; thus

$$X_i = \text{weekly earnings (£s) of the } i\text{th woman}$$

For example, X_2 stands for the earnings of the second woman (i.e. £40 per week), X_5 the earnings of the fifth woman (i.e. £70 per week), and so on. We also let n stand for the number of observations – in this example $n = 5$.

It is usual to denote the mean value of any variable by placing a bar over the symbol used to denote the variable in question. Thus in this example $\bar{X}$ is used to indicate the mean value of X. Therefore, mean weekly wage is given by

$$\bar{X} = \frac{X_1 + X_2 + X_3 + X_4 + X_5}{5} \tag{4.1}$$

In most practical situations we shall be confronted with many more observations than five. In general with n observations we can write

$$\bar{X} = \frac{X_1 + X_2 + \ldots + X_n}{n} \tag{4.2}$$

However, a shorter, more concise way of writing (4.2) makes use of what is known as the *summation operator*. We use the Greek letter sigma Σ to signify 'sum or add together all the values of'. Thus $\Sigma_{i=1}^{n} X_i$ means add together all the values taken by the variable X_i from $i = 1$ (the first value) to $i = n$ (the nth value). In our example X_i denotes the wage of the ith woman, and with five women we have

$$\sum_{i=1}^{n} X_i = X_1 + X_2 + X_3 + X_4 + X_5$$

and the general expression (4.2) can be written more concisely as

$$\bar{X} = \frac{\sum_{i=1}^{n} X_i}{n} \tag{4.3}$$

When the range over which the summation is to be performed is clear (as it usually will be) we can drop the lower and upper summation units $i = 1$ and $i = n$ and simply write ΣX_i.

Comparison of (4.3) with (4.2) illustrates the convenience of the summation notation. We do not want to devote too much time to the algebra of summation; most applications will be obvious at the time we use them. However, we can take note of the following rules (where $i = 1, 2, \ldots, n$):

Summation of sum or difference	$\Sigma(X_i \pm Y_i) = \Sigma X_i \pm \Sigma Y_i$	
Square of summation	$(\Sigma X_i)^2 = \Sigma X_i^2 + 2\sum_{i<j} X_i X_j$	(4
Multiplication by constant	$\Sigma k X_i = k \Sigma X_i$	
Summation of constant	$\Sigma k = (k + k + k + \ldots + k) = nk$	

We can now proceed to consider the calculation of the mean of ungrouped frequency table data.

The mean of an ungrouped frequency distribution

So far we have not explicitly considered the case of data which have been organised into a frequency table but are not grouped. Table 4.2 is an example of such a table and relates to the number

Table 4.2 Number of bedrooms in public-sector rented accommodation

Class (number of bedrooms X_i	Frequency (per cent of all dwellings) f_i	f_iX_i
1	13.2	13.2
2	31.3	62.6
3	52.5	157.5
4	2.7	10.8
5	0.2	1.0
	$\Sigma f_i = 100.0$	$\Sigma f_iX_i = 245.1$

Source: *General Household Survey*, 1973.

of bedrooms in rented accommodation in the public sector in Britain (ignore column three for the moment).

In other words 13.2 per cent of all public-sector accommodation has only one bedroom, 31.3 per cent has two bedrooms, and so on. Now these data are organised into a frequency table but are not grouped, or rather are grouped but into classes each only 1 unit wide. What we want to know is the mean number of bedrooms in public-sector accommodation in Great Britain. Notice that we have used X_i to denote the values in the class column. In Chapter 2 we indicated that X_i is used to denote class mid-points, but when the classes are not grouped, as is the case here, the class mid-points are the same as the class values; the mid-point of the first class is 1, of the second class 2, and so on.

We now define the mean of an ungrouped frequency distribution as

$$\bar{X} = \Sigma f_iX_i/n \tag{4.5}$$

where Σf_iX_i means 'sum together all the products of f_iX_i', i.e.

$$\Sigma f_iX_i = f_1X_1 + f_2X_2 + \ldots + f_nX_n$$

But in a frequency table $n = \Sigma f_i$; therefore, (4.5) becomes

$$\bar{X} = \Sigma f_iX_i/\Sigma f_i \tag{4.6}$$

Note that we cannot cancel Σf_i in the numerator with that in the denominator; Σf_iX_i must be considered as a compound term which cannot be split up in this way. To apply (4.6) we need to calculate the value of the term Σf_iX_i.

The third column of Table 4.2 contains the values of the products f_iX_i. For example, the first value is $f_1X_1 = 13.2 \times 1 = 13.2$, the second $f_2X_2 = 31.3 \times 2 = 62.6$, and so on. The total of the third column gives $\Sigma f_i X_i = 245.1$. The sum of the second column gives $\Sigma f_i = 100$. Thus from (4.6) we can calculate the mean number of bedrooms, i.e.

$$\bar{X} = \frac{245.1}{100} = 2.451$$

Thus the mean number of bedrooms in public-sector rented accommodation is 2.451.

The mean of a grouped frequency distribution

Calculation of the mean from a grouped frequency distribution requires little extra effort than is required for the ungrouped case. Let us return to the example on age of heads of household by colour and calculate the mean age of white heads of household. As it happens, we can use (4.6) with grouped data; X_i still denote the mid-point of the ith class. Thus we need to determine the values of Σf_i and $\Sigma f_i X_i$ for white heads of household. Table 4.3 reproduces the two columns of grouped data shown originally in Table 4.1. The third column gives the class mid-points X_i for white heads of household. The final column gives the values of $f_i X_i$. Note that we have assumed a lower limit of 20 and an upper limit of 84 for the lowest and highest classes respectively.

From Table 4.3 we see that $\Sigma f_i = 100$ and $\Sigma f_i X_i = 5130.2$. Thus from (4.6) we have

$$\bar{X} = 5130.2/100 = 51.302$$

Table 4.3 Calculation of mean age of white heads of household

Class (age of head of household)	Frequency (per cent of all households) f_i	Class mid-point X_i	$f_i X_i$
*20–24	4.2	22.0	92.4
25–29	6.9	27.0	186.3
30–44	25.3	37.0	936.1
45–59	29.2	52.0	1518.4
60–64	10.2	62.0	632.4
65–69	8.8	67.0	589.6
70–79	11.7	74.5	871.6
80–84*	3.7	82.0	303.4
	$\Sigma f_i = 100$		$\Sigma f_i X_i = 5130.2$

* Values assumed.

Our original guess of 50 years was thus not too far out but did not offer the sort of accuracy which is usually required. An identical calculation for coloured heads of household gives a mean age of 40.175, on average more than ten years younger than for whites.

Calculation of the mean from grouped data should not present any problems other than those connected with assuming values for open-ended classes.

There are two other properties of the mean which are of some interest. To investigate these we define a new variable x_i such that

$$x_i = X_i - \bar{X} \tag{4.7}$$

In other words, x_i is found by subtracting the mean value of X from each observation on X. We must be careful to distinguish between X_i and x_i because they have quite different meanings and values. Each x_i represents the difference (or deviation) between $\bar{X}$ and the corresponding X_i. The two properties we wish to establish relate to the x_i and these are:

(a) The sum of the deviations of a variable X_i about its mean $\bar{X}$ equals zero, i.e.

$$\Sigma(X_i - \bar{X}) = \Sigma x_i = 0$$

where

$$\Sigma(X_i - \bar{X}) = (X_i - \bar{X}) + (X_2 - \bar{X}) + \ldots + (X_n - \bar{X})$$

using the summation of differences from (4.4).

(b) The sum of the squares of the deviations of a variable X_i about its mean $\bar{X}$ is smaller than the sum of squares of the deviations about any other value, i.e.

$$\Sigma(X_i - \bar{X})^2 < \Sigma(X_i - \tilde{X})^2$$

or

$$\Sigma x_i^2 < \Sigma \tilde{x}_i$$

where

$$\Sigma(X_i - \bar{X})^2 = (X_1 - \bar{X})^2 + (X_2 - \bar{X})^2 + \ldots + (X_n - \bar{X})^2$$

using the results in (4.5), and where $\tilde{X}$ is some value of X *other* than the mean $\bar{X}$, and $\tilde{x}$ represents the deviations of each observation, around this other value.

We can demonstrate these properties using the example of women's earnings, which we present in tabular form in Table 4.4. Thus the sum of the deviations of the observations about their mean is zero. In other words, the sum of deviations on one side of

Table 4.4 Properties of Σx_i and Σx_i^2

X_i	$x_i = (X_i - \bar{X})$	x_i^2
20	$20 - 46 = -26$	676
40	$40 - 46 = -6$	36
40	$40 - 46 = -6$	36
60	$60 - 46 = 14$	196
70	$70 - 46 = 24$	576
	$\Sigma x_i = 0$	$\Sigma x_i^2 = 1520$

the mean (negative values) is equal to the sum of deviations on the other side of the mean (positive values).

As far as the Σx_i^2 term is concerned, if we had chosen any other value for $\bar{X}$, say 20, or 50, or whatever, we would have obtained a value for Σx_i^2 of *more* than 1520. This property, referred to as *least squares*, is of great importance; we will encounter it again later.

We now want to move on to consider the second measure of location.

4.3 The median

The median of ungrouped data

For a simple array of n numbers the median is the middle number if n is odd and the mean of the two middle numbers if n is even. Formally, denoting the median by Md, for a set of n numbers $X_1, X_2, \ldots, X_n$ we have, if n is odd,

$$Md = X_{(n+1)/2} \quad (4.8)$$

and if n is even

$$Md = \frac{X_{n/2} + X_{1+n/2}}{2} \quad (4.9)$$

In general, therefore, if there are n observations, then the median is the value attached to that observation which leaves $n/2$ observations below it and $n/2$ observations above it. For instance, to return to the example of the weekly wage of five women workers considered earlier, which were

20 40 40 60 70

In this case $n = 5$, i.e. n is odd, and using (4.8) the median is the value of $X_{(5+1)/2} = X_3$, i.e. the third observation. Thus the median is 40.

Suppose now that there is a sixth woman, so that the array of earnings is now

20 40 40 50 60 70

In this case $n = 6$ is even, and using (4.9) the median is the value of

$$(X_{6/2} + X_{1+6/2})/2 = (X_3 + X_4)/2 = (40 + 50)/2 = 45$$

Using these formulae is perhaps to make things seem more complex than they actually are. We can see directly from the array of numbers exactly what the middle observation is in the case of the five values and what the mean of the two middle numbers is in the case of the six values.

The median of a frequency distribution

Let us consider once again the data contained in Table 4.1, relating to age of heads of household by colour. The procedure is a little more difficult than for the mean, but basically the method is to find the class containing the middle observation and determine the median value by interpolating within the observations in this class to locate the value of this middle observation. Again we have to assume that the frequencies are distributed evenly through each class.

To illustrate the method let us calculate the median age of white heads of households. For simplicity the assumption is usually made that there are an odd number of observations; we want to locate the position of the $(n + 1)/2$th, i.e. the imaginary 50.5th observation.

In the first three classes there is a total of 36.4, i.e. 4.2 + 6.9 + 25.3, observations, which leaves us 14.1, i.e. (50.5 – 36.4), observations 'short' at the beginning of the fourth class. At this point, the value is the boundary value between the third and fourth classes, i.e. 44.5. In the next class, (45–59), there are a further 29.2 observations, of which we want only the first 14.1. In other words, we want (14.1/29.2)ths of the next class, which is fifteen wide. Thus we want (14.1/29.2) x 15 of the (45–59) class to add to the boundary value of 44.5 to provide the median value. This means that the median age of white heads of household is equal to

$$44.5 + (14.1/29.2)15 = 51.743 \text{ years}$$

Compared with the average age as measured by the arithmetic mean the median value is a little higher. In this context it is interesting to note that one person's understanding of an average value might well differ from that of another. We shall return to discuss this point shortly. Notice also that, as is *not* the case with the mean, calculation of the median value does not usually require us to make assumptions about the limits of open-ended classes. We can formalise the procedure required to calculate the median by presenting the following equation (assuming an even distribution of frequency within the median class):

$$\text{Median} = L + \left(\frac{\frac{n+1}{2} - f_{\text{cum}}}{f} \right) w \tag{4.9}$$

where L is the lower boundary of the class containing the $(n + 1)/2$th observation, n is the number of observations, w is the width of this class, f is the frequency of this class and f_{cum} is the total frequency of all those classes up to but not including the median class.

To calculate the median age of coloured heads of households we first determine that the 50.5th observation lies somewhere in the third class, i.e. (30–44), from which we can ascertain that

$$L = 29.5 \quad w = 15 \quad f = 54.5 \quad f_{\text{cum}} = 18.5$$

Applying (4.8) gives

$$\text{Median} = 29.5 + \left(\frac{50.5 - 18.5}{54.5} \right) 15 = 38.307$$

This value is a little lower than the average value as calculated by the mean, but again we reserve comment on the reason for this difference (opposite in effect to the result for white heads of household) until a little later.

Quartiles

It is possible to think of the frequency distribution as being subdivisible into four equal parts, i.e. one-quarter of the observations in each part. The observations which achieve this are logically referred to as the first, second and third quartiles and may be denoted by Q_1, Q_2 and Q_3. Figure 4.2 illustrates the general idea

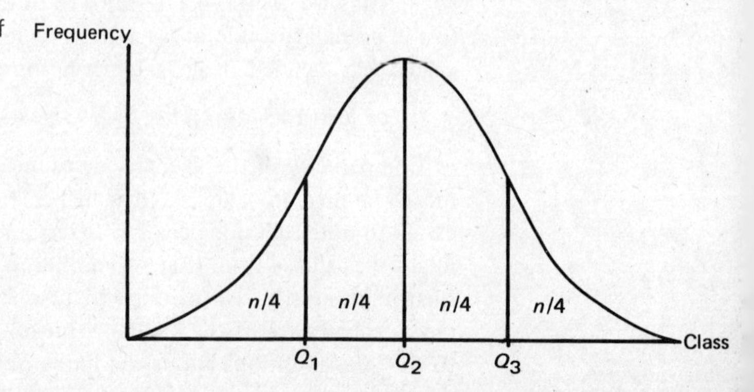

Figure 4.2 Division of frequency distribution into quartiles

in which each segment of the frequency distribution contains $n/4$ observations.

Notice that Q_2, the second quartile, has $n/2$ observations on each side of it and is thus nothing other than the median. Calculation of the quartile values Q_1 and Q_3 presents no new problems. To calculate the first quartile Q_1 we locate the class containing the $(n + 1)/4$th observation and apply a modified version of (4.9), i.e.

$$Q_1 = L + \left(\frac{(n+1)/4 - f_{\text{cum}}}{f}\right)w \tag{4.10}$$

Where L, f_{cum}, w and f refer to the first quartile class, and similarly for the third quartile:

$$Q_3 = L + \left(\frac{3(n+1)/4 - f_{\text{cum}}}{f}\right)w \tag{4.11}$$

Incidentally, the intersection of the 'less than' and 'more than' ogives in any particular example will yield the median (or second-quartile) value. There is no reason why we should not divide the number of observations in the frequency distribution into tenths or hundreds to produce deciles or percentiles; the calculation of such values would be via the application of (4.9), with the $(n + 1)/2$ term suitably modified.

Finally, we want to consider a third measure of location.

4.4 The mode

The final measure of average, location or central tendency (whatever we want to call it) we wish to consider is the *mode*.

Quite simply, it is defined as the most commonly occurring observation(s) in a set of observations. For example, with our five women workers with earnings of

20 40 40 60 70

the mode or modal value is 40, since this occurs twice whereas all the other values occur once only. With raw data such as these there is no difficulty in finding the exact value of the mode, though there may well be more than one value (a set of observations with two modes is called 'bi-modal') or there may be no value which occurs more than once (in which case we have either as many modes as there are observations or none at all, depending upon how we choose to interpret such a result).

More usually we shall be dealing with data in frequency distribution form, in which case calculations of value of the mode is somewhat more difficult. For our purposes we can make do with what is known as the *crude* mode, which is the value of the mid-point of the class with the highest frequency density (again there may sometimes be more than one).

In our heads of household example the highest frequency density for whites is 2.04 (for the 60–64 class) and for coloured is 3.63 (for the 30–44) class). The corresponding mid-points are 62 and 37 respectively.

* * *

The values of the average age of heads of household as computed by the three methods described above are summarised in Table 4.5. Two of the three measures produce reasonably similar values

Table 4.5 Average age of heads of household using three different measures

Measure of average	Age of head of household	
	White	Coloured
Mean	51.2	40.2
Median	51.7	38.3
Mode	62.0	37.0

for white and coloured heads of household, but all three confirm the age difference between the two. Further analysis would be required if this difference were to be explained.

The question we might now logically ask ourselves is which measure of average of those considered should be used?

4.5 Comparison of mean, median and mode

Consider once again the weekly wages of the five women, together with various measures of location:

20 40 40 60 70
mean = 46 median = 40 mode = 40

If we now change just one of the observations, we have

20 40 40 60 200
mean = 72 median = 40 mode = 40

We see that the mean is sensitive to extreme values and in this case gives a misleading impression of the central tendency, since four of the values are below 72 and only one exceeds it.

The median is unchanged and, particularly in econometric work, when we might wish to disregard extreme values as being unlikely to occur again, this may be an advantage. However, the mean uses *all* the information present in the data and preserves knowledge relating to the deviation of each observation from the mean. In this regard the mean is to be preferred, though some difficulties may be presented if we have open-ended classes or if there are some observations at the ends of the frequency distribution about which we have no knowledge; in this case it may be impossible to calculate the mean, and instead the median will have to be used. The mode is little used in economics but may find application when a peak or maximum value has to be allowed for, e.g. supplying peak demand for electricity, or of transport services, etc.

When the frequency distribution is symmetric, then all three measures will produce the same value. In this case the mean should be used if possible since it lends itself to further mathematical development and is a *good sampler*. By this we mean that if we view our set of data as a sample taken from a population (e.g. our five women workers might be a sample taken from a much larger work-force which is considered as the 'population'), then two samples will produce two mean values in general closer together than two median values.

When the distribution is not symmetric, then we face a problem. As we have seen, the mean is sensitive to extreme values and the extreme values in the long tail of an asymmetric distribution will drag the value of the mean in that direction. Thus the mean might give a misleading impression of central tendency, and when applied to a skewed distribution, such as that of income data, will produce a value for average annual income which is higher than that enjoyed by the majority of income-earners.

In these cases the median might more truly reflect the value around which most observations tend to be located. The fact that the mean is dragged out towards extreme values in an asymmetric distribution causes the three measures of location to take on a different order depending on the direction of the skew. We illustrate this feature in Figure 4.3. In Figure 4.3(a) the distribution is left-skewed and the value of the mean is less than the median, which in turn is less the modal value. In Figure 4.3(b) the distribution is symmetric and all three values are identical. In Figure 4.3(c) the distribution is right-skewed and the mean now has a higher value than the median.

Thus, if we know the value of the mean and median for a uni-modal distribution we will know also whether and in what direction the distribution is skewed. Examination of the relative size of the three measures in Table 4.5 above reveals that the age

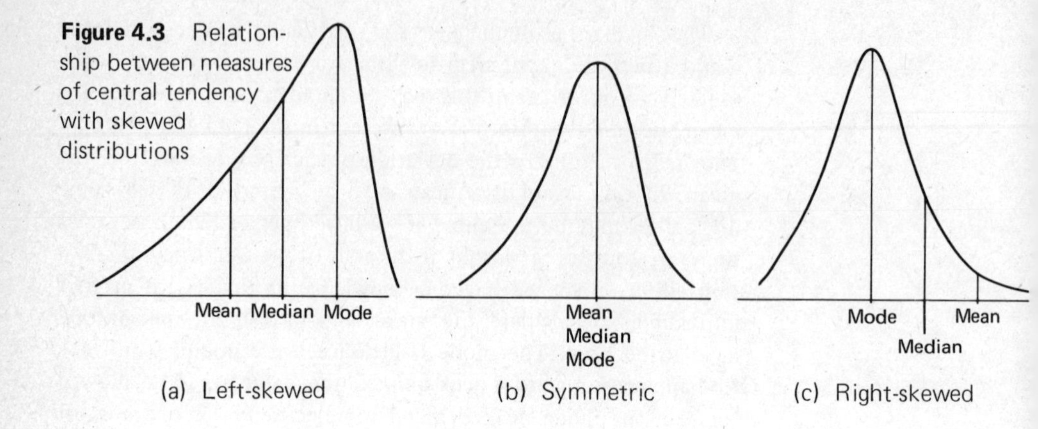

Figure 4.3 Relationship between measures of central tendency with skewed distributions

distribution of white heads of household is left-skewed while that of coloured heads is right-skewed.

4.6 The geometric mean

In our discussion of the mean in section 4.2 we discussed only the arithmetic mean. This is certainly by far the most commonly occurring mean but there is another one which is very useful in calculating average rates of growth. This is the *geometric mean*, which we can denote using the symbol X_G. It is defined as follows:

$$X_G = \sqrt[n]{X_1 X_2 \ldots X_n} = (X_1 X_2 \ldots X_n)^{1/n} \tag{4.12}$$

As an example of its use consider the numbers unemployed in the United Kingdom between 1975 and 1978 (in thousands):

1975	*1976*	*1977*	*1978*
866	1332	1450	1446

Suppose we wish to calculate the average percentage increase in unemployment over the four-year period. The percentage increases year to year are:

1975–6	53.81 per cent
1976–7	8.86 per cent
1977–8	–0.28 per cent

Because each year uses a different base, the arithmetic mean (which has a value of 20.80 per cent) is not a satisfactory measure.

Instead we can calculate X_G using (4.12), i.e.

$$X_G = \sqrt[3]{1.5381 \times 1.0886 \times 0.9972} = (1.6697)^{0.3333}$$

i.e.

$$X_G = 1.1863$$

In other words, the rate of growth in unemployment between 1975 and 1978 was 18.63 per cent. Note that this is lower than the arithmetic mean.

In fact, if we are seeking just the rate of growth, the computation can be made easier by using the following expression, which gives it directly and uses only the first and last values of the actual data:

$$r = \sqrt[n]{P_n/P_0} - 1 \tag{4.13}$$

Where r is the required rate of growth over n time periods and P_0 and P_n are the first and nth values of the original data. For example, in the above case,

$$P_0 = 866; \qquad P_n = 1446$$

and substituting in (4.13) gives

$$\begin{aligned} r &= \sqrt[3]{1446/866} - 1 \\ &= (1446/866)^{0.3333} - 1 \\ &= 1.1863 - 1 \\ &= 0.1863 \end{aligned}$$

i.e. the rate of growth is 18.63 per cent, as before.

4.7 Summary

In this chapter we have examined three common measures of average. The most useful and the most widely used is the arithmetic mean, and for a distribution which is symmetric the mean is to be preferred. For asymmetrical distributions the median will often prove more suitable, though (as we have seen) the median does not employ all the information contained in the data.

However, the measure of location of a variable is only one of the two important summary measures which we can use to describe the essential features of a set of observations. The second measure relates to the spread or dispersion of these observations around the mean. In the chapter which follows we examine several ways of measuring the dispersion of a variable.

EXERCISES

4.1 Calculate the values of the mean, median and mode for each of the following sets of measurements:

(a)	−3	5	0	−1	6	2	3	5
(b)	1	2	2	3	3	3	6	
(c)	4	3	4	4	6	3		
(d)	24	3	4	4	6	3		

Show that the value of Σx_i in each case is equal to zero.

4.2 Calculate the mean and median values for the following set of data:

2 3 5 7 3 4

Recalculate the mean and median values after

(a) adding 2 to each score
(b) subtracting 2 from each score
(c) multiplying each score by 2
(d) dividing each score by 2

Comment on your results.

4.3 Using the data in exercise 2.1, calculate the mean and median hourly earnings of male workers. Suppose the workers in all industries receive an increase in earnings of 10 per cent. What will the mean and median earnings be now?

4.4 Table 4.6 contains information on the number of two-parent families (in thousands) receiving Family Income Supplement in 1978, by number of children. Calculate the mean number of children in such families.

Table 4.6

Number of children	Number of families
1	9
2	14
3	13
4	8
5	4
6 or more	3

Source: *Annual Abstract of Statistics*, 1980, table 3.25.

4.5 Use the data in the frequency table of exercise 3.2 (Table 3.6) to calculate the mean, median and modal ages of males in the United Kingdom in 1901 and 1971. Comment on your results. What do these values reveal about the symmetry of the age distributions?

4.6 The data in Table 4.7 refer to the size of manufacturing establishments by number of employees in Great Britain in 1961. Calculate the three quartile values and comment on your results.

Table 4.7

Size (number of employees)	Number of establishments
11–24	12571
25–49	14704
50–99	12774
100–249	8714
250–499	3499
500–999	1693
1000–1999	777
2000–4999	351
5000 or more	78

Source: *British Labour Statistics, Historical Abstracts 1886–1968*, table 207.

4.7 The number of school-leavers (in thousands) in the United Kingdom with at least one 'A' level (or its equivalent) for selected years between 1967 and 1977 is given below:

1967	*1970*	*1971*	*1972*	*1974*	*1975*	*1976*	*1977*
105.7	120.8	118.2	131.5	134.3	134.3	140.1	146.7

Calculate and compare the rates of growth for the periods (a) 1967–72, (b) 1972–7, (c) 1967–77.

5 Measures of Dispersion

OBJECTIVES

After reading this chapter and working through the examples students should understand the meaning of the following terms:

range
inter-quartile range
mean deviation
standard deviation
variance
normal distribution
standard normal distribution

Students should be able to:

calculate the range of a set of data
calculate the inter-quartile range
calculate the mean deviation
calculate the standard deviation
calculate the variance
standardise the score on a normal variable
use the z-table to determine areas under the normal curve

5.1 Introduction

At the beginning of Chapter 4 we stated that there are two important summary measures we can use to describe the principal characters of a set of observations on some variable. The first of these summary measures was the measure of location, discussed in Chapter 4 (the most important of which was the arithmetic mean). In this chapter we consider the second type of summary measure which relates to the way the data are spread out or *dispersed* around the mean. In our initial discussion of the frequency distribution of the age of heads of household shown in Table 4.1 we noted that the age distribution of coloured heads of household appeared to be much less spread out than that for white heads of household; in the former case three-quarters of all the observations were concentrated in only two age groups compared with just over a half of the observations in the same two groups for white heads of household. But this was only a general subjective impression, and in this chapter we develop more rigorous measures of dispersion. As we shall see, there are several measures of dispersion to be considered but only one of these has a more than limited role, and this measure, the *standard deviation* (or its square, the *variance*) has important ramifications which extend through the remainder of this book.

5.2 The range

We first encountered the *range* at the beginning of Chapter 2. Recall that for raw data it is simply the difference between the largest and smallest values. For data arranged in the form of a frequency distribution, the range is the difference between the lower boundary of the first class and the upper boundary of the last class. The range is not often used in the economic or social sciences (it is more popular, for example, in quality control). However, it does provide a quick method of gauging the spread of data, though it pays no attention to intermediate values and hence wastes a lot of information. As an example of the range consider the following data which relate to expenditure on take-overs and mergers of firms in the food industry between 1966 and 1969 (£ million):

11 29 40 107

The range is thus equal to 107 – 11 = £96 million. The range of the frequency distribution of the age of white heads of household shown in Table 4.1 is equal to 84.5 – 19.5 = 65 years. For coloured heads of household it is necessarily the same.

5.3 The inter-quartile range

One measure of dispersion which avoids this dependence on two extreme values is the inter-quartile range. This is defined as the difference between the first and third quartile values, i.e.

$$\text{Inter-quartile range} = Q_3 - Q_1 \tag{5.1}$$

Although we avoid sensitivity due to extreme values, the inter-quartile range (like the range) does not use all the information contained in the set of data. Furthermore, as is the case with the median, it does not lend itself to further mathematical development and so is of limited use. However, it does have some popularity because, compared with the standard deviation (which we consider shortly), computation of its value, particularly from raw data, is relatively easy. For this reason it is preferred by some statisticians, for example Tukey (1977) and other 'exploratory' authors. Let us calculate the inter-quartile range of the age for both white and coloured heads of household, the frequency distribution of which is given in Table 4.1. We use equations (4.9) and (4.10) to establish the following results.

For white heads of household

$Q_1 = 37.9$ years; $Q_3 = 64.5$ years

Inter-quartile range $= Q_3 - Q_1 = 27.6$ years

For coloured heads of household

$Q_1 = 31.4$ years; $Q_3 = 46.5$ years

Inter-quartile range $= Q_3 - Q_1 = 15.1$ years

As we anticipated at the beginning of this section, the spread of ages as measured by the inter-quartile range is considerably less for coloured than for white heads of household. Computation of the quartile values does not make use of all of the information in the frequency distribution, and for this reason the value of inter-quartile range is not particularly sensitive to changes in the data and therefore offers a robust measure of dispersion.

5.4 The mean deviation

The rationale behind the mean deviation as a measure of dispersion is that if the observations are clustered fairly tightly around the mean, the difference between the value of each observation and the value of the mean, i.e. the quantities $(X_i - \bar{X})$ or x_i, will in general be small and the sum of all these smallish values, i.e. $\Sigma(X_i - \bar{X})$ or Σx_i, will also tend to be small. By comparison, if the observations are widely spread around the mean, then the difference between each observation and the mean, i.e. the quantities $(X_i - \bar{X})$, will now tend to be large and the sum of all such largish values will be correspondingly large. So by measuring the size of $\Sigma(X_i - \bar{X})$, i.e. $(X_1 - \bar{X}) + (X_2 - \bar{X}) + \ldots + (X_n - \bar{X})$, we should be able to use the result to indicate something about the dispersion of the data in question. Unfortunately there is a snag.

To see what this difficulty is consider Table 5.1, which relates to the overtime worked by the ten workers in two ficticious factories *A* and *B*. It is certainly true that the size of the individual x_is, i.e. the $(X_i - \bar{X})$s is larger for factory *B* than for factory *A*, reflecting the greater dispersal around the common mean of 13, but we cannot use the sum of these values, Σx_i, since, as we have demonstrated above in connection with the mean, the sum of these values, i.e. the term Σx_i, always equals zero!

Table 5.1 Calculation of mean deviation for hours of overtime

Factory *A*		Factory *B*	
x_i	$x_i = (X_i - \bar{X})$	X_i	$x_i = (X_i - \bar{X})$
13	0	2	−11
15	−2	5	−8
16	−3	24	11
14	−1	10	−3
12	1	21	8
11	2	9	−4
13	0	18	5
12	1	14	1
11	2	17	4
13	0	10	−3
$\bar{X} = 13$	$\Sigma x_i = 0$	$\bar{X} = 13$	$\Sigma x_i = 0$

However, we can overcome this by ignoring the positive and negative signs attached to the deviation terms and adding up the *absolute* values, i.e. computing the value of the term $\Sigma \mid x_i \mid = \Sigma \mid X_i - \bar{X} \mid$. (The two vertical lines mean 'take the absolute value', i.e. ignore the sign attached to the values and consider only their magnitude.) Doing this we have

Factory *A* $\quad \Sigma \mid x_i \mid = 12$

Factory *B* $\quad \Sigma \mid x_i \mid = 58$

The wider dispersion of the scores for factory *B* is reflected in the fact that $\Sigma \mid x_i \mid$ for factory *B* is approximately five times larger.

We have one final step to take, and that is to adjust the measure for the number of observations, since the size of $\Sigma \mid x_i \mid$ (as described above) is likely to be influenced by the size of n, and this is clearly undesirable.

This leads us to define a measure of dispersion known as the *mean deviation* as follows:

$$\text{Mean deviation} = \frac{1}{n} \Sigma \mid X_i - \bar{X} \mid = \frac{1}{n} \Sigma \mid x_i \mid \tag{5.2}$$

The value of the mean deviation tells us how far away from the mean of a set of data an observation is typically likely to be.

For the two-factory example above, the mean deviations for the number of hours overtime are

$$\text{Factory } A \quad \text{mean deviation} = \frac{1}{10} \times 12 = 1.2 \text{ hours}$$

$$\text{Factory } B \quad \text{mean deviation} = \frac{1}{10} \times 58 = 5.8 \text{ hours}$$

Thus on average each observation for factory A will typically be only 1.2 hours different from the mean of 13 hours, but for factory B the typical distance from the mean is 5.8 hours. In other words, the spread of data is nearly five times greater in the case of factory B than in that of factory A.

We do not intend to make further use of the mean deviation in this book. It does have application in the more advanced reaches of statistics but we cannot consider those. We have considered it here only because it introduces the notion of the deviation of an observation from a mean, which in itself is an important and useful idea to get hold of and leads logically to a much more important measure of dispersion which we consider next.

5.5 Standard deviation and variance

Mean deviation avoids the problem of $\Sigma(X_i - \bar{X}) = \Sigma x_i = 0$ by ignoring the signs and summing absolute values of the deviations. A different way of overcoming the zero sum is to square each deviation term instead, since squaring a negative value provides a positive result. We still divide by n to ensure that the value is not dependent on the number of observations. Since it is difficult to ascribe a meaning to something which is expressed in squared units (for example, what meaning can be ascribed to something measured in hours squared?), we take the square root of the final answer to produce a result which is in the same units as the original data.

The measure of dispersion we thus obtain is known as the *standard deviation*. The standard deviation of a population of raw scores (i.e. not arranged in a frequency distribution) is denoted by the Greek symbol σ, i.e.

$$\sigma = \sqrt{\frac{1}{n}\Sigma(X_i - \bar{X})^2} = \sqrt{\frac{1}{n}\Sigma x_i^2} \tag{5.3}$$

(In theory taking the square root provides us with two answers, but we can ignore the negative result as being unrealistic.)

The standard deviation (unlike the mean deviation) lends itself to further mathematic development and thus has a role to play in what is to come. It is by far the most important measure of dispersion (although we shall increasingly tend to use its square, i.e. *variance*). Let us calculate the standard deviation for factory *A* in the example above. The calculations are set out in Table 5.2.

Table 5.2 Calculation of standard deviation for hours of overtime worked in factory *A*

X_i	$x_i = (X_i - \bar{X})$	x_i^2
13	0	0
15	−2	4
16	−3	9
14	−1	1
12	1	1
11	2	4
13	0	0
12	1	1
11	2	4
13	0	0
$\bar{X} = 13$	$\Sigma x_i = 0$	$\Sigma x_i^2 = 24$

From the table we have $\Sigma x_i^2 = 24$. Therefore, using (5.3)

$$\sigma = \sqrt{\frac{1}{10} \times 24} = 1.55 \text{ hours}$$

A similar calculation for factory *B* yields $\sigma = 6.68$ hours, which confirms that the dispersion or spread of the observations around the mean is much greater for factory *B* than for factory *A*.

Interpretation of the meaning of this result is a little more difficult than for mean deviation and we shall deal with this shortly.

In many cases computation of the standard deviation will be performed using a programmed calculator or computer but in the event that we have to solve it 'manually', equation (5.3) is more conveniently replaced by equation (5.4), which can be derived mathematically from (5.3), i.e.

$$\sigma = \sqrt{\frac{\Sigma X_i^2}{n} - \bar{X}^2} \tag{5.4}$$

Use of (5.4) removes the necessity to calculate x_i, and this reduces the amount of labour involved in calculation.

The square of the standard deviation, i.e. σ^2, is known as the *variance*. It too can thus be used as a measure of dispersion, but since it produces a measurement which is expressed in terms of the square of the original units of measurement it is correspondingly more difficult to interpret. For this reason we will be inclined to use standard deviation rather than variance when we want to discuss the spread of a set of data around the mean. However, as we shall see, variance is more often used in more advanced statistics and the term will occur with increasing frequency later in this book.

5.6 Standard deviation from a grouped frequency distribution

The standard deviation of a set of observations arranged in a frequency distribution is given by the following expression:

$$\sigma = \sqrt{\frac{\Sigma (X_i - \bar{X})^2 f_i}{n}} \tag{5.5}$$

where X_i = the mid-point of the ith class, f_i = the frequency of the ith class, $\bar{X}$ = the mean of the grouped data (from equation 4.6) and n = the number of observations ($= \Sigma f_i$).

If we have to calculate the standard deviation of grouped data manually, the use of (5.6) will reduce the labour involved, i.e.

$$\sigma = \sqrt{\frac{\Sigma X_i^2 f_i}{\Sigma f_i} - \left(\frac{\Sigma X_i f_i}{\Sigma f_i}\right)^2} \tag{5.6}$$

The variance σ^2 of grouped data is simply the square of the result obtained from (5.6).

We can illustrate the use of these equations by calculating the standard deviation of the age distribution of white heads of household shown in Table 4.1, the first two columns of which are reproduced in Table 5.3. The last three columns are required to complete the calculation of σ.

If we examine the two frequency distributions in Table 4.1, it is not difficult to see that the distribution of observations is wider in the case of white heads of household than it is with coloured heads. In the latter case two groups alone contain 67.5 per cent

Table 5.3 Calculation of standard deviation for white heads of household

Class (age)	Frequency (per cent of all households) f_i	Mid-point X_i	X_i^2	$X_i^2 f_i$
*20–24	4.2	22.0	484	2032.8
25–29	6.9	27.0	729	5030.1
30–44	25.3	37.0	1369	34635.7
45–59	29.2	52.0	2704	78956.8
60–64	10.2	62.0	3844	39208.2
65–69	8.8	67.0	4489	39503.2
70–79	11.7	74.5	5550	64937.9
80–84*	3.7	82.0	6724	24878.8
	$\Sigma f_i = 100$			$\Sigma X_i^2 f_i = 289183.5$

* Assumed limits.

of all household heads, whereas it requires three groups to provide a similar coverage for whites. Of course, this is only a subjective *qualitative* assessment, and the purpose of a standard deviation is to provide a qualitative measure so that more rigorous comparisons can be made.

From Tables 4.3 and 5.3 we have

$$\Sigma f_i = 100$$

$$\Sigma f_i X_i = 5130.2$$

$$\Sigma X_i^2 f_i = 289\,183.5$$

Therefore substitution into (5.6) gives

$$\sigma = \sqrt{\frac{289\,183.5}{100} - \left(\frac{5130.2}{100}\right)^2}$$

$$\sigma = 16.12 \text{ years}$$

A similar calculation applied to the age distribution of coloured heads of household yields a value for the standard deviation of $\sigma = 11.76$ years. So the spread of the observations about the mean is in fact about 40 per cent greater for white than for coloured heads of household.

* * *

Table 5.4 summarises the values of the various measures of location and dispersion of the distribution for the age of heads of household data of Table 4.1.

Table 5.4 Measures of location and dispersion for age of heads of household

	White	Coloured
Mean	51.30	40.20
Median	51.74	38.30
Mode*	62.00	37.00
Range†	65.00	65.00
Inter-quartile range	27.6	15.1
Standard deviation	16.12	11.76

* Taken as the mid-point of class with highest frequency density.
† See Table 4.3 for assumed upper and lower boundaries.

5.7 Interpreting standard deviation

We have calculated that the age distribution of white heads of household has a mean of 51.30 years and that the spread of the observations is measured by the value of 16.12 years obtained for the standard deviation. But what exactly does this value of 16.12 years indicate? How are we to interpret it? So far we have only been able to make the comparative statement that the distribution of ages for white heads of household is greater than that for coloured heads (because 16.12 is greater than 11.76). As it happens, we can say something about the value of 16.12 itself because of the property of a normal frequency distribution. This is that if a frequency curve is normal in shape, then about two-thirds (68.26 per cent) of the area under the curve and hence two-thirds of the observations lie no further than one standard deviation either side of the mean value. In other words, applying this to our example enables us to state that, assuming a *normal* distribution, 68.26 per cent of all white heads of household are aged between the mean minus one standard deviation and the mean plus one standard deviation, i.e. between (51.30 – 16.12) and (51.30 + 16.12) or between 35.18 years and 67.42 years. This idea is illustrated in Figure 5.1.

We must emphasise that these values will *only* be accurate if the frequency distribution has this special normal shape. The further away from normal the curve is, the less will these area properties hold. An inspection of the values in Table 4.1 suggests that the age distribution of white heads of household is at least reasonably symmetric (this is confirmed by the closeness of the three measures of location in Table 5.4) and that the age distribution of coloured heads of household is not as symmetric (again supported by the values in Table 5.4). It is in fact right-skewed, as the frequency density histograms of Figure 5.2 show.

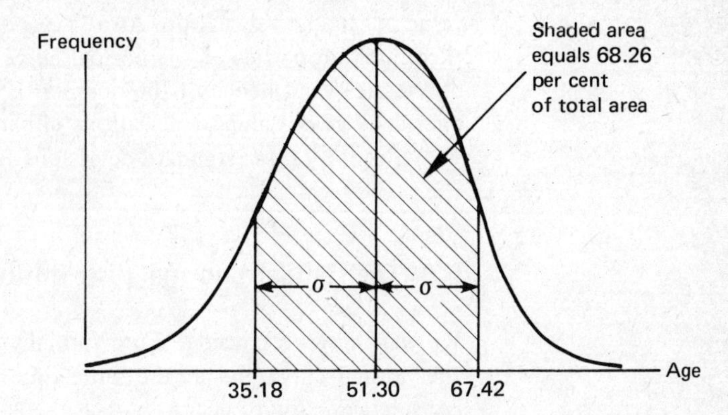

Figure 5.1 Properties of the area under a normal curve

The standard deviation for coloured heads of household is 11.76, which means that *about* two-thirds of all coloured heads of household were aged between 28.35 years and 51.87 years. Thus a span of 23.52 years covers approximately two-thirds by age of all coloured heads of household. The corresponding figure for whites is 32.24 years. These values confirm our first impressions of the dispersion of ages around the respective means. For comparison purposes we have sketched the approximate frequency curves on to the histograms (Figure 5.2).

We can see that although neither curve is normal, that for whites is 'more normal' than that for coloureds and the area properties would therefore be less true for the latter. None the less the ideas outlined do enable us to place some meaning on the

Figure 5.2 Frequency density histograms for the age distribution of heads of household

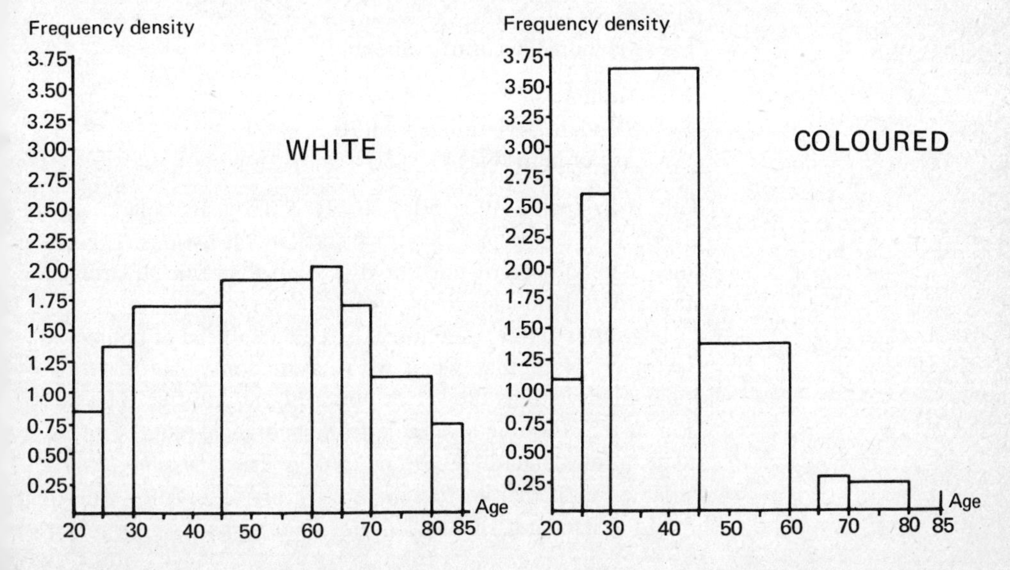

value of standard deviation. As a final point we should note that these area properties of the normal curve mean that 95 per cent of the area (and hence of the number of observations) is encompassed by *two* standard deviations either side of the mean and 99 per cent by *three* standard deviations either side.

5.8 The standard normal distribution

We want now to consider more formally these area properties of the normal curve and at the same time introduce a concept of considerable importance.

Suppose we have one white and one coloured head of household who are aged 60 and 50 respectively and we ask the question: '*Relative* to the mean age of their *own* group, who is the older of the two?' We can answer this question by expressing their age not in years but in terms of the number of standard deviations each age is from their respective means. The calculation proceeds as follows. We let X_i denote the age of the ith individual. Then, for white heads of household,

Mean age $\bar{X}$ = 51.30
Standard deviation σ = 16.12
Age of individual X_i = 60

Therefore, this individual is $60 - 51.30 = 8.7$ years older than the mean. In terms of standard deviations this is $8.7/16.12 = 0.54$ standard deviations. In other words, this white head of household is a little over one-half of a standard deviation older than the mean age of all whites.

For coloured heads of household:

Mean $\bar{X}$ = 40.11
Standard deviation σ = 11.76
Age of individual X_i = 50

This individual is thus $50 - 40.11 = 9.89$ years older than the mean. This is $9.89/11.76 = 0.84$ standard deviations. Therefore, this coloured head of household is more than three-quarters of a standard deviation above the mean.

In *relative* terms, therefore, the coloured head of household is the older. Note that when we measure something in terms of standard deviations the values obtained are unit-free. The operation of transforming a variable from its original units of measurement into standard deviation units is called *standardising* the variable, and if we were to standardise every observation we would obtain a standardised distribution. As it turns out, the principal reason for standardisation is to make use of the area properties of a normal curve. Thus we shall be mainly concerned with standardising from normal (or approximately normal) distributions, and this produces the *standard normal distribution*, also known as the z-distribution or (for reasons which will shortly become apparent) the unit-normal distribution.

The operation described above of subtracting from a raw score its mean and dividing by its standard deviation to produce the z-distribution is summarised by equation (5.7):

$$z = (X_i - \bar{X})/\sigma \tag{5.7}$$

The standard normal distribution given by (5.7) is normal in shape and has a mean of 0 and a standard deviation of 1.0. We can show this by standardising the distribution values shown in Figure 5.1 (which relates to white heads of household).

To standardise $X_i = 35.18$ we have, using (5.7),

$$z = \frac{35.18 - 51.30}{16.12} = \frac{-16.12}{16.12} = -1.0$$

To standardise $X_i = \bar{X} = 51.30$ we have

$$z = \frac{51.30 - 51.30}{16.12} = 0$$

and finally for $X_i = 67.42$ we have

$$z = \frac{67.42 - 51.30}{16.12} = \frac{16.12}{16.12} = 1.0$$

The standardised version of Figure 5.1 is shown in Figure 5.3. The fact that the z-distribution is unit-free means that we can compare any variables regardless of their units of measurement by first converting them into their equivalent z-scores.

For example, suppose that the mean level of unemployment rate in the countries of the EEC is 4.0 per cent, with a standard

Figure 5.3 Properties of the standard normal distribution

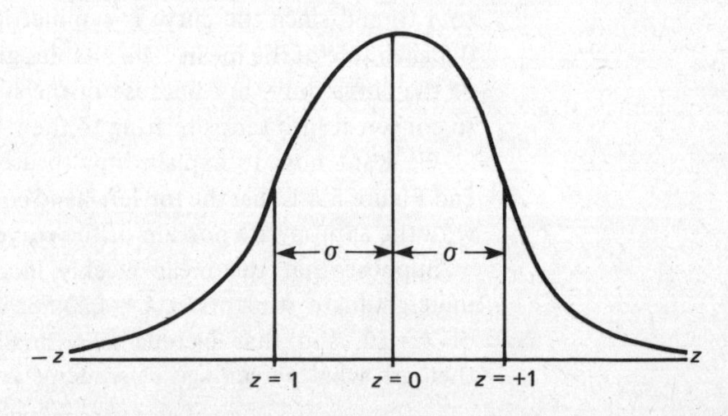

deviation of 0.5 per cent, and that the mean level of annual government spending on unemployment benefit *per capita* is £20, with a standard deviation of £4. If some individual country with 3.5 per cent unemployed spends £18 *per capita*, on which measures does it perform relatively 'better'?

For unemployment we have

$$z_U = \frac{3.5 - 4.0}{0.5} = -1.0$$

For benefit we have

$$z_B = \frac{18 - 20}{4} = -0.5$$

The results show that relatively speaking the country in question has a lower relative level of unemployment (the negative sign attached to $z = -1.0$ indicates a level of unemployment *below* the mean) than it has of spending on unemployment benefit. We have been able to compare two variables with completely different units of measurement.

Using the z-tables

The z-distribution is a very versatile technique which we shall make much use of. The reason for this is that there exists a table showing the area under a standard normal curve for any given value of z. Such a table is given in Table A.1 in the appendix to this book (and will be found at the back of most statistical or econometric texts). The table is usually presented in one of two forms. It will either give the area under the curve *between* $z = 0$ and some chosen value of z, or it will give the area under the curve *beyond* some chosen value of z. The table in the appendix uses the former method of presentation, which is the more common, but care should be used when using an unfamiliar z-table for the first time. The total area under the standard normal curve is equal to 1.0, and since the curve is symmetrical this means an area of 0.5 each side of the mean. The z-tables give the values for *one*-half of the curve only, but because of the symmetry it is easy enough to convert it into terms relating to the whole curve.

We want now to explain how to use the z-tables, and to this end Figure 5.4 is just the top left-hand corner of a z-table, together with the appropriate portion of the curve.

Suppose that the mean weekly income (£) of all university undergraduate students is $\overline{X} = £20.58$, with a standard deviation of $\sigma = £2$, and that income is normally distributed. We might then ask what percentage of students will have incomes between

Figure 5.4 Showing use of z-tables

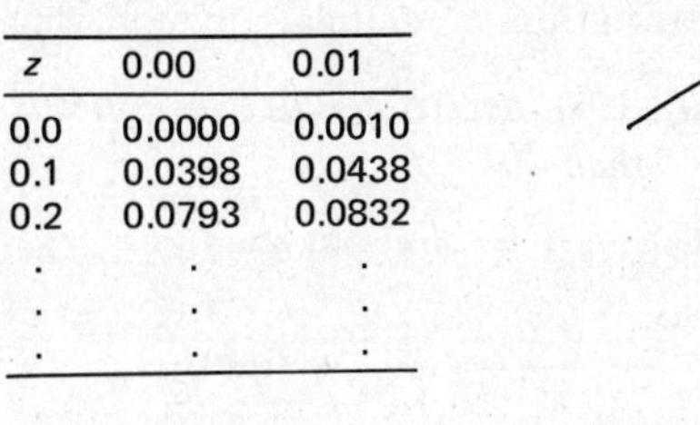

z	0.00	0.01
0.0	0.0000	0.0010
0.1	0.0398	0.0438
0.2	0.0793	0.0832
.	.	.
.	.	.
.	.	.

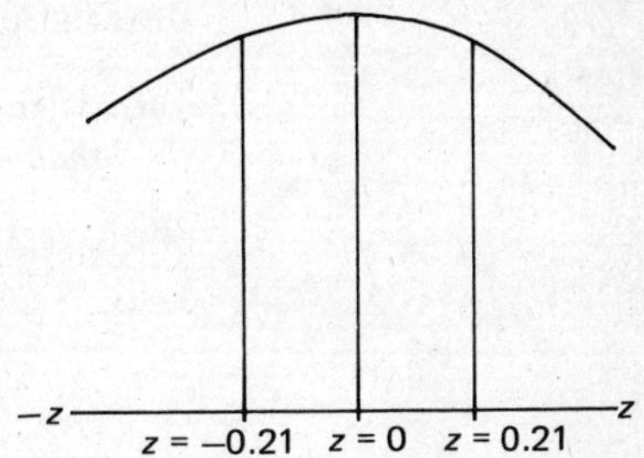

£20.58 and £21. Thus we have

$$\bar{X} = 20.58; \quad \sigma = 2; \quad X_i = 21$$

$$z = \frac{21 - 20.58}{2} = 0.21$$

By looking first of all down the left-hand column of the z-tables to the 0.2 row and then along this row to the 0.01 column (to give a value of $0.2 + 0.01 = 0.21$) we find the value 0.0832. This means that 0.0832 of the total area under the standard normal curve of 1 lies between $z = 0$ and $z = 0.21$ (which corresponds to $X_i = 21$). Therefore, we can say that 8.32 per cent of students will have incomes between £20.58 and £21. Alternatively, we might ask what percentage of students have incomes greater than £22? In which case

$$z = \frac{22 - 20.58}{2} = 0.71$$

From the z-tables we see that the area between $z = 0$ and $z = 0.71$ is 0.2611, i.e. this means that the area to the right of $z = 0.71$ is $0.5 - 0.2611 = 0.2389$. In other words 23.89 per cent of students have incomes greater than £22. The situation is illustrated in Figure 5.5.

To consolidate our understanding of the use of the z-tables let us answer the following three questions relating to the age distribution of heads of household (where we assume that these

Figure 5.5 Areas under the z-curve

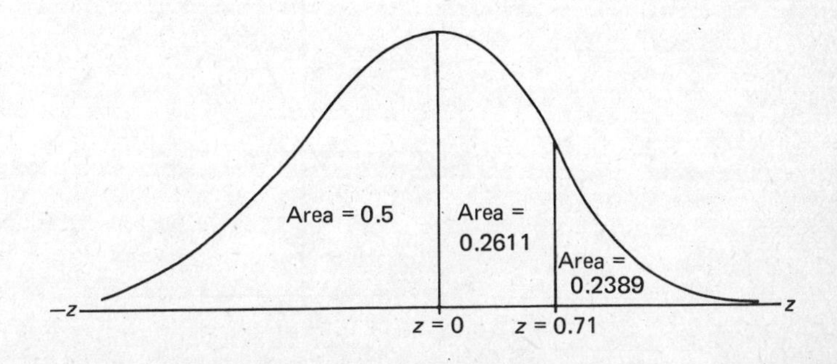

distributions are normal (or approximately so):

(a) *What percentage of white heads of household are younger than 40?*

We have $\bar{X} = 51.3$, and $\sigma = 16.12$. Therefore,

$$z = \frac{40 - 51.3}{16.12} = -0.701$$

The minus sign means that we are working in the left-hand half of the curve. From the z-tables the area between $z = 0$ and $z = -0.701$ is 0.2583 (by interpolation between $z = 0.7$ and $z = 0.71$). This means that the area to the *left* of $z = 0.701$ is $0.5 - 0.2583 = 0.2417$. Therefore, 24.17 per cent of white heads of household are aged less than 40. The original distribution measured in years and the corresponding standard normal distribution is shown in Figure 5.6.

(b) *What percentage of coloured heads of household are aged between 30 and 50?*

This calculation takes two parts: the calculation of the percentage between 30 years and the mean of 40.18 years, and calculation of the percentage between the mean 40.18 years and 50 years.

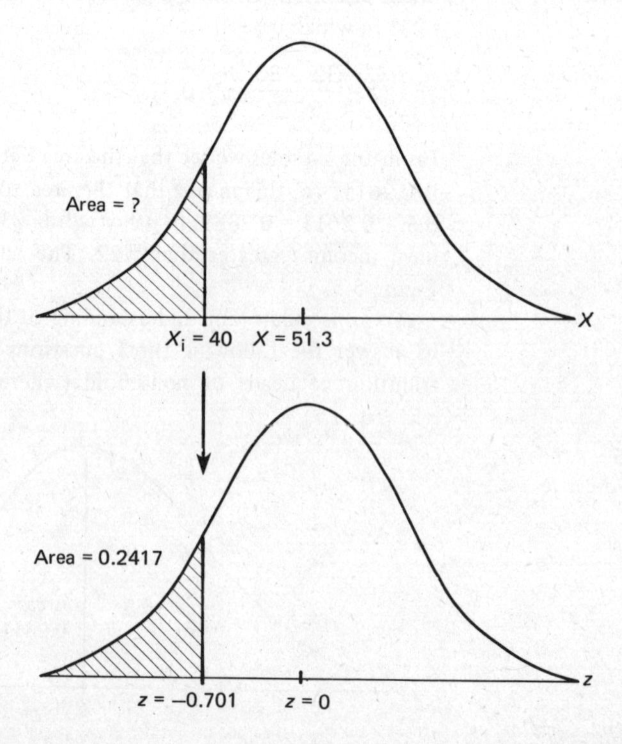

Figure 5.6 Areas under normal curves for white heads of household aged under 40

We have $\bar{X} = 40.11$, and $\sigma = 11.76$. Therefore,

$$z_{30} = \frac{30 - 40.11}{11.76} = -0.86$$

$$z_{50} = \frac{50 - 40.11}{11.76} = 0.84$$

The z-tables reveal that 0.3051 of the area lies between $z = 0$ and $z = -0.86$ and that 0.2995 lies between $z = 0$ and $z = 0.84$. The total area between $z = -0.86$ and $z = 0.84$ is thus 0.3051 + 0.2995 = 0.6046, i.e. 60.46 per cent of coloured heads of household are aged between 30 and 50. This is illustrated in Figure 5.7.

(c) *What is the age which 20 per cent of white heads of household are younger than?*

This problem presents itself in reverse. We start by knowing the area under the z-curve, i.e. 0.20 in the left-hand tail. Thus we look in the main *body* of the z-tables to find the corresponding value of z. But we have to be careful because our z-tables give us the area between $z = 0$ and some value of z, and do not give the area in the tail beyond z. If the area in the tail is 0.2, then the area between $z = 0$ and the appropriate value of z is $0.5 - 0.2 = 0.3$. Thus we look up 0.3 in the body of the z-table and find the value of z to be $z = -0.84$.

We now rearrange (5.7) to the form

$$X_i = z\sigma + \bar{X} \tag{5.8}$$

Thus $X_i = (-0.84 \times 16.12) + 51.30 = 37.76$ years.

Thus our answer is that 20 per cent of white heads of household are younger than 37.76 years.

All of the above presupposes that our assumption of normality in the distribution in question is reasonable (or at least not

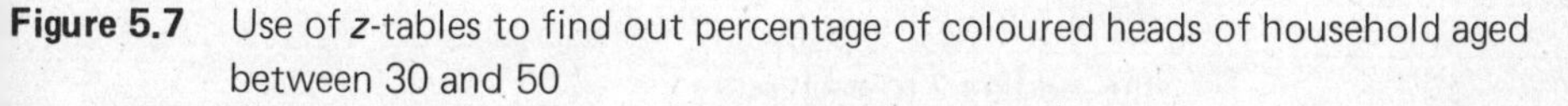

Figure 5.7 Use of z-tables to find out percentage of coloured heads of household aged between 30 and 50

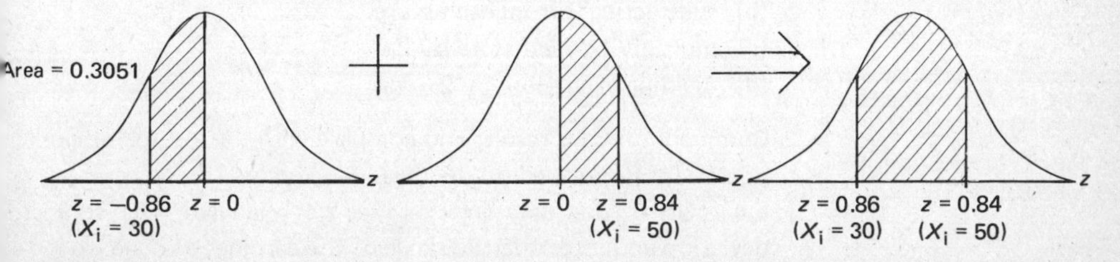

unreasonable). If we know that it is not (or suspect so), we must view with caution the results of any calculations involving the z-distribution.

5.9 Summary

There are two principal summary descriptive measures of data. We discussed the first, the measure of location, in Chapter 4. In this chapter we have discussed the second, i.e. measures of dispersion of the data. Although we have examined four such measures, only two of these, the inter-quartile range and the standard deviation, are of more than passing interest. The standard deviation and its square, the variance, are measures of great importance and we shall be much concerned with them in the material which is to follow.

Also of great importance for future work is the concept of the standard normal distribution and how it may be used to describe the area properties of normally distributed variables. In the next chapter we shall discuss some elements of probability theory; we shall see how the standard normal distribution can be extended to deal with the ideas of probability.

EXERCISES

5.1 Use the summation rules of (4.4) to prove that

$$\sigma = \sqrt{\frac{1}{n}\Sigma x_i^2} = \sqrt{\frac{\Sigma X_i^2}{n} - \bar{X}^2}$$

and that

$$\sigma = \sqrt{\Sigma x_i^2 f_i / \Sigma f_i} = \sqrt{\Sigma X_i^2 f_i/\Sigma f_i - (\Sigma X_i f_i/\Sigma f_i)^2}$$

5.2 Calculate the mean, range and standard deviation of the following sets of data:

(a) 2 3 5 7 3 4
(b) 12 3 5 7 3 4

5.3 For the data in (a) in exercise 5.2, recalculate the standard deviation after

(a) adding 2 to each score
(b) subtracting 2 from each score
(c) multiplying each score by 2
(d) dividing each score by 2

Comment on your results and compare with the results for similar operations performed on the mean in exercise 4.2.

5.4 Using the data in exercise 2.1, calculate the standard deviation and inter-quartile range of the earnings of male workers.

Using knowledge of the value of the mean earnings of this group calculated in exercise 4.3, interpret the result for the standard deviation.

5.5 Calculate the standard deviation and inter-quartile range for the data in exercise 3.2, and interpret your results in the light of knowledge of the mean and median values for the same data calculated in answer to exercise 4.5.

5.6 The data in Table 5.5 refer to the weekly income (£) of households with male and with female heads in the United Kingdom in 1978 by percentage of the total number of households. Calculate the mean and standard deviation weekly household income for the two groups and comment on your results.

Table 5.5

Household income (£ per week)	Per cent of all households	
	Male heads	Female heads
Under 20	0.5	4.8
20– 29	2.8	29.5
30– 39	5.5	13.7
40– 49	5.4	10.7
50– 59	4.6	7.5
60– 69	5.6	5.8
70– 79	5.6	4.8
80– 89	6.4	4.5
90– 99	6.4	3.7
100–109	7.3	2.8
110–119	6.9	2.6
120–129	5.8	2.2
130–149	10.5	2.2
150–169	8.1	2.1
170–199	7.9	1.9
200 or more	10.7	1.1

Source: *Family Expenditure Survey*, 1978, table 3.

5.7 Assume that the mean age of a group of 100 young offenders is 16.5 years, with a standard deviation of 2.3 years. Use the z-table to answer the following questions:

(a) How many are aged 17 and over?
(b) How many are aged 15 and under?
(c) How many are aged between 14 and 18?
(d) What ages would divide the group into
  (i) the youngest 20 per cent
  (ii) the middle 50 per cent
  (iii) the oldest 10 per cent?

PART II

PROBABILITY AND PROBABILITY DISTRIBUTIONS

6 Probability

OBJECTIVES

After reading this chapter and working through the examples students should understand the meaning of the following terms:

statistical inference
population
parameter
sample
simple events
sample space
compound events
intersection of events
union of events
mutually exclusive events
expected relative frequency
probability of an event
the addition rule
the multiplication rule
sampling without replacement
independent events
discrete random variables
probability distribution
probability histogram
continuous random variables
probability density
expected value
joint probability distribution
covariance

Students should be able to:

calculate the probability of a simple event
use the addition and multiplication rules to calculate the probability of a compound event
construct the probability distribution for a random variable
draw a probability histogram
calculate the mean and variance of a random variable
calculate a joint probability distribution for two random variables

---

6.1 Introduction

The first part of this book has been concerned with the efficient organisation and presentation of data and with the determination of summary measures (i.e. measures of location and dispersion) to describe the data concisely. This is the *first* of the two major purposes of statistics. The *second* purpose, with which the remainder of this book is concerned, relates to the process known as *statistical inference*.

Broadly speaking, statistical inference is the process of guessing or estimating the value of certain features or *parameters* of a population on the basis of the evidence gathered from a sample. Suppose we have a box of 100 apples and wish to know whether they are sweet or sour. We could choose one apple, eat it, and on the basis of this evidence decide whether or not the remainder of the apples were sweet or sour.

In this simple example the box of 100 apples represents the *population* under examination, the single apple we take and eat represents the *sample*, the decision as to whether the remaining apples are sweet or sour the process of *statistical inference*. A second aspect of statistical inference is related to judging how accurate this inferential decision is in terms of describing the population on the basis of the sample information. In other words, we choose an apple, it is sweet, we decide therefore that all the apples are sweet. Can we be 100 per cent certain that this is the right decision, i.e. that this decision accurately describes the sweetness or sourness of all the apples?

We can see intuitively that this accuracy will depend upon the size of the sample. We would be foolish to base our guess on the taste of *one* apple out of 100. We would feel more confidence if we took a sample of, say, ten apples and made our decision after examining these. We might feel even more confident with a sample of twenty apples. In other words, the larger the sample size, the more accurate is our inference likely to be.

A knowledge of probability theory is crucial in appraising how accurate any decision we make on the basis of sample information is likely to be. This chapter discusses some appropriate techniques. We start first with a discussion of the meaning of probability, followed by an examination of the formal rules of probability theory. We then investigate what are known as random variables and their associated probability distributions, and note the parallel between these and the frequency distributions encountered in Chapter 2. In this context we use the concept of the expected value operator to obtain expressions for the mean and variance of random variables. Finally, we encounter the idea of joint probability distributions and the concept of *covariance* as a measure of the degree of closeness of the relationship between two variables.

6.2 The meaning of probability

There are several different approaches to the question of probability, and although it is a concept which occurs frequently in many branches of the naturai and social sciences (not to mention everyday life) it remains difficult to define unambiguously and is the subject of several competing theories. We shall not spend any time discussing the various approaches since for our purposes the *relative frequency* concept is the most appropriate and is quite satisfactory. Before we explain the idea behind the relative frequency approach we need to establish the meaning of a few terms which commonly occur in any discussion on probability.

Much of elementary probability theory is explained using the analogy of games of chance (e.g. tossing coins, rolling dice, choosing cards). Although this may seem far removed from the world of economics and the other social sciences, it does provide an easily understood framework via which the elementary concepts and rules of probability can be understood. Once we have dealt with this we can extend our views of probability to include more realistic applications.

In all of these games of chance we assume that several possible outcomes are possible. For instance, if we toss a coin there are two possible outcomes, heads or tails; if we roll a dice there are six possible outcomes; and so on. Each of these outcomes happens purely by *chance* alone, and is not influenced by external events. Tossing a coin, rolling dice, choosing a card are all referred to as *experiments*, and each separate distinct outcome from any such experiment is called a *simple event*.

If we denote an event by the letter E, then the experiment of tossing a coin once has two possible simple events associated with it, i.e. E_1 (head) and E_2 (tail). Rolling a dice once has six simple events, E_1, E_2, E_3, E_4, E_5 and E_6 (corresponding to the values on the dice of 1, 2, 3, 4, 5 and 6). And so on. It may help to think of the collection of all the simple events resulting from an experiment as comprising a *sample space*, in which each simple event is represented by a point in that space.

We can depict sample space graphically; Figure 6.1 shows the sample space corresponding to the above two experiments. Note that the geometric displacement of the points in a sample space has no significance; the diagrams are only symbolic or schematic representations. The important feature is that every single distinct outcome or simple event must be represented in the sample space.

It is important to note that the events must correspond to *distinct* outcomes. For example, if the dice we use has the face with 1 on it changed into a 6, then the sample would now contain only five points, i.e. 2, 3, 4, 5, 6, and not 6, 2, 3, 4, 5, 6 – since in the latter case the two sixes cannot be distinguished as separate outcomes.

Figure 6.1 Sample space for two experiments

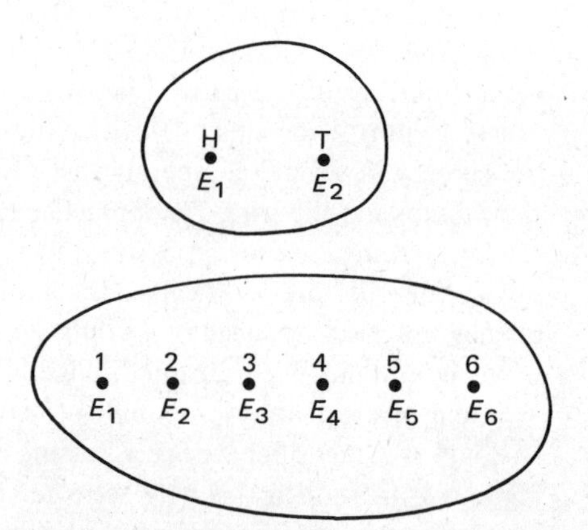

Let us now move on to consider *compound events*. For example, in the dice-rolling experiment, suppose we were betting on the outcome being an odd number. Then, as far as we are concerned, the experiment now produces only two distinct outcomes: 'odd' or 'even'. However, each of these events is a composite of a number of simple events. If we use A to denote a compound event, then A_1 can indicate an odd number and A_2 an even number. Thus

$$A_1 = (E_1 \quad E_3 \quad E_5) \quad \textit{odd}$$

$$A_2 = (E_2 \quad E_4 \quad E_6) \quad \textit{even}$$

We can depict the sample space corresponding to these compound events by enclosing those simple events corresponding to the compound event in question. In Figure 6.2 we show how this is done for this problem.

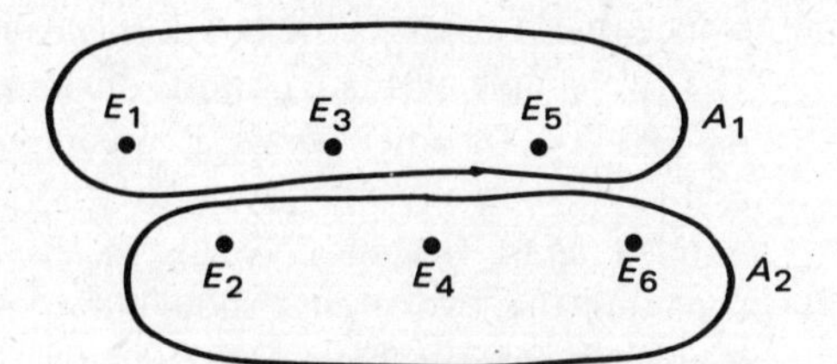

Figure 6.2 Sample space for compound event

We call the compound events in this example *mutually exclusive* since the dice is *either* odd *or* even. There is no other possible result. Notice that in this case the sample spaces corresponding to A_1 and A_2 do not overlap, i.e. there are no points which are common to both spaces. However, this will not always be so. For example, suppose the experiment consists of a single toss of two distinguishable coins. There are four possible outcomes or simple events, namely:

E_1 head on first coin, head on second coin (HH)

E_2 head on first coin, tail on second coin (HT)

E_3 tail on first coin, head on second coin (TH)

E_4 tail on first coin, tail on second coin (TT)

The sample space for this experiment is shown in Figure 6.3.

Let us define two compound events as follows:

A_1 getting at least one head

A_2 getting at least one tail

A_1 comprises the simple events E_1, E_2 and E_3

A_2 comprises the simple events E_2, E_3 and E_4

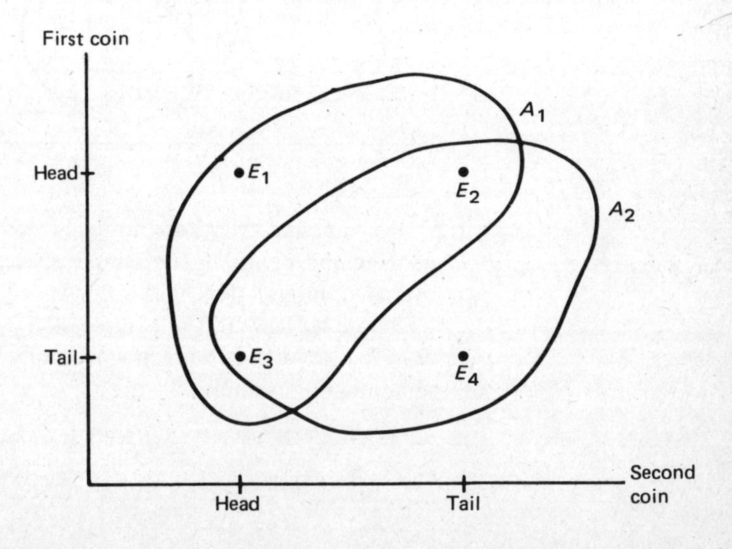

Figure 6.3 Sample space corresponding to tossing two coins

Notice that A_1 and A_2 are not mutually exclusive in the example. The two simple events E_2 and E_3 belong to the sample spaces of A_1 *and* A_2. In other words, two compound events can only be said to be mutually exclusive if they do not contain any *common* simple events. Graphically, this means that the sample spaces depicting the two events should not overlap or *intersect*. The overlapping area in Figure 6.3 contains the two points E_2 (corresponding to HT) and E_3 (TH) and both these points satisfy the requirements of both A_1 *and* A_2. The area or set containing the points in the overlapping region is called the *intersection* of the two events A_1 and A_2. The intersection of the two sets A_1 and A_2 is written (A_1 and A_2). In this example

$$(A_1 \text{ and } A_2) = (E_2, E_3)$$

Finally, we want to define the *union* of two events as being those points which are (i) in one event, or (ii) in the other event, or (iii) in both. We denote the union of two events as (A_1 or A_2). In this example all four points are in A_1 or A_2, and so

$$(A_1 \text{ or } A_2) = (E_1, E_2, E_3, E_4)$$

We illustrate the general idea of the intersection and union of two events in Figure 6.4.

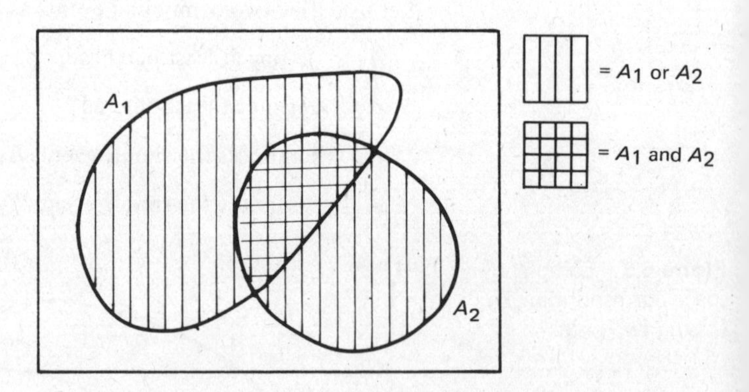

Figure 6.4 Intersection and union of two events

If compound events are mutually exclusive, then there will be no common points in the sample spaces and hence no area of overlap or intersection. This is shown in Figure 6.5.

Before we leave this brief discussion on terminology we should note that there are commonly occurring alternatives to the above notation. For example,

$$(A_1 \text{ or } A_2) \text{ can be written } \quad (A_1 \cup A_2) \text{ or } (A_1 + A_2)$$

$$(A_1 \text{ and } A_2) \text{ can be written } \quad (A_1 \cap A_2) \text{ or } (A_1 \times A_2)$$

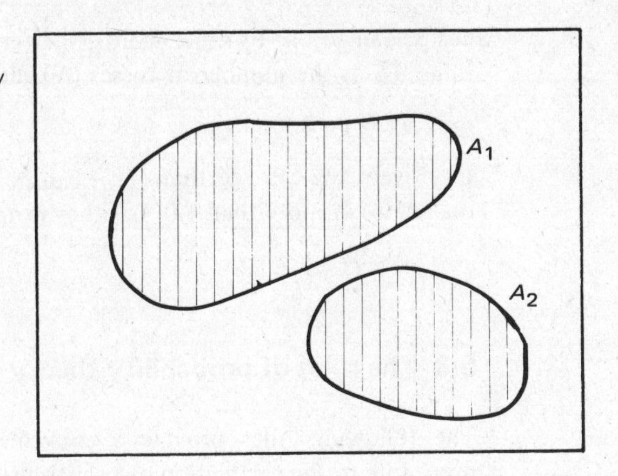

Figure 6.5 Sample spaces for two mutually exclusive events

Definition of probability

Having established the meaning of some terms common in probability theory we can now attempt a more formal definition of probability.

Suppose an experiment results in n simple events E_i where

$$E_i = E_1 \text{ or } E_2 \text{ or } \ldots \text{ or } E_n$$

Suppose we repeat the experiment N times and observe that a particular outcome, say E_i, occurs N_i times. Then the probability of the event E_i, which we denote as $P(E_i)$, is defined thus:

$$P(E_i) = \lim_{N \to \infty} (N_i/N) \tag{6.1}$$

where the right-hand side is read as 'the expected limiting value of the term (N_i/N) as N approaches infinity', or as the value which we expect the term (N_i/N) to attain if we were to repeat the experiment a very large (ultimately an infinite) number of times.

The whole of the right-hand side of (6.1) is called the expected relative frequency of event E_i. Probability is thus defined in terms of relative frequency.

To illustrate the use of (6.1) suppose we have the familiar experiment of tossing a coin once, for which we have seen there are two simple events, head (E_1) and tail (E_2).

The probability of event E_1, i.e. getting a head, is given, according to (6.1), by

$$P(E_1) = \lim_{N \to \infty} (N_1/N) \tag{6.2}$$

The question we have to answer is: 'What value would we expect N_1 to take (i.e. how many heads would we expect to get) if we were to toss the coin 100 times?' The answer would be 'about 50'. If we tossed the coin 500 times we would expect to get about 250 heads, if we tossed it 10 000 times we would expect 5000 heads, and so on. In other words, the term N_1/N approaches the value 0.5 as the number of tosses (N) gets larger and larger. Thus

$$P(E_1) = 0.5$$

Intuitively we can see that $P(E_i)$ can never be negative (i.e. less than zero) or more than 1.0. In other words

$$0 \leqslant P(E_i) \leqslant 1$$

6.3 The rules of probability theory

The following rules provide a convenient and quick solution procedure to the calculation of probabilities.

The addition rule

The addition rule for two events A_1 and A_2 is concerned with the probability of at least one of them occurring. It is written

$$P(A_1 \text{ or } A_2) = P(A_1) + P(A_2) - P(A_1 \text{ and } A_2) \qquad (6.3)$$

which is read as 'the probability of A_1 or A_2 (or both) occurring is equal to the probability of A_1 occurring plus the probability of A_2 occurring minus the probability of A_1 and A_2 occurring'. It is called the addition rule because the terms on the right-hand side of (6.3) are added.

The points in the sample space favourable to the occurrence of A_1 and A_2 have to be subtracted so that they are not counted twice (once in A_1 and once in A_2). This of course assumes that the events are not mutually exclusive. If they are, then there will be no common points and the term (A_1 and A_2) will equal zero. In these circumstances the addition rule for mutually exclusive events is

$$P(A_1 \text{ or } A_2) = P(A_1) + P(A_2) \qquad (6.4)$$

We now illustrate the use of the addition rule with two examples.

Let us return to the experiment of tossing two coins once only. There are four points in the sample space, HH, HT, TH and TT, and these are shown in Figure 6.6.

Example 1

Let us calculate the probability of getting no heads, denoted as the event A_1, or a tail on the first coin, event A_2, or both. The sample spaces corresponding to a favourable outcome of the experiment are shown in Figure 6.6. Notice that A_2 space encloses two points, TH and TT, while A_1 space encloses one point, TT.

These two events are not mutually exclusive: they could both happen simultaneously and so have points in the sample space in

Figure 6.6 Sample space for tossing two coins: A_1 is no heads, A_2 is tail on first coin

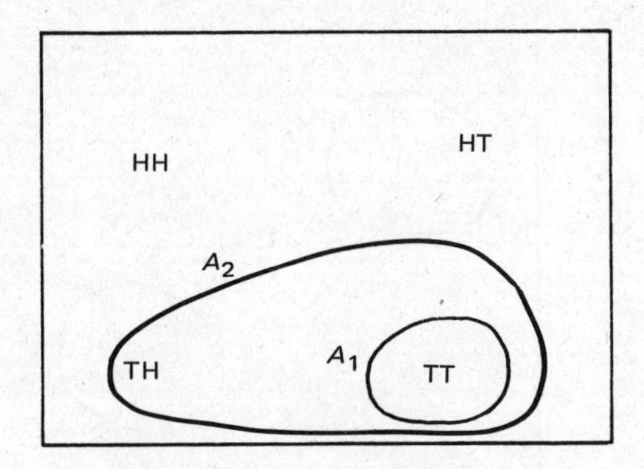

common (or rather one point, TT). We can use the sample space and its points to guide us in determining probabilities.

To calculate $P(A_1 \text{ or } A_2)$ using (6.3) we need to calculate the values of all the terms on the right-hand side of that expression. Let us take them one at a time.

The event A_1 has only one point, TT, in the sample space. Since there are four points in all (and each equally likely) then the probability of A_1 occurring is

$$P(A_1) = \tfrac{1}{4} = 0.25$$

There are two points corresponding to event A_2. Thus

$$P(A_2) = \tfrac{1}{2} = 0.5$$

The events (A_1 and A_2) will have points in the sample space which are in the space for A_1 and the space for A_2. There is only *one* such point (TT), and so

$$P(A_1 \text{ and } A_2) = \tfrac{1}{4} = 0.25$$

Thus substituting into (6.4) gives

$$P(A_1 \text{ or } A_2) = \tfrac{1}{4} + \tfrac{1}{2} - \tfrac{1}{4} = \tfrac{1}{2} = 0.5$$

This result means that if we toss two coins, the probability of getting either no heads or a tail on the first coin (or both) equals 0.5.

Example 2

In the same experiment suppose the event of interest is two heads (A_1) or two tails (A_2) or both. The events are shown in Figure 6.7.

Fairly obviously the two events cannot both occur simultaneously (we either have two heads *or* two tails) and so there are no points in common in the sample spaces. In other words, the events are *mutually exclusive*. Thus we use (6.4) to determine $P(A_1 \text{ or } A_2)$, i.e.

$$P(A_1 \text{ or } A_2) = P(A_1) + P(A_2)$$

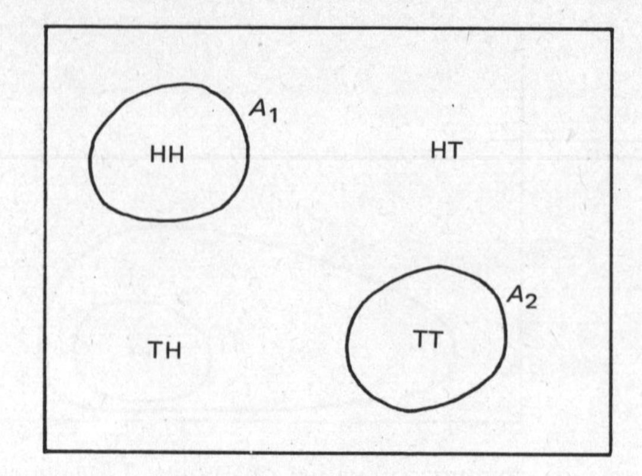

Figure 6.7 Tossing two coins with A_1 (two heads) and A_2 (two tails)

From Figure 6.6 we can see that A_1 is represented by only one point, HH, and so is A_2, the point TT. There are four points in all and so

$$P(A_1) = \tfrac{1}{4} = 0.25$$

$$P(A_2) = \tfrac{1}{4} = 0.25$$

Thus

$$P(A_1 \text{ or } A_2) = 0.25 + 0.25 = 0.5$$

The probability of getting either two heads or two tails or both, when the experiment consists of tossing two coins once, is 0.5.

The multiplication rule

This states that $P(A_1 \text{ and } A_2)$, the probability of events A_1 and A_2 occurring simultaneously, is given by

$$P(A_1 \text{ and } A_2) = P(A_1)P(A_2 \mid A_1) \tag{6.5}$$

It is called the multiplication rule because the terms on the right-hand side are multiplied together.

The term $P(A_2 \mid A_1)$ is read as 'the probability of the event A_2 occurring *given that* A_1 has already occurred', and is known as the *conditional probability* of A_2 on A_1.

We can illustrate the use of this rule with an example. Suppose we take an ordinary deck of cards from which we choose two cards at random, but that we do *not* replace the first card before the second card is taken (this is known as *sampling without replacement*). What is the probability of drawing two kings?

We proceed as follows:

let A_1 = the event that the first card is a king
and A_2 = the event that the second card is a king

Then

$$P(A_1) = 4/52$$

since there are four kings in a pack of fifty-two cards.

Now if we have already drawn one king, the pack will have fifty-one cards and contain three kings. Therefore, the probability of drawing a second king *given* that we have already drawn one is given by

$$P(A_2 \mid A_1) = 3/51$$

We can now substitute into (6.5) to give

$$P(A_1 \text{ and } A_2) = 4/52 \times 3/51 = 0.0045$$

In other words, if we were to repeat the experiment 10 000 times in about forty-five cases we would draw two kings.

Independent events

In the above example the probability that the second card is a king is dependent on whether or not the first card drawn was a king or not. If it were, the probability of the second card being a king is reduced.

Not all events, however, are dependent. In the experiment of tossing two coins, for example, the two coins do not influence one another. The fact that one comes down heads, say, in no way affects what side the other coin will show. In the same way successive spins of a roulette wheel are separate and *independent* events. A black this spin does not tell us anything about what we can expect on the next spin. It follows that if two events A_1 and A_2 *are* independent then

$$P(A_2 \mid A_1) = P(A_2)$$

and (6.5) becomes

$$P(A_1 \text{ and } A_2) = P(A_1)P(A_2) \tag{6.6}$$

and this is the multiplication rule for independent events. We illustrate the use of this rule with an example.

Example

A roulette wheel has thirty-six segments, eighteen black and eighteen red. Calculate the probability of getting two successive reds.

Let A_1 = the event getting a red first spin
and A_2 = the event of getting a red on the second spin

Now these events are independent, so we use (6.6) to calculate

$P(A_1 \text{ and } A_2)$:

$$P(A_1) = 18/36$$

$$P(A_2) = 18/36$$

Thus

$$P(A_1 \text{ and } A_2) = P(A_1)P(A_2) = \frac{18}{36} \times \frac{18}{36} = 0.25$$

Equation (6.6) can be extended to include any number of independent events. Thus the probability of getting three successive reds on a roulette wheel is

$$18/36 \times 18/36 \times 18/36 = 0.125$$

* * *

Let us consolidate our understanding of the above rules of probability with another example.

Example

A room contains 100 economists. Suppose that 20 per cent are women of whom 60 per cent could be labelled liberal (e.g. state interventionists) while the remaining 40 per cent are conservative. Of the male economists 40 per cent are liberal and 60 per cent are conservative. The information is summarised in Table 6.1. If an economist is chosen at random,

(a) What is the probability that a woman will be selected (event A_1)?
(b) what is the probability that a man will be selected (event A_2)?
(c) what is the probability of selecting a liberal (event A_3)?
(d) what is the probability of selecting a conservative (event A_4)?
(e) what is the probability of selecting a male conservative (event (A_2 and A_4))?
(f) what is the probability of selecting a man or a conservative (or both) (event (A_2 or A_4))?
(g) If *two* economists are chosen with replacement, what is the probability of selecting first a woman (event A_1) and second a woman (event A_1')?

Table 6.1

	Male	Female	Total
Liberal	32	12	44
Conservative	48	8	56
Total	80	20	100

The solutions are as follows,

(a) $P(A_1) = 20/100 = 0.2$.
(b) $P(A_2) = 80/100 = 0.8 = (1 - A_1)$.
(c) $P(A_3) = 44/100 = 0.44$.
(d) $P(A_4) = 56/100 = 0.56 = (1 - A_3)$.
(e) $P(A_2 \text{ and } A_4) = P(A_2)P(A_4 \mid A_2)$. Now $P(A_4 \mid A_2) = 48/80 = 0.6$. Thus $P(A_2 \text{ and } A_4) = 0.8 \times 0.6 = 0.48$.
(f) $P(A_2 \text{ or } A_4) = P(A_2) + P(A_4) - P(A_2 \text{ and } A_4) = 0.8 + 0.56 - 0.48 = 0.88$.
(g) $P(A_1 \text{ and } A_1') = P(A_1)P(A_1' \mid A_1)$. A_1 and A_1' are independent since sampling is with replacement. Thus $P(A_1' \mid A_1) = P(A_1') = 20/100 = 0.2$. Thus $P(A_1 \text{ and } A_1') = P(A_1)P(A_1') = 0.2 \times 0.2 = 0.04$.

6.4 Probability distributions

Discrete random variables

So far we have discussed the probabilities associated with the outcomes of idealised experiments. In economics we are in practice much more concerned with examining the probabilities that a variable will take certain values. It will still be useful for the moment to use idealised experiments as a vehicle for the discussion which follows, but we want now to introduce the concept of a *discrete random variable*. This is defined as a variable all possible values of which are defined by the points in a sample space (the word *random* means that we have no knowledge of which event or sample point, and therefore what value, the variable will take before the experiment is performed).

Since the points in a sample space can each be associated with a certain probability and each point is also associated with the value of a discrete random variable, then we can assign probabilities to each value that the variable may take.

The following example may help in understanding this idea. Suppose an experiment consists of tossing two coins (see Figure 6.3). There are four simple events associated with this experiment, and thus the sample space consists of the following four points:

E_1 HH

E_2 HT

E_3 TH

E_4 TT

We could, for example, define a random variable X as being equal to the number of heads associated with each sample point (note that we do not know *before* we toss the coins how many heads there will be and therefore what value X will take). The value of X for each sample point is shown in Table 6.2.

Table 6.2 Values of X when tossing two coins if X equals number of heads

Sample point	Number of heads	X
E_1	2	2
E_2	1	1
E_3	1	1
E_4	0	0

Now each event, i.e. each sample point, has a probability of occurrence which we can calculate, using in this example the multiplication rule for independent events, equation (6.6). Thus

$$P(E_1) = P(H \text{ and } H) = P(H)P(H) = 0.25$$

$$P(E_2) = P(H \text{ and } T) = P(H)P(T) = 0.25$$

$$P(E_3) = P(T \text{ and } H) = P(T)P(H) = 0.25$$

$$P(E_4) = P(T \text{ and } T) = P(T)P(T) = 0.25$$

We can combine this information on the values of X and the probability of each value to produce Table 6.3, and this table can be rearranged as shown in Table 6.4.

Table 6.3 Values of X and $P(X)$

Sample point	Number of heads	X	$P(X)$
E_1	2	2	0.25
E_2	1	1	0.25
E_3	1	1	0.25
E_4	0	0	0.25

Table 6.4 Probability distribution of random variable X when X is the number of heads from tossing two coins

X_i	$P(X_i)$
2	0.25
1	0.50
0	0.25
	1.00

Table 6.4 is known as a *probability distribution*. It shows the values which can be taken by a random variable together with their respective probabilities. Notice that the sum of the probabilities, $P(X_i)$, is equal to 1.0, as we might have expected.

(The idea of a probability distribution is very important and underpins much of what is to follow. It is vital to understand Table 6.4 and how it was determined before proceeding.)

We can generalise these ideas by offering the following formal definition of a probability distribution:

> if X is a random variable which can take any one of n values, $X_1, X_2, \ldots, X_n$, with associated probabilities, $P(X_1), P(X_2), \ldots, P(X_n)$, then a listing of the pairs of values,
>
> $$\begin{array}{ll} X_1 & P(X_1) \\ X_2 & P(X_2) \\ \cdot & \cdot \\ \cdot & \cdot \\ \cdot & \cdot \\ X_n & P(X_n) \end{array}$$
>
> where $\Sigma P(X_i) = 1.0$, is known as the probability distribution of X.

A probability distribution is analogous to a relative frequency distribution. In a relative frequency distribution each possible outcome of a variable is paired with a frequency corresponding to the number of times that the outcome *actually* occurred in the observed data. In a probability distribution each possible outcome is paired with the *probability* of it occurring. In a probability distribution the probabilities sum to 1.0, in a relative frequency distribution the frequencies also sum to 1.0 (or to 100 per cent).

Since probability distributions are paralleled by relative frequency distributions, we can extend the principles of graphic display of frequency distributions (described in Chapter 3) to describe probability distributions. In particular we can draw the probability histogram analogous to the frequency histogram. For the probability distribution shown in Table 6.4 the corresponding probability histogram is shown in Figure 6.8.

As a second example consider the experiment of tossing three coins. The sample space consists of the following eight points (all equally likely and each thus having a probability of 1/8):

HHH, HHT, HTH, THH, TTH, THT, HTT, TTT

Suppose we define X as follows:

$X = 1$ if no tails

$X = 2$ if 1 or 2 tails

$X = 3$ if 3 tails

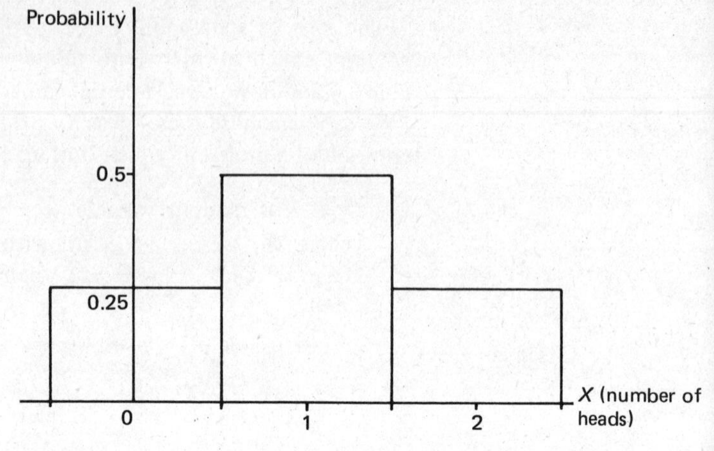

Figure 6.8 Probability histogram corresponding to probability distribution of Table 6.4 (tossing two coins)

Now see Table 6.5. This can be rearranged into a probability distribution and the probability histogram can be drawn (see Figure 6.9).

Finally, let us consider an example, the underlying structure of which will be of interest later. This consists of tossing a single

Table 6.5

Point or event	Number of tails	Value of X	Probability of X
HHH	0	1	$\frac{1}{8}$
HHT	1	2	$\frac{1}{8}$
HTH	1	2	$\frac{1}{8}$
THH	1	2	$\frac{1}{8}$
TTH	2	2	$\frac{1}{8}$
THT	2	2	$\frac{1}{8}$
HTT	2	2	$\frac{1}{8}$
TTT	3	3	$\frac{1}{8}$
			$\Sigma = 1.0$

X_i	$P(X_i)$
1	$\frac{1}{8}$
2	$\frac{6}{8}$
3	$\frac{1}{8}$
	$\Sigma = 1.0$

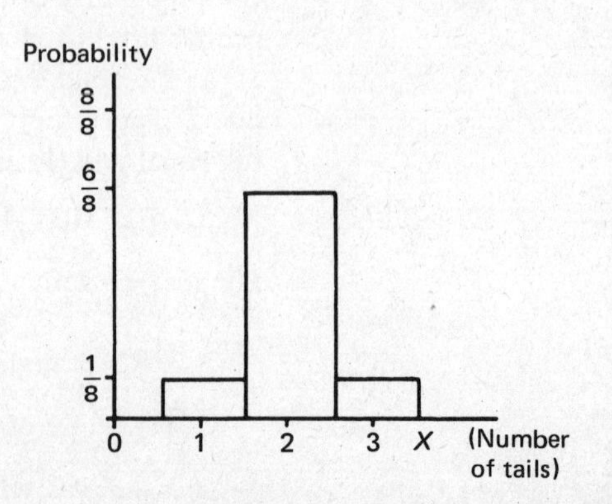

Figure 6.9 Probability distribution and histogram for tossing three coins experiment

coin three times and we define a random variable X as being equal to the number of heads which occur in three tosses. There are eight possible outcomes to this experiment, all equally likely and each thus having a probability of occurrence of $\frac{1}{8}$. This is the same sample space as in the previous example, but we have defined X differently and this produces a different result (as we would expect).

The sample points, the number of heads and the values and probabilities of X are shown in Table 6.6 and we can consolidate these results into the probability distribution and corresponding probability histogram shown in Figure 6.10.

Table 6.6

Point or event	Number of heads	X_i	$P(X_i)$
HHH	3	3	$\frac{1}{8}$
HHT	2	2	$\frac{1}{8}$
HTH	2	2	$\frac{1}{8}$
THH	2	2	$\frac{1}{8}$
TTH	1	1	$\frac{1}{8}$
THT	1	1	$\frac{1}{8}$
HTT	1	1	$\frac{1}{8}$
TTT	0	0	$\frac{1}{8}$
			$\Sigma = 1.0$

As we shall soon see, it is not always necessary to write out the whole probability distribution in detail as we have done in each of the above examples. Some distributions occur so often in practice that algebraic formulae have been established which describe these more common distributions completely. We shall encounter two such distributions later. Before that we turn to consider the probability distribution of continuous random variables.

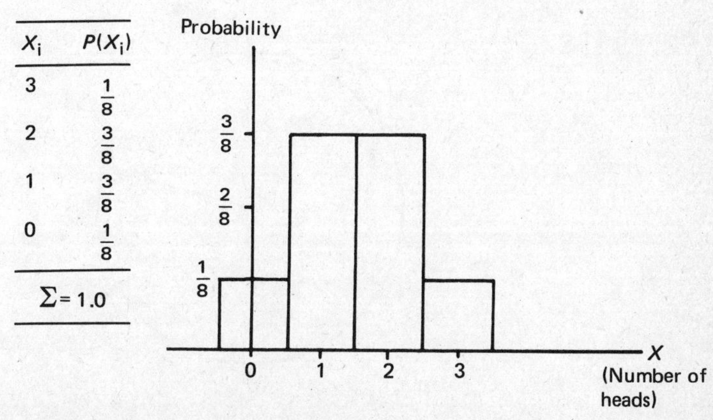

X_i	$P(X_i)$
3	$\frac{1}{8}$
2	$\frac{3}{8}$
1	$\frac{3}{8}$
0	$\frac{1}{8}$
	$\Sigma = 1.0$

Figure 6.10 Probability distribution and histogram for number of heads in three tosses of coin

Continuous random variables and probability density

When dealing with discrete random variables we describe the possible outcome of experiments using sample points separated by finite distances in the sample space. Each point corresponds to a value taken by the random variable and each value is assigned a probability. Rolling a single dice gives us the discrete points 1, 2, 3, 4, 5 or 6. It is not possible to obtain a non-discrete point such as 2.5 or 5.62.

However, and particularly so in economics, 'experiments' do not always yield a sample space consisting of discrete points but consist rather of an entire interval or set of intervals. For example, in the experiment of weighing oneself on a weighing-machine, the moving hand on the dial may stop at any point on the dial, e.g. at 70kg or at 70.2kg, and our estimate of weight is limited only by the gradations of the scale with which the dial is marked. Viewed in this way, there are in fact an infinite number of places where the pointer may stop, and the probability of it stopping at any particular point must thus be zero.

It will not be very helpful, therefore, to plot a graph of probability against the (continuous) possible range of values of X since the probability of X assuming each value is zero.

On the other hand, there is a definite probability that the hand will stop in any given interval, and this probability can be calculated. In a dial marked in 1kg divisions up to a maximum of 150kg, there is a probability of 1/150 that the pointer will come to rest in any given interval, say between 60 and 61kg or between 142 and 143kg, etc., and this presents us with a way of depicting probability distributions for continuous random variables.

For example, suppose an experiment consists of spinning a free hand on a clock or faro-dial (see Figure 6.11). The hand could stop at an infinite number of places anywhere between 0 and 12, and (as we have seen) the probability of it stopping at any particular point must therefore be zero. However, the proba-

Figure 6.11

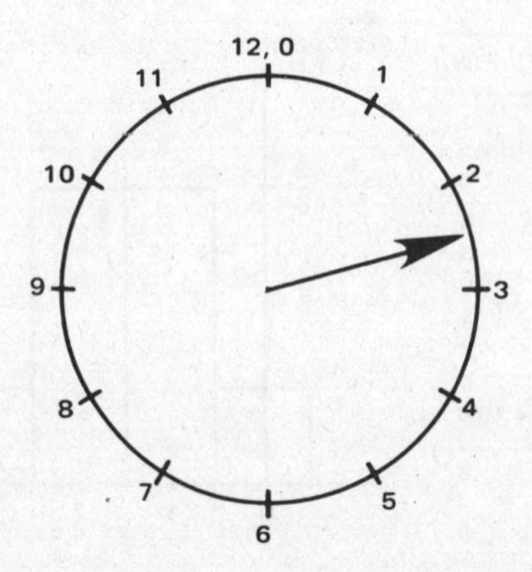

Figure 6.12 Probability density function for clock experiment

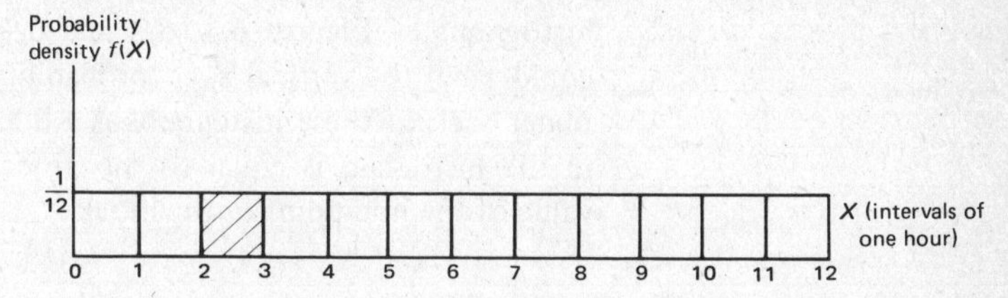

bility of it stopping in any hourly interval, say between 2 and 3, is 1/12 (since there are twelve such intervals), and we can depict this graphically for all intervals, as shown in Figure 6.12.

The probability of the hand stopping between 2 and 3 (or in any hourly interval) is given by the *area* under the curve corresponding to the hour in question, and in this case is $1/12 \times 1 = 1/12$.

Notice that we have labelled the vertical axis of Figure 6.12 as '*probability density, f(X)*', and we define this as follows:

> The probability density function of a continuous variable X, denoted $f(X)$, is represented by a curve, and the probability that X assumes a value in the interval from one given point to another is given by the area under the curve between these two points. The total area under the curve must necessarily be 1.0 since this is the sum of all the interval probabilities.

The probability density function shown in Figure 6.12 is a straight line, but in general probability density functions may take any shape. Even a probability density function as complicated as that shown in Figure 6.13 enables us to calculate any desired probability simply by measuring the area beneath the curve at the appropriate point.

If we want the probability histograms of *discrete* random variables to have these useful area properties, we must ensure that the widths of each separate block are equal to 1.0. Only in this case will the probability densities and the corresponding probabilities be the same. Notice that this happens to be so in the probability

Figure 6.13 A general probability density function

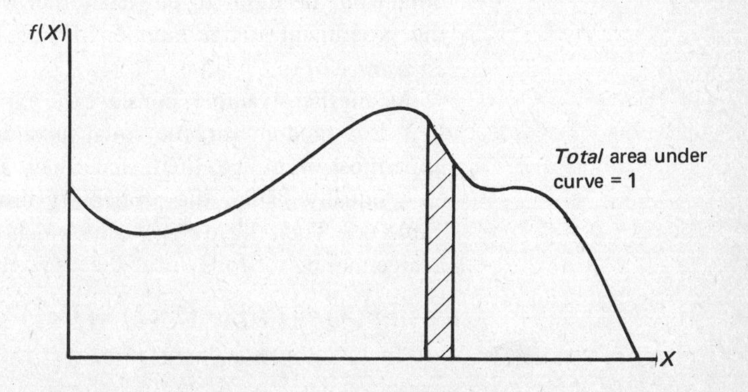

histograms of Figures 6.8, 6.9 and 6.10, and in these cases we could label the vertical axes 'probability density'. The total area under each of these histograms is 1.0 and the area of each block in the histogram is equal to the probability of X assuming the value of the mid-point of the block.

6.5 The mean and variance of random variables – expected value

We have seen that the probability distributions of random variables are directly analogous to the frequency distributions of non-random variables discussed in Chapter 2. We can similarly define the measure of location and dispersion for random variables as we did for the variables discussed in Chapters 4 and 5. We shall confine our remarks to the mean as a measure of location and variance (the standard deviation squared) as a measure of dispersion. To do so we need to introduce a new idea, that of the *expectation operator*, denoted by E.

First we define the *mean* or *expected value* of a *discrete* random variable X as follows:

$$E(X) = \Sigma X_i P(X_i) \tag{6.7}$$

i.e. the value which we *expect* X to take is equal to the sum of all the possible values which X can take, each multiplied by the probability of it taking that value – in other words, the sum of the weighted values of X_i where the weights are the corresponding probabilities.

Thus for the experiment in which X equals the number of heads in tossing two coins, the probability distribution of which is given in Figure 6.3, the mean or expected value of X is given by

$$E(X) = (2 \times 0.25) + (1 \times 0.50) + (0 \times 0.25) = 1.0$$

In other words when we toss two coins we expect that we will obtain one head. Or to put it another way, if we were to repeat the experiment a large number of times, we would expect to get an *average* of one head per toss.

As another example, consider the experiment of rolling a dice. If X is a random variable equal in value to the number of dots uppermost on a dice, then, since there are six possible outcomes each equally likely, the probability distribution is as shown in Table 6.7. Then the average value which we expect X to take over a large number of rolls of the dice is given, using (6.7), by

$$E(X) = (1 \times \tfrac{1}{6}) + (2 \times \tfrac{1}{6}) + (3 \times \tfrac{1}{6}) + (4 \times \tfrac{1}{6}) + (5 \times \tfrac{1}{6}) + (6 \times \tfrac{1}{6}) = 21/6 = 3.5$$

Table 6.7

X_i	$P(X_i)$
1	$\frac{1}{6}$
2	$\frac{1}{6}$
3	$\frac{1}{6}$
4	$\frac{1}{6}$
5	$\frac{1}{6}$
6	$\frac{1}{6}$
	$\Sigma = 1.0$

Of course, there can never be 3.5 spots uppermost on the dice; 3.5 is the *average* value taken by X when the number of throws becomes very large (ultimately infinite).

Second, we define the variance of a discrete random variable X as follows:

$$\text{var}(X) = \Sigma[X_i - E(X)]^2 P(X_i) \tag{6.8}$$

i.e. the variance of a discrete random variable is the weighted average of the squared deviations from the mean, $E(X)$, where the weights are the respective probabilities.

We can use the algebra of expectation to write (6.8) in a different form, i.e.

$$\text{var}(X) = E[X - E(X)]^2 \tag{6.9}$$

The calculations necessary to determine the variance of the number of spots uppermost on a dice using equation (6.8) are shown in Table 6.8. The standard deviation is thus

$$\text{s.d.}(X) = \sqrt{2.92} = 1.71$$

Table 6.8 Calculation of $E(X)$ and var(X) for the number of dots uppermost on a single dice

X_i	$P(X_i)$	$X_iP(X_i)$	$X_i - E(X)$	$[X_i - E(X)]^2$	$[X_i - E(X)]^2P(X_i)$
1	$\frac{1}{6}$	$\frac{1}{6}$	$-\frac{15}{6}$	$\frac{225}{36}$	$\frac{225}{216}$
2	$\frac{1}{6}$	$\frac{2}{6}$	$-\frac{9}{6}$	$\frac{81}{36}$	$\frac{81}{216}$
3	$\frac{1}{6}$	$\frac{3}{6}$	$-\frac{3}{6}$	$\frac{9}{36}$	$\frac{9}{216}$
4	$\frac{1}{6}$	$\frac{4}{6}$	$\frac{3}{6}$	$\frac{9}{36}$	$\frac{9}{216}$
5	$\frac{1}{6}$	$\frac{5}{6}$	$\frac{9}{6}$	$\frac{81}{36}$	$\frac{81}{216}$
6	$\frac{1}{6}$	$\frac{6}{6}$	$\frac{15}{6}$	$\frac{225}{36}$	$\frac{225}{216}$
Σ		$\frac{21}{6}$			$\frac{630}{216}$

$E(X) = \frac{21}{6} = 3.5$ $\qquad$ $\text{var}(X) = \frac{630}{216} = 2.92$

We can see that the probability of X taking a value within one standard deviation of its mean, i.e. between (3.5 – 1.71) and (3.5 + 1.71), i.e. between 1.79 and 5.21, is 4/6 = 0.67. Thus if we rolled the dice 100 times we would expect that the average of the number of spots uppermost would be 3.5, and in *about* sixty-seven throws the number of spots would be no more than 1.71 away from this value, in either direction.

While we are dealing with expected values, it is convenient to note the following further rules of the algebra of expectation operators which will be useful for later work.

First, if Y is some function of the random variable X, i.e. $Y = h(X)$, then

$$E[h(X)] = \Sigma h(X) P(X) \tag{6.10}$$

Second, if X is a random variable and a and b are constants, then

$$E(aX + b) = aE(X) + b \tag{6.11}$$

noting that the expected value of a constant can only be the constant itself.

Also,

$$\text{var}(aX + b) = a^2 \text{var}(X) \tag{6.12}$$

Finally, the mean and variance of a *continuous* random variable are defined exactly as for discrete random variables: the mean is the weighted average of the values of the variable, the weights being their respective probabilities; the variance is the weighted average of the squared deviations of the values of the variable from its mean, the weights being their respective probabilities.

Unfortunately, the algebraic expressions for these terms are not quite as straightforward as in the discrete-variable case, involving as they do integral calculus. However, since we do not specifically require to use these algebraic forms, we can pass over this difficulty.

6.6 Joint probability distributions

Joint probability distributions are analogous to joint frequency distributions discussed in Chapter 2. Consider the experiment of rolling two dice and define two random variables X and Y as shown in Table 6.9. We describe the joint probability that $X = X_i$ and $Y = Y_i$ simultaneously by $P(X = X_i, Y = Y_i)$, or more simply by $P(X_i, Y_i)$.

Now $X = 1$ and $Y = 1$ only if we roll a 1 or 2 on the first dice

Table 6.9

1st dice	X	2nd dice	Y
1	1	1	1
2	1	2	1
3	2	3	1
4	2	4	2
5	2	5	2
6	3	6	2

and a 1, 2 or 3 on the second dice:

$$P(X=1, Y=1) = P(1, 1) = P(1 \text{ or } 2 \text{ on 1st dice } \textit{and } 1, 2 \text{ or } 3 \text{ on 2nd dice})$$

$$= P(1 \text{ or } 2 \text{ on 1st dice})\, P(1, 2 \text{ or } 3 \text{ on 2nd dice})^{1}$$

$$= 2/6 \times 3/6 = 6/36$$

Similarly,

$$P(X=1, Y=2) = P(1, 2) = 6/36$$

and so on.

We use these results to construct a joint probability distribution, as shown in Table 6.10. Notice that analogous to joint frequency

Table 6.10 Joint probability distribution of X and Y in dice-rolling experiment

X	Y = 1	Y = 2	Total
1	$\frac{6}{36}$	$\frac{6}{36}$	$\frac{12}{36}$
2	$\frac{9}{36}$	$\frac{9}{36}$	$\frac{18}{36}$
3	$\frac{3}{36}$	$\frac{3}{36}$	$\frac{6}{36}$
	$\frac{18}{36}$	$\frac{18}{36}$	$\frac{36}{36}$

distributions, the row totals are just the probability distribution of X by itself and the column totals the probability distribution of Y by itself.

Finally, we may note that two random variables are independent if

$$P(X = X_i;\ Y = Y_i) = P(X = X_i)\, P(Y = Y_i)$$

i.e. if

$$P(X_i, Y_i) = P(X_i)\, P(Y_i) \tag{6.13}$$

[1] Using the multiplication rule for independent events.

for all values of X and Y. Thus, for example,

$$P(X_i = 1, Y_i = 1) = 6/36$$

and

$$P(X_i = 1) P(Y_i = 1) = 2/6 \times 3/6 = 6/36$$

thus

$$P(X_i = 1, Y_i = 1) = P(X_i = 1) P(Y_i = 1)$$

If this equality can be demonstrated for *all* values of X_i and Y_i (which it can in this example), then the two variables are independent. If the converse can be demonstrated for at least one joint value, then the two variables are dependent (checking dependency is thus easier this way).

Covariance of two random variables

We now describe the covariance between two variables as the measure of the strength of the association between two variables: the greater the value of the covariance, the stronger the association. For example, high values of one variable tend to be associated with high values of the other (positive covariance), or high values of one tend to be associated with low values of the other (negative covariance). In this context 'high' means values above the mean, 'low' means values below the mean.

Thus if there is a positive relationship between two random variables X and Y, then $[X_i - E(X)]$ and $[Y_j - E(Y)]$ will tend to be of the same sign and their product therefore positive. If there is a negative association, then $[X_i - E(X)]$ and $[Y_j - E(Y)]$ will tend to have opposite signs and their product will therefore be negative. This idea enables us to define the covariance between two random variables X and Y, where X can take the values $X_i (i = 1 \text{ to } n)$ and Y the values $Y_j (j = 1 \text{ to } m)$:

$$\text{cov}(X, Y) = E[X_i - E(X)][Y_j - E(Y)] \tag{6.14}$$

An alternative form of (6.14) is

$$\text{cov}(X, Y) = \sum_{i=1}^{n} \sum_{j=1}^{m} [X_i - E(X)][Y_j - E(Y)]\, P(X_i Y_j) \tag{6.15}$$

Note the use of the double summation sign because the two variables are summed over different ranges. Note also that if two random variables are independent, then

$$\text{cov}(X, Y) = 0 \tag{6.16}$$

Finally, we might note that while independence *necessarily* implies zero covariance, the reverse is not necessarily true. Variables that have a zero covariance are said to be *uncorrelated*. The difference between independence and uncorrelation is that the former requires there to be *no* form of relationship between two variables, while the latter requires only that there be no *linear* relationship. We shall discuss this point further in connection with correlation in Chapter 11.

To illustrate these points let us calculate the covariance between X and Y as they are defined in the dice-rolling example above and using (6.15). These calculations are set out in Table 6.11. This result might have been anticipated since we already know that X and Y are independent.

Table 6.11 Calculation of cov(X, Y) in dice-rolling experiment

X_i	$P(X_i)$	$X_iP(X_i)$	Y_i	$P(Y_i)$	$Y_iP(Y_i)$
1	2/6	2/6	1	3/6	3/6
2	3/6	6/6	2	3/6	6/6
3	1/6	3/6			
Σ	6/6	11/6		6/6	9/6
		$E(X) = 11/6$			$E(Y) = 9/6$

* * *

X_i	Y_i	$P(X_i, Y_i)$	$X_i - E(X)$	$Y_i - E(Y)$	$[X_i - E(X)][Y_i - E(Y)]\ P(X_i, Y_i)$
1	1	6/36	−5/6	−3/6	90/1296
1	2	6/36	−5/6	3/6	−90/1296
2	1	9/36	1/6	−3/6	−27/1296
2	2	9/36	1/6	3/6	27/1296
3	1	3/36	7/6	−3/6	−63/1296
3	2	3/36	7/6	3/6	63/1296
Σ		36/36			0

i.e. $\text{cov}(X, Y) = 0$

We can now add some further rules to the algebra of expectations in the context of joint probability distributions for two random variables X and Y.

First,

$$E(X + Y) = E(X) + E(Y) \tag{6.17}$$

i.e. the expected value of the sum of two random variables is equal to the sum of their expected values.

Second,

$$\text{var}(X + Y) = \text{var}(X) + \text{var}(Y) + 2\ \text{cov}(X, Y) \tag{6.18}$$

i.e. the variance of the sum of two random variables is the sum of their variances plus twice their covariance. In the event that X and Y are independent, then $\text{cov}(X, Y) = 0$, and (6.18) becomes

$$\text{var}(X + Y) = \text{var}(X) + \text{var}(Y) \tag{6.19}$$

Third,

$$E(XY) = E(X)E(Y) + \text{cov}(X, Y) \tag{6.20}$$

i.e. the expected value of the product of two random variables is the product of the expected values plus their covariance.

In the event that X and Y are independent, then $\text{cov}(X, Y) = 0$ (as we have seen) and (6.20) becomes

$$E(XY) = E(X)\,E(Y) \tag{6.21}$$

Equations (6.17) to (6.21) can be extended to any number of variables.

6.7 Summary

In this chapter we have examined the meaning of probability, the rules of probability theory and the probability distribution of both discrete and continuous random variables.

We saw that frequency distributions and probability distributions are directly analogous and that the means and variances of the latter can be derived using the idea of the expectation operator. Finally, we discussed joint probability distributions and the notion of the association between them as measured by covariance.

The contents of this chapter provide us with some very important concepts and techniques which we shall need in the rest of this book.

EXERCISES

6.1 A seminar group contains two men and three women. The tutor decides to pick one of the students at random to lead the discussion. What is the probability

(a) that a woman will be picked?
(b) that a man will be picked?

6.2 Forty students (of whom twenty-five are men) attend a lecture. The lecturer picks two students at random (without replacement) to distribute a handout. What is the probability

(a) that two men will be picked?
(b) that two women will be picked?

(c) that two students of different sex will be picked?
(d) that two students of the same sex will be picked?
(e) that a man will be picked first followed by a woman?
(f) that a woman will be picked first followed by a man?

6.3 Repeat exercise 6.2 with replacement.
6.4 In exercise 6.2 twenty of the men and ten of the women do not smoke cigarettes. One student is to be picked at random. What is the probability that this student will be

(a) either a male or a non-smoker?
(b) either a female or a non-smoker?

6.5 If now two students are to be picked from the group described in exercise 6.4 (with replacement), what is the probability

(a) that at least one of the students smokes?
(b) that at least one of the students is a woman?
(c) that both students are male?

6.6 Repeat exercise 6.5 without replacement.
6.7 The height of new male recruits to a police force is normally distributed with a mean of 5.75 feet and a standard deviation of 0.15 foot. One recruit is picked out at random. What is the probability that this recruit will be

(a) taller than 6 feet?
(b) shorter than 5.7 feet?
(c) between 5.7 feet and 6 feet?

6.8 An experiment consists of rolling two dice. The random variable X is defined as follows:

(a) $X = 1$ if the sum of the uppermost spots is odd.
(b) $X = 2$ if the sum is even (excluding a pair).
(c) $X = 3$ if the dice show a pair.

Describe the probability distribution of X and draw the corresponding probability histogram.
6.9 Calculate the mean and variance of X for the experiment described in exercise 6.8.
6.10 An experiment consists of rolling two dice. The random variable X is defined as being equal to half the sum of the number of spots uppermost. Describe the probability distribution of X and draw the corresponding probability histogram.
6.11 An experiment consists of rolling two dice, and the random variable X is equal to 1 if the first dice is odd and equal to 2 if the first dice is even. The random variable Y is equal to 1 if the number of spots on the second dice is 1, 2 or 3 and equal to 2 if

the number of spots on the second dice is 4, 5 or 6. Construct the joint probability distribution for X and Y.

(a) Are X and Y uncorrelated?
(b) Are X and Y independent?

6.12 Calculate the covariance of X and Y for the experiment described in exercise 6.11.

6.13 For the experiment described in exercise 6.11 calculate

(a) $E(X + Y)$
(b) $\text{var}(X + Y)$
(c) $E(XY)$

PART III

STATISTICAL INFERENCE

7 Sampling Distributions

OBJECTIVES

After reading this chapter and working through the examples students should understand the meaning of the following terms:

random sample	**permutation**
sampling distribution	**normal distribution**
standard error	**discrete variables**
population parameter	**continuous variables**
sample characteristic	**finite population correction factor**
binomial distribution	**central limit theorem**
combination	**probability statement**

Students should be able to:

calculate the standard deviation of a sample
calculate the probability or sampling distribution of a random variable
calculate the mean and variance of a sampling distribution of sample proportions and of sample means
apply a finite population correction factor
make probability statements about sample values using the *z*-table

7.1 Introduction

At the beginning of Chapter 6 we explained briefly and in very simple terms what was meant by statistical inference and how important a concept it was in terms of the material in the rest of this book. We need the probability theory of Chapter 6 to enable us to judge the accuracy of any decision which we make about a feature of a population on the basis of some sample information.

Intuitively we saw that the larger the size of the sample, the more confidence we could have in the accuracy of our inference. Moreover, and just as crucial, is a knowledge of how the sample information varies when repeated samples of the same size are taken from a population. To see what we mean by this, suppose, for some reason, that we were interested in determining the mean number of matches per box in boxes of matches of a certain brand. The 'population' in this example consists of *all* boxes of this brand of matches which have been manufactured. Clearly, even if we had unlimited funds and time, we could not examine every such box since some will have been consumed, some lost, etc. Suppose, therefore, that we take a random sample of ten boxes and determine from this that the mean number of matches per box is 48.1. (We define a *random sample* as one taken in such a way that every member of the population being sampled has an equal likelihood of being chosen for the sample. In this way the sample should accurately reflect the characteristics of the population.)

We could then infer that this was the mean number of matches in all the boxes. Being cautious we might, however, decide to take a second random sample of ten boxes, and this might produce a sample mean of 49.2 matches. A third sample might yield a sample mean of 47.8 matches. The closeness of these three sample means encourages us to believe that the population mean number of matches is not far from these values, or conversely, because the sample values are so close, that we could have confidence that a single sample mean would not be too far away from the population mean. If, however, the three samples yielded means of 60.5, 38.2 and 49.8 we would have much more difficulty in assessing the value of the population mean and we would have much less confidence, if we were able to take only *one* sample (which, unfortunately, will usually be the case in economics), that the result could be relied on to give a value close to the population mean. In other words, if repeated samples yield sample means all reasonably close together, we would have some confidence that the result of any single sample is likely to be correspondingly close to the population mean. If repeated samples yield widely disparate sample mean values, then we would have little confidence that any one of them was close to the population mean.

In brief, we would like to know how likely or probable it is that the sample mean will assume particular values. Since we have no knowledge of the number of matches in any box picked at random before we open it, then this number is a random variable. Hence the sample mean number of matches is a random variable. This then amounts to knowing the probability distribution of the sample means, more usually known as the *sampling distribution* of the sample means. It is very important to distinguish between the distribution of values in a single sample and the sampling distribution of some characteristic obtained from a number of different samples. To illustrate this difference let us return to the matchbox example and the sampling distribution of the mean number of matches per box. Suppose we take fifty samples each consisting of five boxes chosen at random. If we let X_i denote the number of matches in the ith box and assume that the first sample yields the following values:

$$X_1 = 50;\; X_2 = 48;\; X_3 = 51;\; X_4 = 46;\; X_5 = 47$$

then the mean of this first sample, denoted by $\bar{X}_1$, using (4.4), is

$$\bar{X}_1 = \frac{\Sigma X_i}{n} = \frac{242}{5} = 48.4$$

and the standard deviation of this first sample, which we denote by s_1, is

$$s_1 = \sqrt{\frac{1}{n-1}\Sigma(X_i - \bar{X}_1)^2} \tag{7.1}$$

Note that this is slightly different from (5.3) which, repeated here for convenience, is

$$\sigma = \sqrt{\frac{1}{n}\Sigma(X_i - \bar{X})^2} \tag{5.3}$$

We can now reveal that (5.3) is used when we calculate the standard deviation of a *population*, when we have a value for X_i for every single member of the population. In these circumstances the standard deviation is denoted by σ. When we are dealing with a *sample*, then (7.1) is appropriate and the standard deviation is denoted by s. We shall explain later why there is a difference between (7.1) and (5.3) (discussion on this on pp. 171–2).

Thus, by substituting into (7.1), we can calculate that

$$s_1 = 2.0736$$

Suppose a second sample yields the following values of X_i:

$$X_1 = 42;\ X_2 = 49;\ X_3 = 46;\ X_4 = 48;\ X_5 = 45$$

from which

Mean $\bar{X}_2 = 46.0$

Standard deviation $s_2 = 2.7386$

If we were to repeat the sampling procedure a further forty-eight times, we might obtain the fifty values of sample mean shown in Table 7.1.

Table 7.1 Fifty sample mean values for matchbox example

$\bar{X}_1 = 48.4$	$\bar{X}_{11} = 44.2$	$\bar{X}_{21} = 49.3$	$\bar{X}_{31} = 50.0$	$\bar{X}_{41} = 47.6$
$\bar{X}_2 = 46.0$	$\bar{X}_{12} = 51.8$	$\bar{X}_{22} = 49.5$	$\bar{X}_{32} = 51.5$	$\bar{X}_{42} = 43.5$
$\bar{X}_3 = 49.5$	$\bar{X}_{13} = 50.9$	$\bar{X}_{23} = 46.9$	$\bar{X}_{33} = 50.7$	$\bar{X}_{43} = 44.8$
$\bar{X}_4 = 47.6$	$\bar{X}_{14} = 47.5$	$\bar{X}_{24} = 48.0$	$\bar{X}_{34} = 47.0$	$\bar{X}_{44} = 50.1$
$\bar{X}_5 = 49.9$	$\bar{X}_{15} = 45.8$	$\bar{X}_{25} = 44.6$	$\bar{X}_{35} = 48.9$	$\bar{X}_{45} = 48.6$
$\bar{X}_6 = 47.1$	$\bar{X}_{16} = 48.7$	$\bar{X}_{26} = 46.8$	$\bar{X}_{36} = 48.5$	$\bar{X}_{46} = 47.4$
$\bar{X}_7 = 48.1$	$\bar{X}_{17} = 47.3$	$\bar{X}_{27} = 48.8$	$\bar{X}_{37} = 45.9$	$\bar{X}_{47} = 45.0$
$\bar{X}_8 = 48.3$	$\bar{X}_{18} = 48.8$	$\bar{X}_{28} = 45.1$	$\bar{X}_{38} = 51.8$	$\bar{X}_{48} = 48.2$
$\bar{X}_9 = 47.8$	$\bar{X}_{19} = 49.5$	$\bar{X}_{29} = 49.4$	$\bar{X}_{39} = 46.4$	$\bar{X}_{49} = 50.2$
$\bar{X}_{10} = 49.1$	$\bar{X}_{20} = 49.6$	$\bar{X}_{30} = 48.5$	$\bar{X}_{40} = 50.8$	$\bar{X}_{50} = 46.2$

We can arrange these fifty sample means into the form of a histogram diagram (Figure 7.1), to which we have added the frequency curve. Table 7.1, then, is a sampling distribution; in this example it is the sampling distribution of the mean number of matches per box. This sampling distribution has a mean and standard deviation which we can calculate in the usual way using

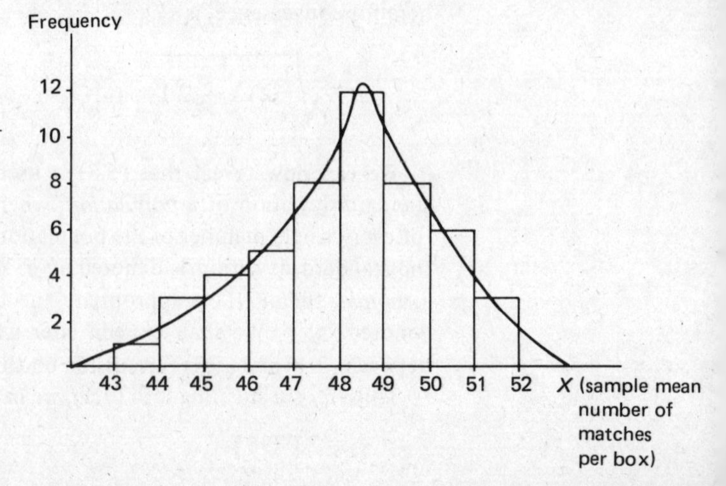

Figure 7.1 Sampling distribution of sample mean number of matches per box; sample size = 5, number of samples = 50

equations (4.3) and (7.1), and which, with $\bar{X}_i$ (the mean of the *i*th sample) as the variable X, become the mean of the sample means, denoted by $\mu_{\bar{X}}$, which is

$$\mu_{\bar{X}} = \Sigma \bar{X}_i / n = 2405.9/50 = 48.1180$$

(notice that n now equals 50 since we are trying to calculate the mean of the *50* sample means) and the standard deviation of the sample means, denoted by $\sigma_{\bar{X}}$, which is

$$\sigma_{\bar{X}} = \sqrt{\frac{1}{n-1} \Sigma(\bar{X}_i - \mu_{\bar{X}})^2} = 2.0082$$

To repeat an important point, we must be careful to distinguish between the mean and standard deviation of a single sample of some characteristic and the mean and standard deviation of a sampling distribution of that characteristic.

We should note that the standard deviation of a sampling distribution is usually known as the *standard error* of that distribution, and in future we shall follow this convention.

7.2 Some terminology

This is a convenient point to set down a notation and terminology. First of all, the characteristics of the population which we are trying to determine are known as the population *parameters*. We will limit our discussion to making inferences only about the population *proportion* (e.g. the proportion of households who have more than one wage-earner, the proportion of the unemployed who are unskilled, etc.), about the population *mean* and about population *variance*. The results which we obtain from a sample (e.g. the sample mean or sample proportion) are known as the *sample characteristics* or sample statistics.

The notation shown in Table 7.2 is widely used. We have already encountered many of the terms in this table and the remainder will be defined shortly in context. We might, however, mention that the word *attribute* is commonly used to mean characteristic or feature, e.g. the possession or non-possession of the attribute, 'red hair' or of the attribute 'car ownership', etc.

We can derive the sampling distribution of some particular characteristic in two ways.

First, we can experiment, by constructing some specific population (which is therefore completely known to us), taking a large number of samples and calculating the value of the sample characteristic in each case. These values enable us to determine the approximate sampling distribution of the characteristic in question

Table 7.2 Notation and terminology in statistical inference

Population (of size N)	A single sample (of size n)	Sampling distribution (repeated samples of size n)
Mean $\mu = \dfrac{\Sigma X_i}{N}$	Mean $\bar{X} = \dfrac{\Sigma X_i}{n}$	Mean $\mu_{\bar{X}} = \mu$
Variance $\sigma^2 = \dfrac{1}{N} \Sigma(X_i - \mu)^2$	Variance $s^2 = \dfrac{1}{n-1} \Sigma(X_i - \bar{X})^2$	Variance $\sigma^2_{\bar{X}} = \sigma^2/n$
Standard deviation $\sigma = \sqrt{\dfrac{1}{N} \Sigma(X_i - \mu)^2}$	Standard deviation $s = \sqrt{\dfrac{1}{n-1} \Sigma(X_i - \bar{X})^2}$	Standard error $\sigma_{\bar{X}} = \sqrt{\sigma^2/n}$
*	*	*
Proportion $\pi = \dfrac{\text{Number with attribute}}{N}$	Proportion $p = \dfrac{\text{number with attribute}}{n}$	Proportion $\pi_p = \pi$ Variance $\sigma_p^2 = \pi q/n$ Standard error $\sigma_p = \sqrt{\pi q/}$

(which is the approach we used in the above example to obtain the sampling distribution of Table 7.1), and while (strictly speaking) the result will apply only to the specific population (with its appropriate characteristics) we have constructed, we hope we can generalise our results somewhat.

The alternative approach is to use the ideas of probability theory to derive a theoretical sampling distribution which would result if we took *all* possible samples of some given size from a population. This approach avoids the necessity for any experimental work, gives a more *precise* result and can be readily generalised (within well-defined limits). The only difficulty is that in some circumstances the theory might be so difficult as to make it impossible for us to obtain an answer.

To sum up, we need to know the sampling distribution of any particular sample characteristic in order to be able to comment on how close any single sample result is likely to be to the population value. We prefer theoretically derived sampling distributions in this context since, as we have seen, they are superior to those empirically determined. In particular we concentrate on two distributions which have been found to be most generally applicable in economic and social sciences: the *binomial* distribution (which is used with discrete random variables), and more importantly the *normal* distribution (which we can use for continuous random variables).

In the remainder of this chapter we describe the derivation of both of these distributions, starting with the binomial distribution and its application to the sampling distribution of sample proportions.

7.3 Sampling distribution of the sample proportion

Let us begin by considering the following example.

Suppose we are interested in whether the employees in a certain factory are trade-union members or not. Let us further suppose that 40 per cent of them are members and 60 per cent are not. Imagine that this information is recorded on each worker's personal record card. We proceed as follows. Shuffle all the record cards together and choose one at random. Note whether that worker is a trade-union member or not. Replace this card and repeat four more times, i.e. we are taking a sample of five. Notice that each separate selection of a record card is an independent event since we are sampling with replacement.

We define a random variable X as follows: X = the number of 'successful' drawings in each sample where being a union member counts as a 'success' and not being a union member counts as a 'failure', and we denote success on the ith draw by S_i and failure by F_i.

Notice that with this population, for each individual drawing in a sample,

$$P(F_i) = \tfrac{6}{10} \text{ and } P(S_i) = \tfrac{4}{10}$$

X can take the values 0, 1, 2, 3, 4 or 5 in any particular sample. The probability that we get no trade unionists on five 'drawings' of record cards is, using the multiplication rule for *independent* events, as follows:

$$\begin{aligned} P(X = 0) \text{ or } P(0) &= P(F_1 \text{ and } F_2 \text{ and } F_3 \text{ and } F_4 \text{ and } F_5) \\ &= P(F_1) \times P(F_2) \times P(F_3) \times P(F_4) \times P(F_5) \\ &= \tfrac{6}{10} \times \tfrac{6}{10} \times \tfrac{6}{10} \times \tfrac{6}{10} \times \tfrac{6}{10} \\ &= 0.6^5 \\ &= 0.0777 \end{aligned}$$

Similarly, the probability of obtaining one trade-union member in drawing five record cards is, using the addition rule for *exclusive* events (a worker is either a member or not), as follows:

$$\begin{aligned}
P(X=1) \text{ or } P(1) &= P(S_1F_2F_3F_4F_5) \text{ or } P(F_1S_2F_3F_4F_5) \text{ or} \\
&\quad P(F_1F_2S_3F_4F_5) \text{ or } P(F_1F_2F_3S_4F_5) \text{ or} \\
&\quad P(F_1F_2F_3F_4S_5)^1 \\
&= P(S_1F_2F_3F_4F_5) + P(F_1S_2F_3F_4F_5) \\
&\quad + P(F_1F_2S_3F_4F_5) + P(F_1F_2F_3S_4F_5) \\
&\quad + P(F_1F_2F_3F_4S_5) \\
&= P(S_1)P(F_2)P(F_3)P(F_4)P(F_5) \\
&\quad + P(F_1)P(S_2)P(F_4)P(F_5) + \ldots \\
&= \tfrac{4}{10} \times \tfrac{6}{10} \times \tfrac{6}{10} \times \tfrac{6}{10} \times \tfrac{6}{10} + \tfrac{6}{10} \times \tfrac{4}{10} \times \tfrac{6}{10} \\
&\quad \times \tfrac{6}{10} \times \tfrac{6}{10} + \ldots \\
&= 0.2592
\end{aligned}$$

In the same way we can show that

$$P(2) = 34560/100000 = 0.3456$$
$$P(3) = 23040/100000 = 0.2304$$
$$P(4) = 1610/100000 = 0.0161$$
$$P(5) = 1024/100000 = 0.0102$$

We summarise these results in Table 7.3, where we also show the calculations for the mean and variance of X (ignore the last

Table 7.3 Sampling or probability distribution of X and calculation of the mean and variance for union membership

X	$P(X)$	$X\,P(X)$	$[X - E(X)]$	$[X - E(X)]^2$	$[X - E(X)]^2 P(X)$	p
0	0.0777	0	−2	4	0.3110	0
1	0.2592	0.2592	−1	1	0.2592	0.2
2	0.3456	0.6912	0	0	0	0.4
3	0.2304	0.6912	1	1	0.2304	0.6
4	0.0768	0.3072	2	4	0.3072	0.8
5	0.0102	0.0512	3	9	0.0922	1.0
Σ	1.0	2.0			1.2	
	$E(X) = 2$			var$(X) = 1.2$		

[1] Strictly speaking the terms on the right-hand-side of the above expression should be written as (for example) $P(S_1$ and F_2 and F_3 and F_4 and $F_5)$.)

column for the moment), using (6.7) and (6.8). Notice that the expected (or mean) value of the random variable X is 2. This implies that, on average, in a sample of size five, we get two workers who are union members, i.e. 2/5 or 40 per cent. This is in fact the same percentage as the population proportion.

This sampling or probability distribution also tells us that of all samples of size five drawn from this population 7.77 per cent of them will contain no union members, 25.92 per cent of the samples will contain one union member, and so on. Finally, notice that in only 1.02 per cent of such samples of size five will there be five union members. If you were to take a sample of five from such a population, having been told that there were only 40 per cent union members in the population and obtained five union members would it not cause us to pause and ask: 'I wonder if the population has been accurately described; perhaps there has been some union recruitment since the 40 per cent figure was determined and we now seem to be sampling from a population which perhaps has a higher percentage of union members.' Of course it is not *impossible* for there to be five members in the sample but the probability of there so being is very small. This is an important point to which we return when we come to consider hypothesis testing in Chapter 10.

In Figure 7.2 we have drawn the probability density histogram and polygon for the sampling distribution of X. Recall that the area under both these curves is 1.0.

Finally, and to anticipate the next section, the sampling or probability distribution of *any* random variable which results from an experiment in which there are only two possible outcomes (success or failure), for which the population proportion of successes is 0.4 and from which we take a sample of five (in such a way that the choices are independent, i.e. with replacement), will have a sampling probability distribution identical to that shown in Table 7.3. Incidentally, an experiment need not be

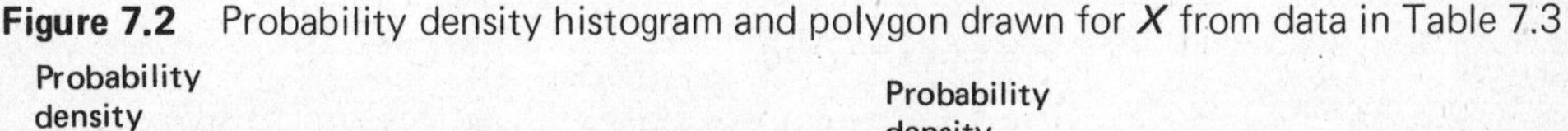

Figure 7.2 Probability density histogram and polygon drawn for X from data in Table 7.3

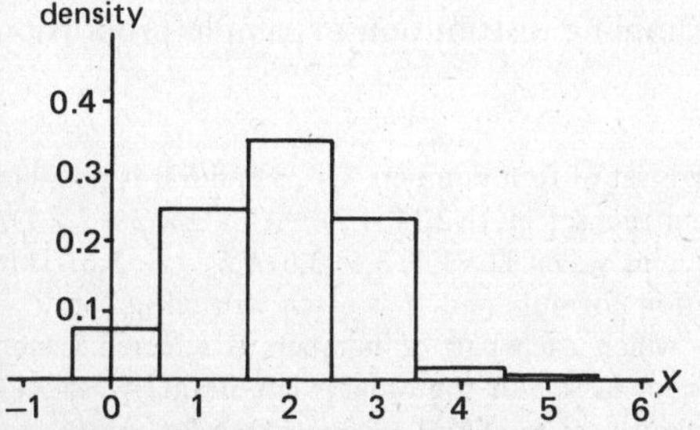

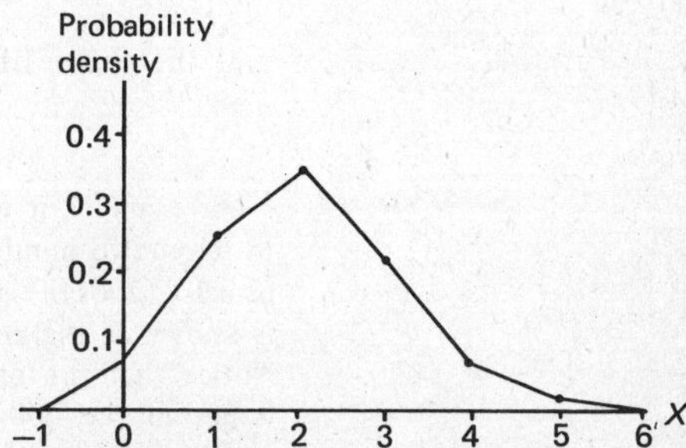

constrained to having only two outcomes to be classified as of the binomial type. For example, in a dice-rolling experiment we may be interested only in getting a six, in which case a six would count as a success, while numbers from one to five would count as a failure.

The binomial distribution

Even for a sample size as small as five, the calculations required to produce the sampling distribution of Table 7.3 are quite lengthy. Fortunately, experiments for which there are only two outcomes, success or failure, belong to a special class which produces what is known as the *binomial distribution*.

The binomial probability or sampling distribution of a random variable X, where X corresponds to the number of successful outcomes, is given by the following equation:

$$P(X = X_i) \text{ or } P(X_i) = \frac{n!}{X_i!(n - X_i)!}\pi^{X_i}q^{n-X_i} \qquad (7.2)$$

where

X_i = number of successes

n = sample size

π = proportion of successes in the population

$q = 1 - \pi$

$\dfrac{n!}{X_i!(n - X_i)!}$ = the total number of different ways that it is possible to select X_i elements from a larger number of elements n.[1]

It can be shown that the mean (or the expected value), the variance and the standard error (s.e.) of the sampling distribution of X are given by

$$E(X) = n\pi \qquad (7.3)$$

$$\text{var}(X) = n\pi q \qquad (7.4)$$

$$\text{s.e.}(X) = \sqrt{n\pi q} \qquad (7.5)$$

and the *shape* of the sampling distribution of sample proportion

[1] For example, if we have a set of four numbers 3, 4, 5 and 6 it is possible to select two numbers from this set in $4!/2!(4 - 2)! = (4 \times 3 \times 2 \times 2 \times 1)/(2 \times 1 \times (2 \times 1))$ = six different ways: i.e. 3,4; 3,5; 3,6; 4,5; 4,6; 5,6. This is known as the *combination* formula and it is often written $_nC_r$ or C_r^n. Notice that the order in which each pair of numbers is selected is not important, i.e. 3,4 is the same as 4,3. If the order *is* important, we should use the *permutation* formula, i.e. $n!/(n - X_i)!$, often written $P_{n,r}$ or P_r^n.

is that of the binomial distribution, i.e. something of the general shape of Figure 7.2, but the exact form will depend on n and π.

In statistics there is a very convenient shorthand which is used to describe the shape and principal features of a distribution. This has the form

$$X \sim (\text{mean, variance})$$

where the sign $\sim$ means 'is distributed'.

The above results relating to the sampling distribution of the number of successes can thus be summarised as

$$P(X) \sim B(n\pi, n\pi q)$$

or as

$$X \sim B(n\pi, n\pi q) \qquad (7.6)$$

In other words, the sampling distribution of X is distributed binomially with mean $n\pi$ and variance $n\pi q$.

If we apply equations (7.3) to (7.5) to the union-membership example, we can confirm the results established in Table 7.3, i.e.

$$E(X) = 5 \times 0.4 = 2$$

$$\text{var}(X) = 5 \times 0.4 \times 0.6 = 1.2$$

$$\text{s.e.}(X) = \sqrt{1.2} = 1.095$$

Examination of (7.2) reveals that a binomial distribution can be described completely (i.e. we know exactly what shape it is and could plot it if necessary) provided only that we know the values of n and π. Conversely, if we know the values of the mean and variance, we can, by a rearrangement of (7.3) and (7.4), determine the values of n and π.

So far we have only considered the sampling distribution of X, the *number* of union members in the sample. We may well be just as interested in the sampling distribution of the *proportion* of successes, which we denote by p. Thus p is the proportion of union members in the sample. Clearly, if X is the number of union members in a sample size n, then the proportion of union members in the sample is equal to X divided by n, i.e.

$$p = X/n \qquad (7.7)$$

The value of p, corresponding to each value of X, is shown in the last column of Table 7.3. The mean, variance and standard error of the sampling distribution of p are given by the following (using the notation of Table 7.2):

$$\text{Mean} \quad E(p) \text{ or } \pi_p = \pi \qquad (7.8)$$

$$\text{Variance} \quad \text{var}(p) \text{ or } \sigma_p^2 = \pi q/n \qquad (7.9)$$

$$\text{Standard error} \quad \text{s.e.}(p) \text{ or } \sigma_p = \sqrt{\pi q/n} \qquad (7.10)$$

and we can use our statistical shorthand to write this as

$$p \sim B(\pi, \pi q/n) \qquad (7.11)$$

i.e. the sampling distribution of p is binomial with mean π and variance $\pi q/n$.

Equation (7.8) is reassuring, indicating as it does that on average the sample proportion is equal to the population proportion. Equations (7.9) and (7.10) imply that the dispersion of p around the population mean of π is inversely proportional to the size of the samples being taken. The larger the value of n, the smaller will the dispersion of p around π be.

Applying these results to our union-membership example gives:

$$E(p) \text{ or } \pi_p = \pi = 0.4$$

$$\text{var}(p) \text{ or } \sigma_p^2 = \pi q/n = 0.4 \times (0.6/5) = 0.048$$

$$\text{s.e.}(p) \text{ or } \sigma_p = \sqrt{0.048} = 0.219$$

We can plot the probability histogram and polygon for p using values from the second and last columns of Table 7.3, noting that $P(X) = P(p)$. These are shown in Figure 7.3. Notice that they are not *density* diagrams, since the horizontal axis is not marked in units of 1, and therefore the area under each curve is thus not equal to 1.0, but in fact is equal to 0.2. To obtain density diagrams we would have to divide each value of probability by 0.2.

Sampling without replacement

In the event that we sample without replacement, i.e. the samples are not independent, then the above equations require some

Figure 7.3 Probability histogram and polygon for p drawn from data in Table 7.3

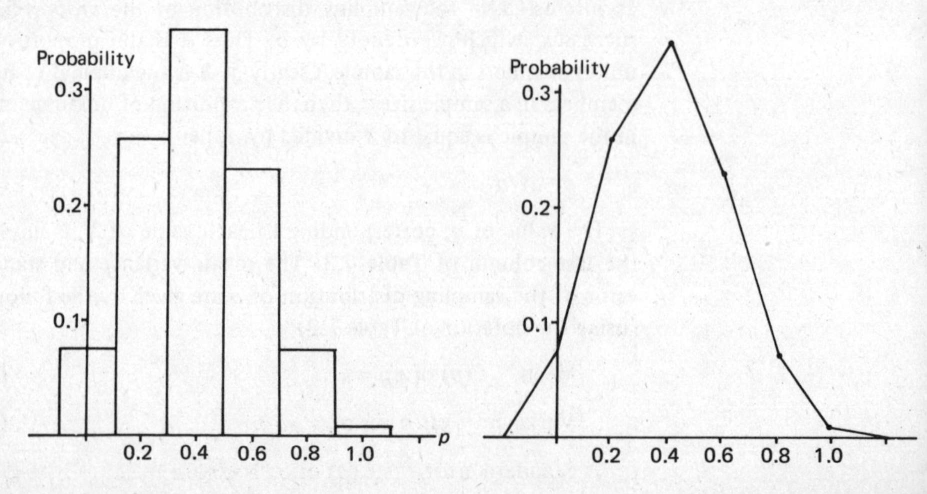

modification. If we take a sample of size n from a population of size N, then we can show that the equations for mean and variance are affected as follows:

$$E(X) = n\pi$$

$$\text{var}(X) = n\pi q \times (N - n)/(N - 1)$$

$$\text{s.e.}(X) = \sqrt{n\pi q \times (N - n)/(N - 1)}$$

and

$$E(p) \text{ or } \pi_p = \pi$$

$$\text{var}(p) \text{ or } \sigma_p^2 = \pi q/n \times (N - n)/(N - 1)$$

$$\text{s.e.}(p) \text{ or } \sigma_p = \sqrt{\pi q/n \times (N - n)/(N - 1)}$$

Notice in fact that the means of the sampling distributions of the sample number of successes and the sample proportions are unaffected but the dispersion of both is smaller, since it is easy to show that

$$(N - n)/(N - 1) < 1$$

The term $(N - n)/(N - 1)$ is called the *finite population correction factor* and in practice we usually find that N is much larger than n which means that

$$(N - n)/(N - 1) \cong 1$$

and we can ignore the effects of sampling without replacement. In the event that the population is infinite in size, then

$$(N - n)/(N - 1) = 1$$

and again it can be ignored. In all that is to follow we assume that the sample size is so small compared with the size of the population that the finite population correction factor equals 1.0.

The normal distribution

Even though equation (7.2) enables us to calculate binomial probabilities relatively quickly, when n becomes large the arithmetic becomes extremely tedious and time-consuming. Fortunately, as the sample size n increases, the binomial distribution tends to become increasingly more and more symmetric irrespective of the value of the population proportion π. Even distributions for which π is close to 1 or 0 which are not at all symmetric or skewed when n is small become symmetric when n is large. Figure 7.4 illustrates this point, showing the binomial distributions for a population with a value of $\pi = 0.1$ for various sizes of n.

In fact as n gets very large, ultimately approaching ∞, the

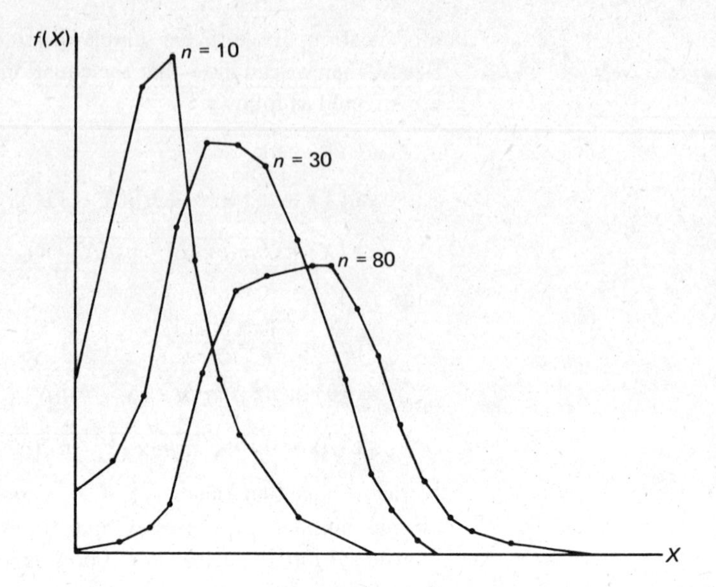

Figure 7.4 Binomial distribution showing increase in symmetry as n gets larger

binomial distribution approaches the shape of the *normal* distribution, i.e.

$$X \sim B(n\pi, n\pi q) \text{ is approximately } N(n\pi, n\pi q) \tag{7.12}$$

and

$$p \sim B(\pi, \pi q/n) \text{ is approximately } N(\pi, \pi q/n) \tag{7.13}$$

Thus (7.13) states that the sampling distribution of sample proportions is distributed *approximately* normal with mean π and variance $\pi q/n$.

Notice that the values of mean and variance are still exact values; it is only the *shape* of the sampling distribution which is approximate.

As we have noted, the approximation is closest for large values of n, i.e. of the order of $n \geqslant 25$, and when the population proportion $\pi = 0.5$. The approximation may still be reasonably close with smaller values of n provided that π is not too close to 0 or 1. A useful rule of thumb is that we can use the approximation when $n\pi$ and $n(1 - \pi)$ both exceed 5.

Essentially, therefore, at least at this stage, we can think of the normal distribution as a useful approximation to the binomial distribution. With this in mind we can now present the following result: the normal distribution of a random variable X has the probability density function

$$f(X_i) = \frac{1}{\sqrt{2\pi}\,\sigma} e^{-1/2[(X_i-\mu)/\sigma]^2} \tag{7.14}$$

where

σ = standard deviation of X

μ = mean of X

π' = the constant 3.14159 (π' is not to be confused with the population mean proportion).

e = the constant 2.71829

$\dfrac{1}{\sqrt{2\pi'}\sigma}$ is a scaling factor which ensures that the area under the curve is 1.

The term

$$e^{-1/2[(X_i-\mu)/\sigma]^2}$$

means raise the constant e to the power

$$-\frac{1}{2}\left(\frac{X_i-\mu}{\sigma}\right)^2$$

an operation which can be performed quite easily on most pocket calculators if necessary.

Because the binomial distribution approaches or approximates the normal distribution as n gets large, then we can use (7.14) as an approximation to the binomial.

Thus, if X represents the *number* of successful outcomes and since the mean of X is $n\pi$ and the standard deviation of X is $\sqrt{n\pi q}$ (using (7.3) and (7.5)), when n is large we can rewrite (7.14) as

$$f(X_i) = \frac{1}{\sqrt{2\pi'}\sqrt{n\pi q}}\, e^{-1/2[(X_i-n\pi)/\sqrt{n\pi q}\,]^2} \tag{7.15}$$

and similarly if p is the *proportion* of successes in a sample of size n, then when n is large we can rewrite (7.14) as

$$f(p_i) = \frac{1}{\sqrt{2\pi'}\sqrt{\pi q/n}}\, e^{-1/2(p_i-\pi)/(\sqrt{\pi q/n})^2} \tag{7.16}$$

since the mean of p is π and its standard deviation $\sqrt{\pi q/n}$.

Notice that we have used the notation $f(X)$ to denote probability density. However, if the variable X is discrete, then $f(X)$ represents probability. This is because the normal distribution is applied to continuous variables. Do not be too intimidated by equations (7.14), (7.15) and (7.16): they look pretty horrendous but in fact are not as bad as they appear. Anyway, we are not going to use them, as we shall see shortly.

Figure 7.5 The normal probability density

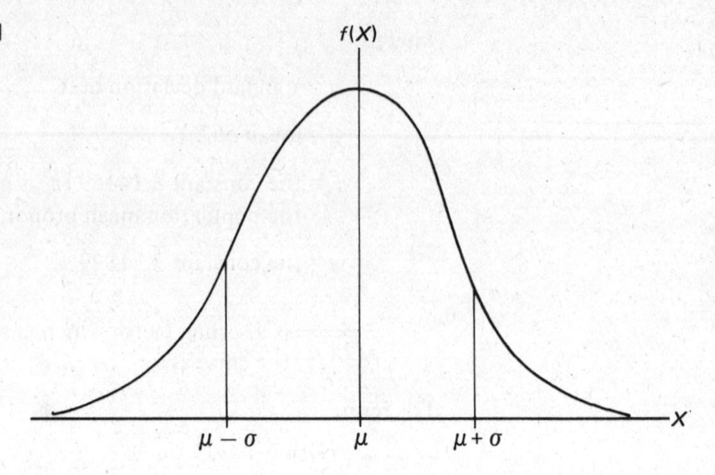

Figure 7.5 is a graphical portrayal of the normal probability density distribution of (7.14). The principal features of the normal probability density distribution are as follows:

(i) The distribution is continuous and symmetric around the mean μ and is bell-shaped.
(ii) It extends from $-\infty$ to $+\infty$.
(iii) The maximum height of the curve occurs at the point $X = \mu$.
(iv) The points of inflection (i.e. where the curve starts to flatten out) occur at the points $X = \mu - \sigma$ and $X = \mu + \sigma$.

In other words, the standard deviation measures the distance from the mean to the point of inflection.

Furthermore, the mean and variance of the normal distribution are denoted

(v) $E(X) = \mu$.
(vi) $\text{var}(X) = \sigma^2$.

The normal distribution is completely described provided we know the values of the parameters μ and σ^2.

The normal distribution is extremely important in economics (as well as in the other social sciences). Besides being a useful approximation to the binomial distribution (particularly when n exceeds 25) it very satisfactorily describes many distributions we might encounter. Moreover, as we have seen in Chapter 5 (pp. 80–3), there exist tables giving the area under the unit normal curve between any two values of the standardised normal variable z. This means that we do not have to work with (7.10), since the tables do the same job much more simply.

As we explained in Chapter 5, the z-table in the appendix to this book has a conventional form: it indicates what proportion of the total area under the curve lies between $z = 0$ and some specific value of z.

In Chapter 5 we were dealing not with probability distributions but with frequency distributions and we were able to use the z-tables in a deterministic way only, e.g. how many white heads of household were between the ages of 30 and 50? Now we have linked the z-tables with probability distributions and hence we can use them in a probabilistic way, e.g. what is the *probability* that a certain number of heads of household are aged between 30 and 50?

Of course, different normal distributions lead to different probabilities, but these differences are only due to different values for the mean and variance, and as we also saw in Chapter 5 (p. 81) we can use the transformation given by equation (5.7), reproduced here for convenience, to convert any scores on a normally distributed variable into a standardised or z-score, i.e.

$$z = \frac{X_i - \mu}{\sigma} \tag{5.7}$$

where the mean is now μ not $\bar{X}$. The procedure with probability distributions is no different from that described in Chapter 5 for frequency distributions, as we shall see in an example in a moment.

First of all let us consider exactly what is involved when we approximate a discrete binomial distribution with a continuous normal distribution. For example, suppose we have defined a *discrete* random variable X as the *number* of successes in a random sample of size 5. Thus X can take any value between 0 and 5. Table 7.4 shows this, together with the *continuous* approximation of X. Also shown is the *discrete* variable p corresponding to the *proportion* of successes and its *continuous* approximation.

Table 7.4 Continuous approximation of discrete random variables X and p with $n = 5$

X_i (discrete)	X_i (continuous)	p (discrete)	p (continuous)
0	−0.5 to 0.5	0	−0.1 to 0.1
1	0.5 to 1.5	0.2	0.1 to 0.3
2	1.5 to 2.5	0.4	0.3 to 0.5
3	2.5 to 3.5	0.6	0.5 to 0.7
4	3.5 to 4.5	0.8	0.7 to 0.9
5	4.5 to 5.5	1.0	0.9 to 1.1

That the approximation between normal and binomial becomes closer as n increases in size is demonstrated by Table 7.5, where we show the continuous approximation to a sample proportion of $p = 0.2$ for various sample sizes, beginning with $n = 5$. Thus, when $n = 5$ we approximate 0.2 with the interval 0.2 ± 0.1, but

Table 7.5 Increasing closeness of binomial and normal approximation as n increase, $p = 0.2$

n	p(continuous)
5	0.1 → 0.3
50	0.19 → 0.21
100	0.195 → 0.205
1000	0.1995 → 0.2005

when $n = 1000$, 0.2 is replaced by the interval of 0.2 ± 0.0005, which is a very close approximation, of the order of 0.25 per cent, to the exact value.

Let us see how well this approximation works in practice in conjunction with the z-distribution, by returning to our union-membership example. Suppose we take a sample of size twenty and calculate the probability that we obtain five union members, first by using the binomial equation (7.2) and then the normal approximation, with the help of the z-table.

Using the binomial distribution we have

$$n = 20$$

$$\pi = 0.4 \quad \text{thus } E(X) = n\pi = 20 \times 0.4 = 8.0$$

$$q = 0.6 \quad \text{and var}(X) = n\pi q = 20 \times 0.4 \times 0.6 = 4.8$$

$$X_i = 5 \quad \text{and s.d.}(X) = \sqrt{4.8} = 2.1909$$

i.e. from (7.2) we have

$$P(5) = \frac{20!}{5!(20-5)!} \times 0.4^5 \times 0.6^{20-5} = 0.0746$$

Now consider Figure 7.6, which shows the continuous normal distribution superimposed on the discrete binomial distribution.

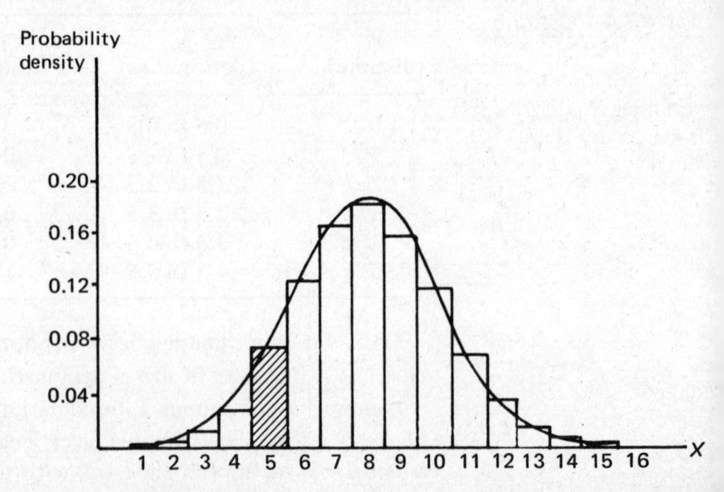

Figure 7.6 Binomial and normal distributions for union-membership example with $n = 20$, $p = 0.4$

The approximation we are discussing means that the area under the normal curve between $X_i = 4.5$ and $X_i = 5.5$ approximates the area under the binomial curve between the same two values. This area is shown shaded.

We want to determine the value of $P(4.5 \leqslant X_i \leqslant 5.5)$, and to do this we must standardise these values of X_i, using (5.7) so that we may use the z-table. Thus, since the mean of X_i is 8.0 and the standard deviation is 2.1909, then, when $X_i = 4.5$,

$$z = \frac{4.5 - 8.0}{2.1909} = -1.598$$

and when $X_i = 5.5$

$$z = \frac{5.5 - 8.0}{2.1909} = -1.1411$$

(the minus signs simply confirm that we are dealing with the left-hand side of the distribution). Thus

$$P(4.5 \leqslant X_i \leqslant 5.5) = P(-1.598 \leqslant z \leqslant -1.1411)$$

The ordinary normal curve of Figure 7.6 has thus been converted into the standardised normal curve, the relevant portion of which is shown in Figure 7.7. We now have to use the z-tables to discover the size of the shaded area.

We can find the area shown shaded by

(i) finding the area between $z = 0$ and $z = -1.598$
(ii) finding the area between $z = 0$ and $z = 1.141$
(iii) subtracting (ii) from (i).

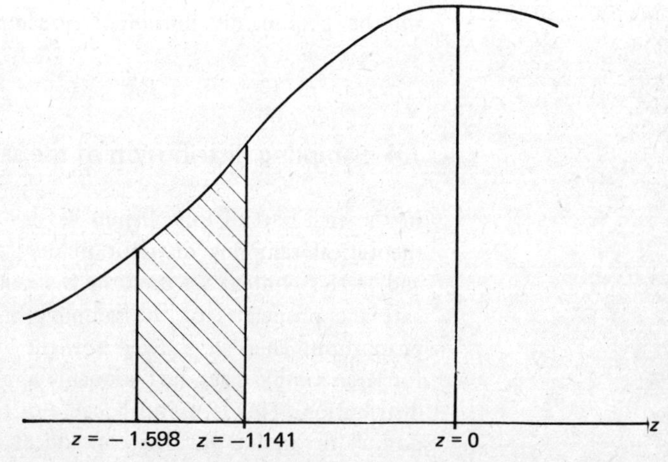

Figure 7.7 Area of standard normal distribution corresponding to values of X of 4.5 and 5.5

Thus

(i) area between $z = 0$ and $z = -1.598 = 0.4450$
(ii) area between $z = 0$ and $z = -1.141 = 0.3730$
(iii) area between $z = -1.141$ and $z = -1.598 = 0.4450 - 0.3730 = 0.0720$

Since the area under a standard normal curve can, as we have explained, be interpreted as probability then we have

$$P(4.5 \leqslant X_i \leqslant 5.5) = 0.0720$$

which is read as 'the probability that the value of X_i will lie between 4.5 and 5.5 is 0.072'.

This compares very well with the binomial probability of 0.0746 – in fact the error is only about 3.6 per cent. We thus see that the normal distribution is an acceptable approximation to the binomial distribution.

* * *

One important feature of the above discussion is that we have been using knowledge of the population proportion to make probability statements about the number of successes we might expect to get in a sampling experiment. There is one difficulty with this concept and that is that in almost all circumstances we do not know what the population proportion is! We are much more interested in a procedure which enables us to determine or estimate the population proportion on the basis of a known number of successes from some given sample. This process is known as *estimation*, and is the subject of the next chapter. We now want to leave discrete random variables and the sampling distribution of proportions using the binomial distribution and turn to the more important subject of continuous random variables and the sampling distribution of the sample mean.

7.4 Sampling distribution of the sample mean

In the first part of this chapter we developed and examined the theoretical sampling distribution of a discrete random variable and in this context we used the binomial distribution to investigate the properties of the sampling distribution of the sample proportion. Then we showed how the binomial distribution can, for large sample sizes, be reasonably approximated by the normal distribution. This is useful because of the existence of the unit-normal or z-table and the consequent ease with which we can make probability statements about the values which the random variable might take.

In all this we were dealing with a discrete random variable and in particular a variable which could assume one of only two values, success or failure.

In this section we want to consider the second of the two theoretical sampling distributions we promised to examine: that is, the sampling distribution of the sample mean. This enables us to consider the distributions of *continuous* random variables and markedly extends the power and scope of the statistical inferential process we are examining. A continuous variable, as we explained in Chapter 6, is one which can assume *any* value in a range of values. For example, if you ask a woman how many children she has had, her answer can only be a discrete number, e.g. 0, 1, 2, 3, etc. If you ask the same woman what the mean age of her children is, her answer could take any value, e.g. 14.2, 8.62, etc., the number of decimal places in her reply being limited only by the arithmetic involved in calculating the mean. The mean age of her children, in other words, is a continuous variable. In economics we are often interested in continuous variables, more so perhaps than those which are discrete. Most of the major economic variables, e.g. income, expenditure, investment, imports, exports, etc., are, to all intents and purposes, continuous, and to consider these we must therefore extend our analysis.

As we shall see, the sampling distribution of the sample mean is a distribution of great importance and plays a major role in much of what is to follow. Most of the basic ideas of theoretical sampling distributions were covered earlier in this chapter, so in this section, therefore, we intend to limit the discussion wherever possible to the essential points.

Assume that we have a population of size N and we use the variable X to denote some characteristic of each member of the population in which we are interested. For example, the population might be all firms in the electronics industry, the characteristic of interest the annual investment by each firm in a particular year, where X_i denotes the investment by the ith firm.

The mean value of X over the whole population is denoted by μ and the dispersion of the individual X_i around this mean, i.e. var(X) is denoted by σ^2.

Let us suppose that we take a large number of independent samples, each of size n and calculate the value of the sample mean in each case ($\bar{X}$ will of course be a random variable). We can then arrange the various sample values of $\bar{X}$ into a probability or sampling distribution (as we did in the matchbox example at the beginning of this chapter). It is possible to show that the mean or expected value and the variance of this sampling distribution of sample means are as follows (using the notation of Table 7.2):

$$E(\bar{X}) \text{ or } \mu_{\bar{X}} = \mu \tag{7.17}$$

$$\text{var}(\bar{X}) \text{ or } \sigma^2_{\bar{X}} = \sigma^2/n \tag{7.18}$$

$$\text{s.e.}(\bar{X}) \text{ or } \sigma_{\bar{X}} = \sqrt{\sigma^2/n} \tag{7.19}$$

Expression (7.17) indicates that on average the mean of the sampling distribution of the sample means is the same as the population mean. In other words, if we were to take a large number of samples, the over-all mean value of all these sample means would on average equal μ. Expressions (7.18) and (7.19) indicate that the dispersion of the sample mean values around $E(\bar{X})$, i.e. around μ, gets narrower as the sample size increases. They also imply that the dispersion of sample means is directly related to the size of the population dispersion of X: the larger is σ^2, the greater will be the dispersion of $\bar{X}$.

To illustrate the use of (7.17) to (7.19) consider a population consisting of just four firms and suppose we are interested in the percentage of female employees in each firm, which we denote by X. The values of X in this population are

12, 20, 24 and 36 per cent

Thus

$\mu = 23$ per cent and $\sigma^2 = 75$ per cent squared

We will work out the mean $\mu_{\bar{X}}$ and the variance $\sigma^2_{\bar{X}}$ of the sampling distribution of the $\bar{X}$ from first principles, i.e. using the probability distribution approach and then confirm our results using equations (7.17) to (7.19). Let us start by considering all possible independent samples of size 2, and in Table 8.6 we let X_1 denote the value of the first of the two firms in the sample and X_2 the second; $\bar{X}$ is also shown. Sampling is with replacement.

The sampling distribution of $\bar{X}$ is shown in Table 7.7, together with the calculations necessary to determine its mean, variance and standard error.

We can confirm these results by substituting in (7.17), (7.18) and (7.19). Thus

Mean $\quad \mu_{\bar{X}} = \mu = 23$ per cent

Variance $\quad \sigma^2_{\bar{X}} = \sigma^2/n = 37.5$

Standard error $\quad \sigma_{\bar{X}} = \sqrt{37.5} = 6.12$ per cent

For samples of size 3 we would have:

Mean $\quad \mu_{\bar{X}} = \mu = 23$ per cent

Variance $\quad \sigma^2_{\bar{X}} = \sigma^2/n = 25$

Standard error $\quad \sigma_{\bar{X}} = \sqrt{25} = 5$ per cent

Table 7.6 Values of $\bar{X}$ for all possible samples with $n = 2$

Sample number	X_1	X_2	$\bar{X}$
1	12	12	12
2	12	20	16
3	12	24	18
4	12	36	24
5	20	12	16
6	20	20	20
7	20	24	22
8	20	36	28
9	24	12	18
10	24	20	22
11	24	24	24
12	24	36	30
13	26	12	24
14	36	20	28
15	36	24	30
16	36	36	36

As we see, the spread of the sampling distribution of the $\bar{X}$s (measured by $\sigma^2_{\bar{X}}$) becomes narrower as n increases from 2 to 3.

Sampling from finite populations

As was the case with sample proportions, if sampling takes place *without* replacement, i.e. if the samples are not strictly independent, then the above formulae relating to the dispersion of the $\bar{X}$s

Table 7.7 Sampling distribution of $\bar{X}$ when $n = 2$

Sample number	Number of samples	$\bar{X}$	$P(\bar{X})$	$\bar{X}\,P(\bar{X})$	$[\bar{X} - E(\bar{X})]$	$[\bar{X} - E(\bar{X})]^2$	$[\bar{X} - E(\bar{X})]^2\,P(\bar{X})$
1	1	12	1/16	0.75	−11	121	7.562
2,5	2	16	2/16	2	−7	49	6.125
3,9	2	18	2/16	2.25	−5	25	3.125
6	1	20	1/16	1.25	−3	9	0.563
7,10	2	22	2/16	2.75	−1	1	0.125
4,11,13	3	24	3/16	4.5	1	1	0.188
8,14	2	28	2/16	3.5	5	25	3.125
12,15	2	30	2/16	3.75	7	49	6.125
16	1	36	1/16	2.25	13	169	10.562
Σ				23			37.5

$E(\bar{X})$ or $\mu_{\bar{X}} = 23$ $\quad$ var($\bar{X}$) or $\sigma^2_{\bar{X}} = 37.5$

s.e.($\bar{X}$) or $\sigma_{\bar{X}} = 6.12$

require some modification using the finite population correction factor.

In these circumstances (7.18) and (7.19) become

$$\text{Variance} \qquad \sigma^2_{\bar{X}} = \sigma^2/n \times [(N-n)/(N-1)] \qquad (7.20)$$

$$\text{Standard error } \sigma_{\bar{X}} = \sqrt{\sigma^2/n \times [(N-n)/(N-1)]} \qquad (7.21)$$

Again, if N is large compared with n, as will often be the case in practice, then

$$(N-n)/(N-1) \cong 1$$

and (7.18) and (7.19) may be used without modification.

Notice, as with sample proportions, that the mean of the sampling distribution is unaffected.

The shape of the sampling distribution of sample means

So far we have said nothing about the *shape* of the sampling distribution of the $\bar{X}$s. We shall do so now and start by considering the case where the variable X in which we are interested is *normally distributed* in the population, with mean μ and variance σ^2, i.e. $X \sim N(\mu, \sigma^2)$.

Now if we select at random one member of this population and denote by X_1 the value which this member takes, then the probability distribution of X_1 must be the same as the frequency distribution of X in the population. We can think of X_1 as standing for all possible values of X that can be obtained. Since we have assumed that X is normally distributed, it follows that X_1 must also be normally distributed.

A sample of size 2 (taken with replacement) will produce values of X_1 and X_2, and following the same argument both X_1 and X_2 will be normally distributed and so on up to samples of size n, in which case $X_1, X_2, \ldots, X_n$ are each normally distributed. Now, we know that

$$\bar{X} = \frac{1}{n}(X_1 + X_2 + \ldots + X_n)$$

i.e.

$$\bar{X} = \frac{1}{n}X_1 + \frac{1}{n}X_2 + \ldots + \frac{1}{n}X_n \qquad (7.22)$$

We now need to make use of the following theorem, which we state without proof:

> If $X, Y, Z \ldots$ are normally and independently distributed random variables and if $a, b, c \ldots$ are constants, then the linear combination of $aX + bY + cZ + \ldots$ is also normally distributed.

Now in (7.22) n and therefore $1/n$ are constants and thus we can think of $\bar{X}$ as being a linear combination of n independent normal variables, $X_1, X_2, \ldots, X_n$. But by the above theorem a linear combination of normally distributed independent random variables is also normally distributed. This implies that if X (and thus $X_1, X_2, \ldots, X_n$) is normal, then $\bar{X}$ *is also normal*, regardless of sample size.

To sum up, and using the results (7.17) and (7.18), if

$$X \sim N(\mu, \sigma^2) \tag{7.23}$$

then

$$\bar{X} \sim N(\mu, \sigma^2/n) \tag{7.24}$$

In other words, if the random variable X in a population is normally distributed with a mean of μ and a variance of σ^2, and if we take a random sample of size n from that population, then the sampling distribution of sample means will also be normally distributed with a mean $\mu_{\bar{X}}$ equal to the population mean μ and variance $\sigma^2_{\bar{X}}$ equal to the population variance σ^2 divided by the sample size n.

As we noted above, the inverse relationship between $\sigma_{\bar{X}}$ and n means that as n increases in size then $\sigma_{\bar{X}}$ gets smaller. We illustrate these results in Figure 7.8. We must always bear in mind that in general we are able to take only *one* sample and clearly we want to have as large a value of n as possible, since (as we can see from Figure 7.8) this is likely to give us a value for $\mu_{\bar{X}}$ closer to μ than would be the case with a smaller n.

The result demonstrated by (7.23) and (7.24) is clearly very important. If X in the population is normally distributed, then $\bar{X}$

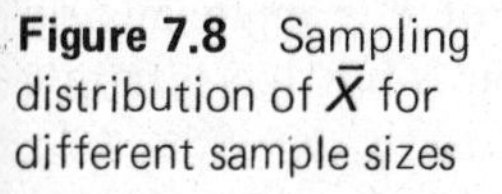

Figure 7.8 Sampling distribution of $\bar{X}$ for different sample sizes

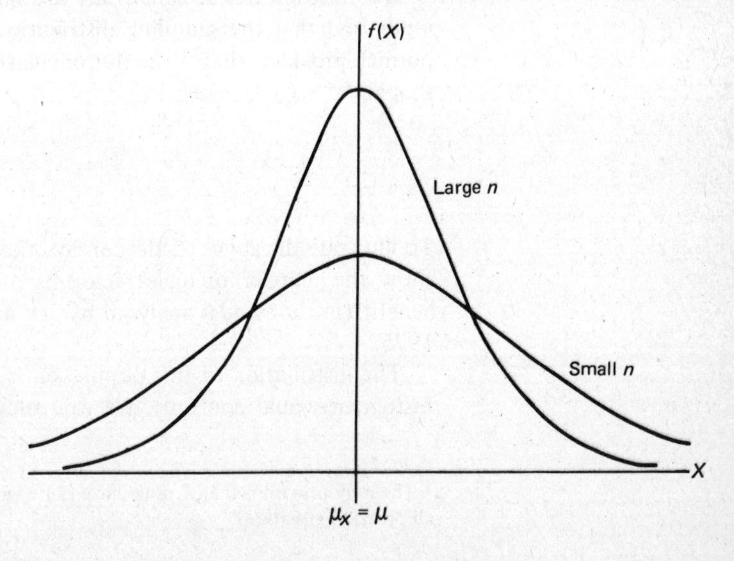

will be normally distributed. It is important to realise that, regardless of the *shape* of $\bar{X}$ in the population, the mean and variance of $\bar{X}$ will *still* be equal to μ and σ^2/n respectively.

The central limit theorem

As important as the results given by (7.23) and (7.24) are, there is a result of even greater importance which we state without proof as follows:

> If X in a population is distributed with a mean of μ and a variance of σ^2, then the distribution of $\bar{X}$ will have a mean of μ and a variance of σ^2/n and its shape will approach the normal distribution as the sample size n increases.

In other words, *whatever* the shape of X in the parent population, the distribution of the sample means will approach normality as the size of the sample increases.[1]

This very important result is known as the *central limit theorem* and is clearly of great significance. We can summarise it as follows:

$$\text{if } X \sim K(\mu, \sigma^2), \text{ then } \bar{X} \sim \text{approximately } N(\mu, \sigma^2/n) \qquad (7.25)$$

where K stands for a distribution of any shape.

In economics we often encounter populations whose distributions are either known to be not normal or where it is unclear what shape their distribution is. The central limit theorem is of great comfort in this context since it assures us that the distributions of sample means will none the less approach normality, and for reasonably large n the approximation may become very close.

We illustrate this schematically in Figure 7.9. Even with small n we find that the sampling distribution of $\bar{X}$ is approximately normal provided that X in the population is not too bizarre a shape.

*　　　*　　　*

To illustrate the above results consider the data in Table 7.8 which show the number of male recipients of sickness and invalidity benefit (in thousands) analysed by age in the United Kingdom in 1975.

The distribution of this population is clearly left-skewed (as a histogram would confirm). We can calculate the mean, variance

[1] The only proviso is that X must have finite variance, which will be true in all practical situations.

Figure 7.9 Illustration of central limit theorem for non-normal populations

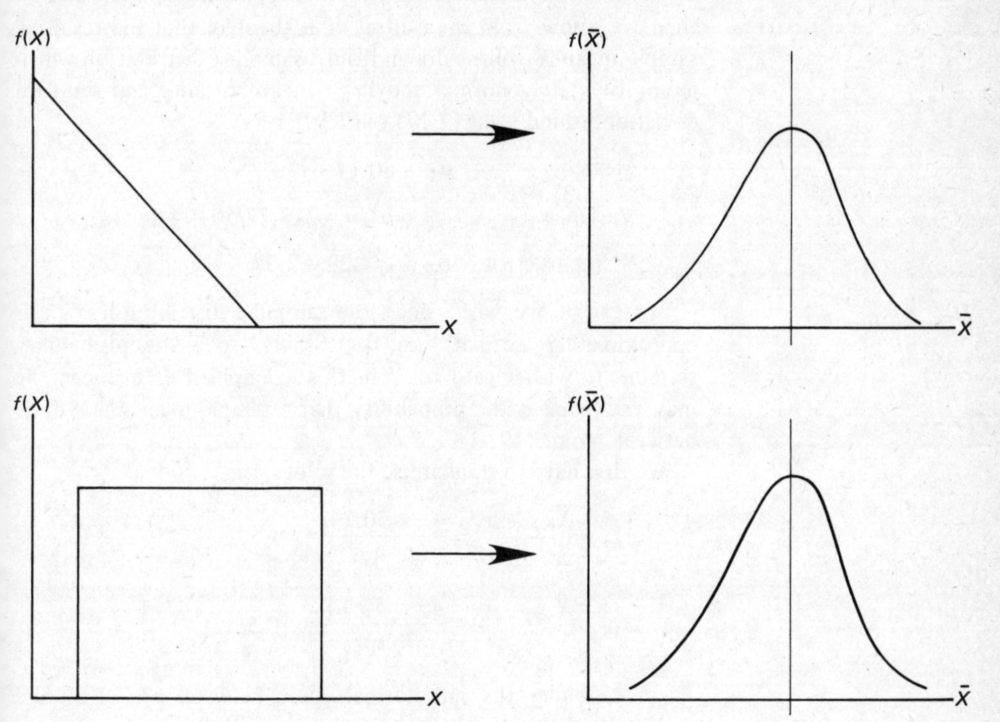

and standard deviation of this population, using (4.6) and (5.6), to be

$$\mu = 50.14 \text{ years}$$

$$\sigma^2 = 213.77$$

$$\sigma = 14.62$$

Table 7.8 Age distribution of male sickness and invalidity beneficiaries, United Kingdom, 1975

Age (years)	Number of recipients (thousands)
0–19	17
20–29	73
30–39	91
40–49	122
50–59	204
60–64	204
65–74*	25

* Assumed value.
Source: *Annual Abstract of Statistics.*

If we take a random sample of, say, fifty from this population, then we know from the central limit theorem that the resulting sample mean $\bar{X}$ will be drawn from a sampling distribution which is approximately normal and has a mean, variance and standard deviation, found using (7.17) to (7.19), of

Mean $\qquad \mu_{\bar{X}} = 50.14$

Variance $\qquad \sigma_{\bar{X}}^2 = \sigma^2/n = 213.77/50 = 4.28$

Standard error $\qquad \sigma_{\bar{X}} = \sqrt{4.28} = 2.07$

We can if we wish, since the sampling distribution of $\bar{X}$ is approximately normal, use the z-tables to make probability statements with regard to $\bar{X}$ in this example. For instance, we may ask 'what is the probability that a sample mean age will lie between 47 and 49?'

We first have to standardise the values, i.e.

$$z_{47} = \frac{\bar{X}_{47} - \mu_{\bar{X}}}{\sigma_{\bar{X}}} = \frac{47 - 50.14}{2.07} = -1.52$$

$$z_{49} = \frac{\bar{X}_{49} - \mu_{\bar{X}}}{\sigma_{\bar{X}}} = \frac{49 - 50.14}{2.07} = -0.55$$

The area under the unit-normal curve between $z = -1.52$ and $z = -0.55$ will be equal to the required probability.

Now, the area between $z = -1.52$ and $z = -0.55$ *equals* the area between $z = 0$ and $z = -1.52$ *minus* the area between $z = 0$ and $z = -0.55$. This is illustrated in Figure 7.10.

From the z-tables we have the area between $z = 0$ and $z = -1.52$

Figure 7.10 Area under $\bar{X}$ and z curves corresponding to $\bar{X} = 47$ and $\bar{X} = 49$

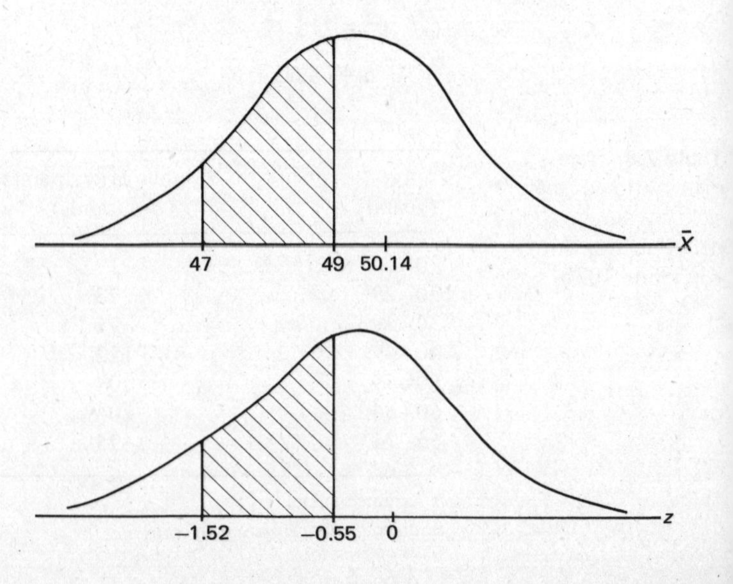

equals 0.4357, and the area between $z = 0$ and $z = -0.55$ *equals* 0.2088. Thus the area between $z = -1.52$ and $z = -0.55$ *equals* 0.4357 – 0.2088, *which equals* 0.2269.

In other words, there is a probability of 0.2269 that a single random sample of size $n = 50$ drawn from this population will yield a sample mean age of between 47 years and 49 years. Or to put it another way, if we were to take a number of such samples of size 50 from this population, just over 22 per cent of them could be expected to yield a sample mean age of between 47 and 49. We can write this probability statement thus:

$$P(47 \leqslant \bar{X} \leqslant 49) = 0.2269$$

We could ask a somewhat different question. For example, 'What values could we expect to find $\bar{X}$ lying between in 95 per cent of such samples of size 50?' The situation is illustrated in Figure 7.11.

Figure 7.11 Boundary values of $\bar{X}$ and z to include 95 per cent of all beneficiaries

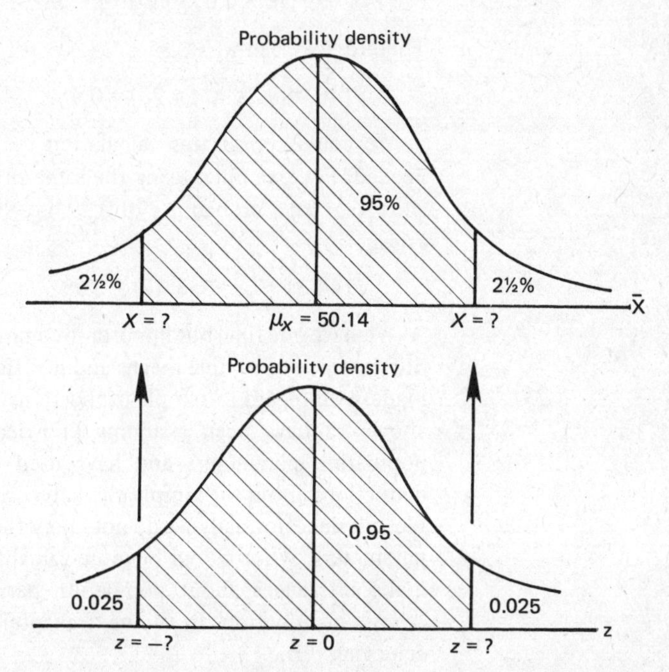

We assume here that the 95 per cent is symmetrical around the mean of the distribution. There are of course an infinite number of intervals containing 95 per cent of samples. The symmetrical interval is the smallest.

We seek the values of z that will make the shaded area equal 0.95 of the total area under the unit-normal curve, i.e. 0.475 in each half since the curve is symmetrical about $z = 0$. We proceed as follows:

1. Find the value of z which provides for an area of 0.475 between

$z = 0$ and $z = ?$ There will be two otherwise identical values of z, one positive and one negative.

2. Use the transformation, i.e. (5.8), to get from the appropriate value of z to the corresponding values of $\bar{X}$. These values of $\bar{X}$ will ensure that 47.5 per cent of the area lies either side of the mean, $\mu_{\bar{X}} = 50.14$ years.

From the z-tables the appropriate values of z are -1.96 and 1.96. Using (5.8) we have

$$\bar{X} = z\sigma + \mu$$

Thus, when $z = -1.96$

$$\bar{X} = (-1.96 \times 2.07) + 50.14 = 46.08 \text{ years}$$

and when $z = 1.96$

$$\bar{X} = (1.96 \times 2.07) + 50.14 = 54.20 \text{ years}$$

Therefore we can write

$$P(46.08 \leqslant \bar{X} \leqslant 54.20) = 0.95$$

We could repeat this calculation for any level of probability needed. For example, using the same procedure as above, we can show that for a probability of 0.99, i.e. 99 per cent of all beneficiaries,

$$P(44.81 \leqslant \bar{X} \leqslant 55.47) = 0.99$$

We have one final but important comment to make on sampling distributions of sample means and it is the same comment that we made with regard to sample proportions. That is, that in all of the above we have been assuming knowledge of the value of the population parameters and have used this knowledge to make deductions about the sample characteristics. Clearly this is unsatisfactory since normally we do not know the value of the population parameters. What would be more valuable would be the ability to make inferences about population parameters on the basis of sample information. In the next chapter we shall discuss ways of achieving this.

7.5 Summary

In this chapter we have examined several important ideas and established some interesting results.

We started by considering the concept of statistical inference, i.e. the idea of investigating population parameters by making inferences from sample characteristics.

We saw how the binomial distribution can be used to describe the probability or sampling distribution of a discrete random variable X which can take one of only two values. In particular we saw that we can describe the sampling distribution of sample proportions in binomial terms. We noted the following results in connection with the sample proportion p:

Mean	$\pi_p = \pi$
Variance	$\sigma_p^2 = \pi q/n$
Standard error	$\sigma_p = \sqrt{\pi q/n}$

i.e. the mean of the sampling distribution of sample proportions is equal to the population proportion and the dispersion of this distribution decreases as sample size increases.

We then proceeded to show how, for reasonably large samples, the normal distribution can approximate the binomial distribution. This is useful because the existence of z-tables relating to the area under the unit-normal curve enables us to make probability statements about the sample proportion p.

The normal distribution is used for continuous variables and we saw that that of principal interest is the sampling distribution of the sample mean $\bar{X}$. In this regard, sampling from a population with mean of μ and variance σ^2 we noted the following results:

Mean	$\mu_{\bar{X}} = \mu$
Variance	$\sigma_{\bar{X}}^2 = \sigma^2/n$
Standard error	$\sigma_{\bar{X}} = \sqrt{\sigma^2/n}$

i.e. the mean of the sampling distribution of sample means is equal to the population mean, while the dispersion of this distribution gets narrower as sample size increases. We then used the z-tables to make probability statements about $\bar{X}$.

The central limit theorem is of great importance. It tells us that regardless of the shape of a parent population the shape of the sampling distribution of sample means will approach normality as sample size increases.

Although this chapter contains much that is interesting and important, we are still not yet in a position to make inferences in the only direction that is of any practical use, i.e. *from* samples *to* populations. In the next chapter we investigate ways of doing this, using the concept of *estimation*.

EXERCISES

7.1 Assume that the fifty-five industries listed in the earnings data in exercise 2.1 constitute the *population* of all industries.

Consider male workers only.

(a) Calculate the mean μ and variance σ of this population. (You may have already performed the appropriate calculations in answering exercises 4.3 and 5.4.)

(b) If we were to take repeated samples of size 5 from this population (with replacement) and calculate the value of the sample mean $\bar{X}$ in each case, what would you anticipate the values of the mean and variance of the sampling distribution of these means to be?

(c) What would they be for samples of size 10?

(d) How are the values in (b) and (c) related to the population mean and variance?

(e) Now take fifty such random samples of size 5 from this population (using, for example, the table of random numbers (Table A.6) at the end of this book). Calculate the value of the mean in each case and arrange these values in a frequency table. Sketch the corresponding histogram and compare it with the histogram of the population male earnings data, paying particular attention to the normality and symmetry of the two figures (you may have already drawn the population histogram in answering exercise 3.1). What is the role of the *central limit theorem* in this respect?

(f) Calculate the mean and variance of the sampling distribution of sample means using the frequency table constructed in part (e) and compare your results with the theoretical values you anticipated in part (b).

(g) Repeat (e) and (f) for samples of size 10 and compare your results with those obtained for the smaller sample size.

(h) Discuss how all these results might be affected if sampling took place without replacement.

7.2 A first-year statistics class contains fifty students whose smoking habits are recorded below (S = smoker, N = non-smoker):

S N N N S N N S S N
N N S S N S N N N N
N S N S N N N S S N
S N S N N N S N N S
N N N N S S N N N S

(a) If this class is considered as a population, calculate the mean π of the proportion of smokers.

(b) If repeated samples of size 10 are taken from this population (with replacement), what would you expect the mean and variance of the sampling distribution of sample proportions to be? How are these values related to the population values?

(c) Now take fifty samples of size 10, and calculate the sample proportion in each case and arrange these values in a frequency table. Hence calculate the mean and variance of the sampling distribution of the sample proportion and compare it with your expected values in part (b).

7.3 A continuous variable, X, representing the IQ of inmates in a prison is distributed with a mean of 95 and a variance of 64. A sample of 100 prisoners is taken and the sample mean $\bar{X}$ is calculated. What is the probability that $\bar{X}$ will be

(a) less than 75
(b) less than 90
(c) between 90 and 100
(d) greater than 90
(e) greater than 100
(f) equal to 100?

8 Estimation of the Population Mean

OBJECTIVES

After reading this chapter and working through the examples students should understand the meaning of the following terms:

estimation	consistency
hypothesis testing	maximum likelihood

estimator
estimate
sampling error
bias
mean square error
efficiency
best linear unbiased
asymptotic property
loss function
point estimator
interval estimator
confidence interval
confidence level
chi-square distribution
degrees of freedom
t-distribution

Students should be able to:

describe the desirable small and large sample properties of estimators
calculate large sample point and interval estimates of a population mean using the z-distribution
calculate an unbiased estimator of population variance
calculate small sample point and interval estimates of a population mean using the t-distribution

---

8.1 Introduction

Once we have taken a random sample from a population and calculated those sample characteristics in which we are interested we can then proceed in either one of two separate (but complementary) directions. These are the two branches into which the process of statistical inferences can be divided and are known as *estimation* and *hypothesis* testing.

The difference between them is that in estimation (*not* in hypothesis testing) we have no preconceived or *a priori* ideas as to the numerical value which a population parameter might take. Estimation is the process by which we hope to be able to determine that value.

In hypothesis testing, on the other hand, we shall have some *a priori* idea or hypothesis as to the value of the population parameter in question; hypothesis testing is the procedure by which we test whether these preconceived ideas are supported by the sample evidence. Figure 8.1 illustrates these concepts schematically.

To further illustrate the difference between the processes consider the following example. For the past ten years, ever since he originated it, Professor Plum has been teaching a course in statistics to economics students. This year he decides to change quite radically his method of teaching, abandoning his traditional 'chalk, talk and walk' approach in favour of programmed learning. At the end of the year the mean score of his students in the

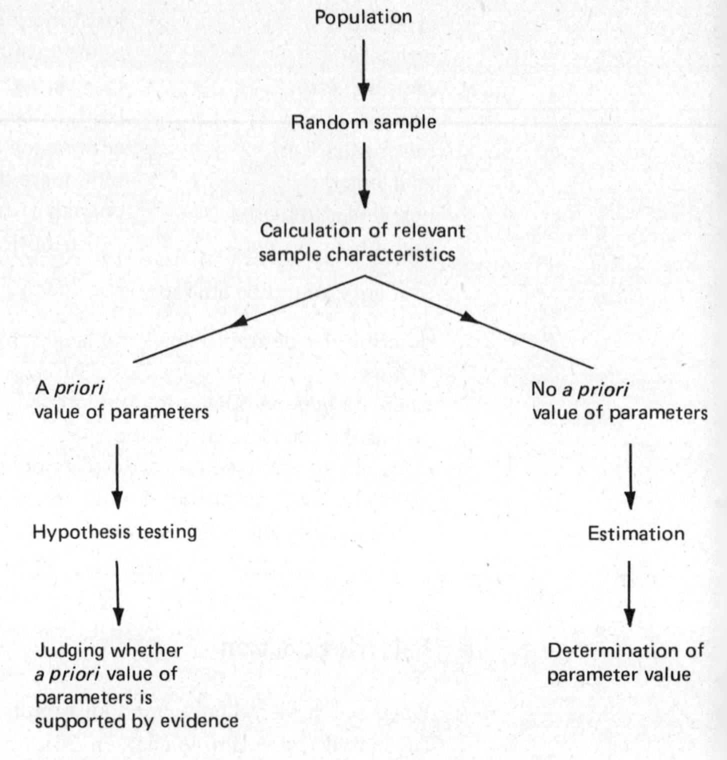

Figure 8.1 Division of statistical inferential process into two branches: estimation and hypothesis testing

examination is 64 per cent. His immediate impression is that this mark is somewhat better than those of previous groups of students taught using traditional methods. To confirm this he gets from the files the marks of all students taking the examination in the previous ten years together with the current year's marks, all of which taken together he considers to be the population. The mean mark thus obtained is 61 per cent. He is about to rush off to the head of his department to be patted on the back when he remembers that the sample mean is unlikely to be the same as the population mean anyway since the sample score is only one mark drawn from a range of possible values known as the sampling distribution. He wonders, therefore, whether the difference of 3 per cent is evidence of a real difference in examination performance or whether it has arisen purely by chance.

Put another way, Professor Plum's dilemma is whether his class is a sample from a population with a mean of 61 per cent (differing from it merely due to chance factors) or a sample from a population with a different mean (one of which is possibly higher). Since in this example Professor Plum has a preconceived or *a priori* idea as to the value of the population parameter (he thinks the mean is 61 per cent), this is a situation in which *hypothesis testing* would help him resolve the problem.

Suppose Professor Plum, his interest aroused, decides he would like to have some idea as to the mean mark of all students in all the first-year examinations in the university. Not having the time to examine the performance of every student in every examination, he takes a random sample of 100 students and determines the population mean mark in this way.

In this example, Professor Plum has no *a priori* idea as to the mean mark in all examinations. He wants to use the sample information as a basis for making an inference as to the value of the population mean, in other words he wants to estimate the value. This situation thus lends itself to the process of *estimation*. In this chapter we discuss *estimation*. We start by considering the two complementary approaches to the problem of choosing suitable estimators, and then follow with a discussion on estimating population means and population proportions.

8.2 Properties of estimators

We have seen that statistical inference is concerned with making informed 'guesses' as to the value of certain population parameters on the basis of information gleaned from sample evidence. We are trying to *estimate* the value of, say, the population mean using knowledge gained from the sample. Usually we will employ some suitable algebraic formula to estimate the value of the population parameter in question. Such a formula is known as an *estimator*, and the numeric result it produces is known as the *estimate*.

Obviously we want estimators that produce 'good' results. For example, if we are trying to estimate the value of the population mean, we want an estimator which can generally be relied upon to produce a value close to the true population mean. In other words, there are certain properties of estimators which we would consider to be desirable and in this context we want to specify what those properties might be.

We start with some general remarks about the properties of estimators and this will give us the opportunity to establish some notation. We can then examine small sample and large sample properties in turn.

Let us consider a population which consists of all possible values of some random variable X and suppose that we are interested in some parameter of this population which we denote as θ (for example, X might be the age of all unemployed males in the United Kingdom and θ the population mean age of this group). Thus we seek an *estimate* of θ, which we can denote as $\widehat{\theta}$, and we determine $\widehat{\theta}$ using some suitable estimating technique, i.e. by means of a suitable *estimator*, applied to sample data.

The sampling distribution of $\widehat{\theta}$ is described by two basic characteristics, its mean and variance, i.e.

$$\begin{aligned} \text{Mean of } \widehat{\theta} &= E(\widehat{\theta}) \\ \text{var}(\widehat{\theta}) &= E[\widehat{\theta} - E(\widehat{\theta})]^2 \text{ (from (6.9))} \\ &= E(\widehat{\theta}^2) - [E(\widehat{\theta})]^2 \end{aligned} \qquad (8.1)[1]$$

Also of importance in this connection are the following:

$$\begin{aligned} \text{Sampling error} &= \widehat{\theta} - \theta \\ \text{Bias} &= E(\widehat{\theta}) - \theta \\ \text{Mean square error} &= E(\widehat{\theta} - \theta)^2 \end{aligned}$$

Sampling error is the difference between the population value of θ and the value of an estimator resulting from a particular sample (sampling error will vary from one sample to another).

Bias is the difference between the mean of the sampling distribution of an estimator $\widehat{\theta}$ and the population mean θ (this will be constant for a given estimator).

Mean square error (often abbreviated to m.s.e.) is a measure of the dispersion of the distribution of the estimator $\widehat{\theta}$ around the population mean θ (this is different from *variance*, which measures the dispersion of $\widehat{\theta}$ around the mean of its sampling distribution $E(\widehat{\theta})$).

It is not difficult (see, for example, Kmenta, 1971, p. 156) to establish the following relationship between m.s.e. and variance:

$$\text{m.s.e.}(\widehat{\theta}) = E[\widehat{\theta} - E(\widehat{\theta})]^2 + [E(\widehat{\theta} - \theta]^2 \qquad (8.2)$$

i.e.

$$\text{m.s.e.}(\widehat{\theta}) = \text{var}(\widehat{\theta}) + \text{bias}^2 \qquad (8.3)$$

Thus the mean square error of $\widehat{\theta}$ equals the variance of $\widehat{\theta}$ plus the bias of $\widehat{\theta}$ squared. The concept of m.s.e. is important when we have to choose between estimators which may be biased and may not have minimum variance (this will be made clearer shortly).

Having established the above concepts and terminology let us now consider some properties of estimators which may be thought desirable. We start by considering properties which hold for *small samples*, i.e. small n. In the section following this we shall consider

[1] $E[\hat{\theta} - E(\hat{\theta})]^2 = E\{\hat{\theta}^2 - 2\hat{\theta}E(\hat{\theta}) + [E(\hat{\theta})]^2 = E(\hat{\theta}^2) - 2E(\hat{\theta})E(\hat{\theta}) + [E(\hat{\theta})]^2\}$
$= E(\hat{\theta})^2 - 2[E(\hat{\theta})]^2 + [E(\hat{\theta})]^2 = E(\hat{\theta})^2 - [E(\hat{\theta})]^2$

properties which only hold (and then approximately) when n is large; we refer to these as *asymptotic* or *large sample* properties.

Small sample properties

Unbiasedness

An unbiased estimator is one that has a sampling distribution the mean of which is equal to the true population parameter θ, whatever the size of the sample. If we use an unbiased estimator, we obtain estimates (from different samples) some of which are below θ and some above (some may indeed coincidentally equal θ), but for an infinite number of samples these values will average out to equal θ (we should stress that unbiasedness tells us nothing about the *difference* between θ and estimates of θ, only that these differences will all cancel out).

Using the expectation notation we can write the property of unbiasedness thus:

$$E(\hat{\theta}) = \theta \tag{8.4}$$

Figure 8.2 illustrates the property of unbiasedness for a symmetric sampling distribution. We should note that for an unbiased estimator

$$\text{m.s.e.}(\hat{\theta}) = \text{var}(\hat{\theta})$$

from (8.3).

The property of unbiasedness, however desirable, is not of much comfort by itself. An estimator may well be unbiased but have a very large variance, in which case the value of $\hat{\theta}$ is likely to be considerably different from θ. This leads us to the next property.

Figure 8.2 An unbiased estimator

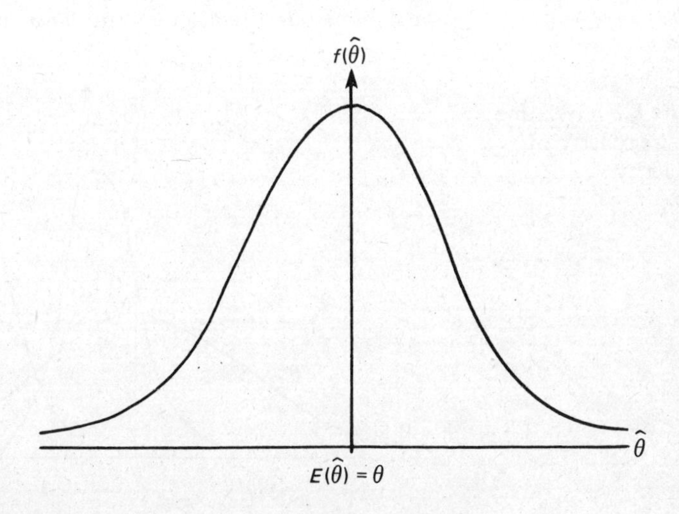

Efficiency

Consider the sampling distributions of three different estimators shown in Figure 8.3. The estimator $\hat{\theta}_1$ is unbiased but has large variance, $\hat{\theta}_2$ has a small bias but also small variance, $\hat{\theta}_3$ has a large bias and small variance. In other words, in general, $\hat{\theta}_2$ will give estimates closer to $\hat{\theta}$ than will $\hat{\theta}_1$ or $\hat{\theta}_3$, despite the fact that $\hat{\theta}_1$ is unbiased. A combination of $\hat{\theta}_1$ and $\hat{\theta}_2$, i.e. one which was unbiased *and* had small variance, would provide even better estimates.

This concept leads us to define a second desirable property which we would like an estimator to have, i.e.

> An *efficient* estimator has the smallest variance among all unbiased estimators.

Put another way, an estimator $\hat{\theta}$ is *efficient* if

$$\left.\begin{array}{ll} E(\hat{\theta}) = \theta & \text{(i.e. } \hat{\theta} \text{ is unbiased)} \\ \operatorname{var}(\hat{\theta}) \leqslant \operatorname{var}(\tilde{\theta}) & \text{(where } \tilde{\theta} \text{ is any other unbiased} \\ & \text{estimator)} \end{array}\right\} \quad (8.5)$$

For obvious reasons an efficient estimator is sometimes also called a *minimum variance* unbiased estimator, or just a *best* unbiased estimator.

The problem with efficiency is that it is difficult to measure, in the sense that we would have to examine *all* unbiased estimators to ensure that any particular one was efficient. It is, however, much easier to compare the variances of only a small number of unbiased estimators, and in this case we can think of an estimator as being *relatively efficient*, i.e. having the smallest variance amongst a small group of unbiased estimators. Even this process may not be easy with certain types of unbiased estimator, but if we confine our attention only to *linear* unbiased estimators (and

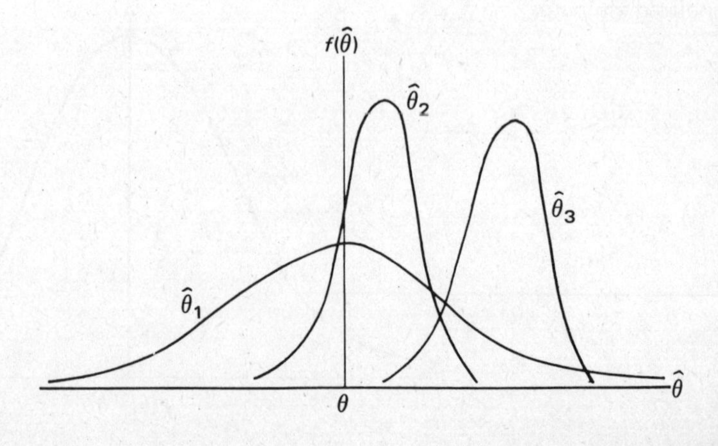

Figure 8.3 Relating to the property of efficiency

we shall see what linear means in a moment), then we can redefine efficiency in a narrower sense thus:

$\hat{\theta}$ is called a *best linear unbiased estimator* of θ provided that

$$\left.\begin{array}{l}\hat{\theta} \text{ is a linear function of the sample observations} \\ E(\hat{\theta}) = \theta \text{ (i.e. } \hat{\theta} \text{ is unbiased)} \\ \text{var}(\hat{\theta}) \leqslant \text{var}(\tilde{\theta}) \text{ (where } \tilde{\theta} \text{ is any other linear} \\ \qquad\qquad\qquad \text{unbiased estimator)}\end{array}\right\} \quad (8.6)$$

(Best linear unbiased estimator is often abbreviated to BLUE, or BLU, estimator.)

The linearity condition means that we must be able to express $\hat{\theta}$ in terms of the sample observations thus:

$$\hat{\theta} = a_1X_1 + a_2X_2 + \ldots + a_nX_n \quad (8.7)$$

where the a_i are constants which remain fixed regardless of the particular sample values of X.

In connection with efficiency we should finally note that it is assumed that the estimator uses *all* of the information in the sample which relates to θ. For example, the sample mean does this, whereas the sample median (which does not consider the actual value of each observation but only their number) does not.

We now turn to look at large sample properties.

Asymptotic or large sample properties

When we talk about small sample and large sample (or asymptotic) properties of estimators perhaps we might more accurately talk about the properties of estimators for which the sample size is *fixed* and about those for which the sample size can be increased to $n = \infty$. Since in economics we cannot envisage situations where we can increase n indefinitely, why should we need to consider asymptotic properties? Two important reasons are, first, there are some estimators for which it is impossible to establish the small sample properties, in which case we can only compare asymptotic properties, and second, it seems more sensible if we have a choice of estimator to choose one whose properties would 'improve' if only we could increase n, rather than one whose properties were incapable of improvement.

There are three asymptotic properties which we want to consider. They closely parallel the small sample properties.

Asymptotic unbiasedness

An estimator is said to be asymptotically unbiased if the bias, i.e. the difference between the asymptotic mean or expected value and the population mean, reduces to zero as the sample size increases to infinity.

Thus the estimator θ is asymptotically unbiased if

$$\lim_{n \to \infty} [E(\hat{\theta}) - \theta] = 0 \tag{8.8}$$

which is read as 'as n approaches infinity the limiting value of $[E(\hat{\theta}) - \theta]$ is zero'.

Estimators which have the small sample property of unbiasedness will also be asymptotically unbiased, but the reverse is not always true.

Consistency

We can say that an estimator is consistent if, as the sample size increases towards ∞, the variance reduces to zero. When the variance becomes zero we say that the distribution collapses around a single point, i.e. it becomes a vertical line.

Thus $\hat{\theta}$ is a consistent estimator if the following condition is satisfied:

$$\lim_{n \to \infty} [\text{var}(\hat{\theta})] = 0 \tag{8.9}$$

Note that it does not necessarily follow that any bias which the estimator might have is similarly reduced to zero. If this is the case, larger samples increase the certainty about something which is incorrect. For this reason many authors prefer to define a *consistent* estimator as one in which both the variance *and* the bias reduce to zero as n increases to infinity. This is a definition which we prefer.

In Figure 8.4 we show an estimator which loses its bias and reduces its variance as n is progressively increased.

Figure 8.4 The property of consistency

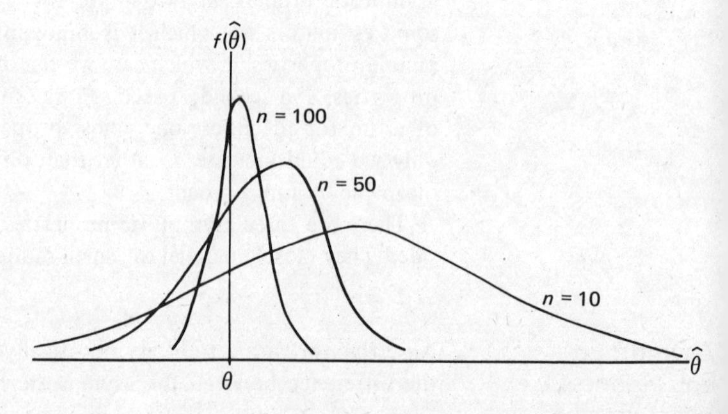

Asymptotic efficiency The concept of efficiency is a little more difficult to deal with, but broadly speaking an estimator is asymptotically efficient if it satisfies the following two conditions:

(i) It is consistent.
(ii) It has a smaller asymptotic variance than any other consistent estimator.

Thus $\hat{\theta}$ is an asymptotically efficient estimator if amongst all consistent estimators it has the smallest variance.

Maximum likelihood At the end of our discussion on small sample properties we established the concept of searching for the best linear unbiased estimator since this has all the desirable properties considered. Analogously we can think of another type of estimator in the context of asymptotic properties – this is an estimator known as the *maximum likelihood estimator* (MLE). The mathematics of the MLE prevent us considering it other than superficially, but the important feature (as far as we are concerned) of maximum likelihood estimators is that they are consistent and asymptotically efficient. Thus if we are presented with a maximum likelihood estimator we do not need to investigate its asymptotic properties – we know what they are (just as, if we were presented with a BLU estimator, we would know its properties). We often do not know what the small sample properties of an MLE are, however, and these would have to be analysed.

Very briefly the principle behind maximum likelihood estimation is examining the sample and asking the question 'which population did this sample probably come from?'

For example, suppose a sample has a mean of $\bar{X} = 15$ and we have three different populations from which the sample might have been drawn: the first with $\mu_1 = 10$, the second with $\mu_2 = 15$ and the third with $\mu_3 = 20$.

We would conclude that the population mean is most *likely* to be $\mu_2 = 15$ (we use the word 'likely' rather than 'probably' because the population mean is either 15 or it is not).

We now want to consider a parallel but rather different approach to the problem of finding suitable estimators.

8.3 Choice of estimator: the loss function

We want to conclude this examination of the desirable properties of estimators with a brief discussion of the problems of choosing the 'best' estimator. In this context the word 'best' is to be interpreted as meaning the estimator which leads to the least serious consequence or loss in the event that the estimator $\hat{\theta}$ differs from the population parameter θ (as it inevitably will).

To see what we mean, suppose a greengrocer has 120 kg of peaches which he wishes to sell. An economist friend of his believes that the demand equation is

$$q = 240 + \beta p$$

where q = quantity sold (kg), p = price (p/kg)

Suppose the true population value of β is –4 but the economist overestimates it and decides $\hat{\beta} = -3$. He therefore determines the price which will sell all the peaches using

$$120 = 240 - 3p$$

thus $\quad p = 40.$

The amount of peaches actually sold will thus be

$$q = 240 - (4 \times 40) = 80 \text{ kg}$$

The greengrocer will thus have 40 kg of peaches on his hands, which is his loss caused by an overestimation of β.

Losses may not be symmetrical, i.e. the loss caused by underestimation of a parameter may not equal the loss caused by overestimation, and the loss may not be linear – in other words, it may get *progressively* worse as the degree of over- or underestimation increases. For example, we might have a loss function such as that shown in Figure 8.5, where underestimation losses are not as severe or as rapidly increasing as are the losses caused by overestimation.

Let us now consider loss functions in general. As before we assume that θ is the parameter to be estimated and $\hat{\theta}$ is our estimator. We assume that the loss function is generally of the form

$$L = f(\hat{\theta}, \theta) \tag{8.10}$$

Figure 8.5 Non-symmetric loss function

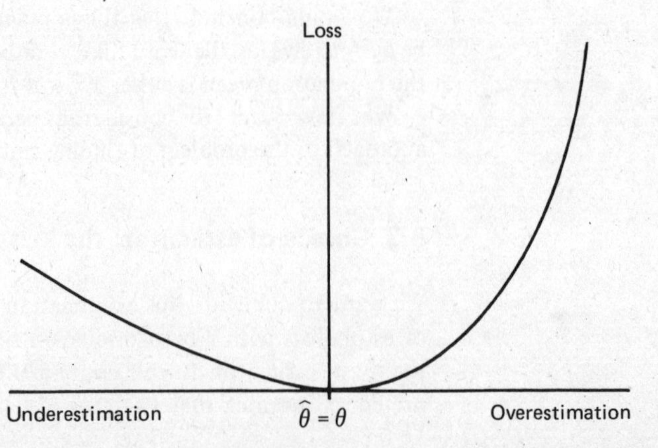

i.e. the loss is a function of both $\hat{\theta}$ and θ.

The specific form of (8.10) will differ from situation to situation but a widely used form is given by

$$L = k(\hat{\theta} - \theta)^2 \tag{8.11}$$

This is a quadratic loss function, the loss depending on the squared difference between $\hat{\theta}$ and θ (and k is some constant). If there is no difference between $\hat{\theta}$ and θ, then the loss is zero, but for differences of either negative or positive sign the loss increases at an increasing rate. This loss function has the general shape shown in Figure 8.6.

Figure 8.6 A symmetric quadratic loss function

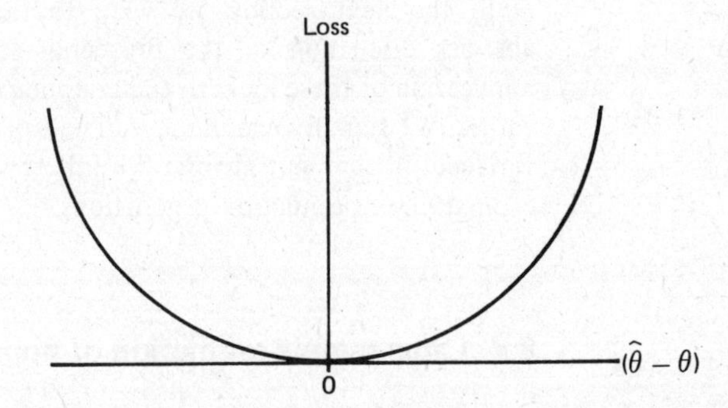

A particular estimator applied to a number of samples will produce a corresponding number of estimates which are of course random variables. Thus for each type of estimator the $\hat{\theta}$ which occurs in (8.10) is not to be thought of as a particular estimate but rather as a random variable. We do not want the estimator which works well with some given sample, rather we want the estimator which does well with all samples of size n. In other words, we are not trying to minimise the loss on a particular sample but the *expected* value of the loss. Thus taking expected values of (8.11) gives

$$E(L) = kE(\hat{\theta} - \theta)^2$$

But $E(\hat{\theta} - \theta)^2 = \text{m.s.e.}$ Thus

$$E(L) = k \times \text{m.s.e.} \tag{8.12}$$

i.e.

$$E(L) = k[\text{var}(\hat{\theta}) + (\text{bias } \hat{\theta})^2] \tag{8.13}$$

In other words, the 'best' estimator is the one for which variance plus squared bias is smallest. We must bear in mind that (8.13) was only arrived at having assumed a loss function of specific form, i.e. a quadratic type as in (8.11), but as we mentioned above this form has proved useful in practice. We have also made the rather large assumption that variance and bias^2 are given equal weights.

* * *

This concludes our general discussion on estimators. In connection with the choice of an estimator we have examined two complementary approaches. We want an estimator which has as many of the desirable properties discussed above as possible. When we have to choose between competing estimators with varying degrees of bias and variance we can make use of the loss function approach (i.e. equation (8.13)) to aid our decision as to which is the best.

In the next section we want to turn from the somewhat abstract discussion of the preceding sections to consider the application of these ideas to the sampling distributions considered earlier. We start by examining estimators of population mean and variance. In the next chapter we follow with a discussion on the estimation of population proportions.

8.4 Large sample estimation of population mean μ

In section 7.4 we discussed the sampling distribution of the sample mean for a population with mean μ and variance σ^2 and established the following results for samples of size n:

$$E(\bar{X}) = \mu \qquad (7.17)$$

$$\text{var}(\bar{X}) = \sigma^2/n \qquad (7.18)$$

$$\text{s.e.}(\bar{X}) = \sqrt{\sigma^2/n} \qquad (7.19)$$

In this section we want to derive an estimator of μ and consider, in the light of the preceding sections, what its properties are. As it happens, there are two distinct types of estimators to consider: *point* estimators and *interval* estimators. A point estimator is essentially our single best unqualified 'guess' as to the value of μ. An interval estimator provides us with some information as to the probability of our guess being within an acceptable distance of the true value of μ.

A point estimator of μ

Equations (7.17) to (7.19) describe the sampling distribution of the sample means in terms of the population mean μ, the variance σ^2 and sample size n (although, as we pointed out at the time, we do not usually know either μ or σ^2, and hence (7.17) to (7.19) are not of much practical use).

The problem which faces us is as follows: what value should be placed on μ to ensure that it is the best guess we can make? We have already seen that

$$E(\bar{X}) = \mu$$

i.e. on *average* the value of the sample mean will equal the population mean, and thus it seems entirely reasonable to choose $\bar{X}$ as a point estimator of μ. But what of its properties? Recall that for any particular sample the sample mean is calculated from

$$\bar{X} = \frac{1}{n}(X_1 + X_2 + \ldots + X_n)$$

i.e.

$$\bar{X} = \frac{1}{n}X_1 + \frac{1}{n}X_2 + \ldots + \frac{1}{n}X_n$$

Thus we see that $\bar{X}$ is a *linear* function of the sample observations.

Moreover (7.17) implies that $\bar{X}$ is an *unbiased* estimator of μ, and (7.18) implies that as n increases the variance of $\bar{X}$ decreases. Taken together (7.17) and (7.18) imply that $\bar{X}$ is a *consistent* estimator of μ. We can state therefore that $\bar{X}$ is a BLU estimator of μ. (It is also possible to show (see, for example, Common, 1976, pp. 138–9) that $\bar{X}$ is also the maximum likelihood estimator of μ provided that the parent population is normally distributed.)

Given that we have $\bar{X}$ as a BLU estimator of μ, we would now like to be able to qualify $\bar{X}$ by attaching to it some probability of it being correct. This leads us to interval estimation.

An interval estimator of μ

We already know that the sampling distribution of the sample means is normal if the parent population is normal and approximately normal (by the central limit theorem) even when the parent population is not normal. To simplify matters and save time we are going to assume from now on that we are sampling from a normal population. Thus

$$\bar{X} \sim N(\mu, \sigma^2/n)$$

We can standardise $\bar{X}$ using

$$z = (\bar{X} - \mu)/\sqrt{\sigma^2/n}$$

but we know that z is also normally distributed with a mean of zero and a variance of 1.0, which means that

$$(\bar{X} - \mu)/\sqrt{\sigma^2/n} \sim N(0, 1) \qquad (8.1)$$

and we can use the z-tables if we want to make probability statements about μ. To see how this is done, let us consider a specific example first and then generalise.

We proceed as follows. Recall that for a variable with a unit-normal distribution 95 per cent of all values lie between $z = -1.96$ and $z = 1.96$, or (put another way) 0.95 of the total area under the curve lies between these two values. Remember that we look up $0.95/2 = 0.475$ in the body of the z-tables because of the symmetry of the normal curve around the mean (0.5 of the area in each half). In other words, 0.475 of the area lies between $z = -1.96$ and $z = 0$ and 0.475 between $z = 0$ and $z = 1.96$. Thus we can write

$$P(-1.96 \leqslant z \leqslant 1.96) = 0.95$$

or, using (8.14),

$$P[-1.96 \leqslant (\bar{X} - \mu)/\sqrt{\sigma^2/n} \leqslant 1.96] = 0.95$$

We can rearrange this by multiplying terms inside the brackets by $\sqrt{\sigma^2/n}$, subtracting $\bar{X}$ and multiplying through by -1.0. We then have

$$P(\bar{X} - 1.96\sqrt{\sigma^2/n} \leqslant \mu \leqslant \bar{X} + 1.96\sqrt{\sigma^2/n}) = 0.95 \qquad (8.15)$$

This expression means that if we took an infinite number of samples of size n from the population in question, and if for each sample we calculated the interval using (8.15), then 95 per cent of those intervals would embrace or enclose the true population parameter μ.

So we can be 95 per cent confident that an interval so constructed will contain μ.

For this reason the interval

$$(\bar{X} - 1.96\sqrt{\sigma^2/n} \leqslant \mu \leqslant \bar{X} + 1.96\sqrt{\sigma^2/n}) \qquad (8.16)$$

is known as the 95 per cent *confidence interval* for μ and provides us with an interval estimate of μ.

It is important to note that we cannot interpret (8.15) or (8.16) as meaning that μ will be in the interval 95 per cent of the time. The population mean μ is a fixed and constant value and therefore either will or will not lie in the interval (which, we do not know). It is the interval which is moving, being dragged around from sample to sample by the different sample means. If in 95 per cent of the samples the interval does enclose μ, this implies that in 5 per cent of the samples it does not. This moving about of the interval from one sample to another is illustrated in Figure 8.6.

For any given population the interval has a fixed width (equal to $3.92\sqrt{\sigma^2/n}$) but the actual position of the interval will depend on the particular sample value $\bar{X}$. In Figure 8.7 we can see that the intervals formed from samples 1 and 2 do embrace μ, but not that formed from sample 3.

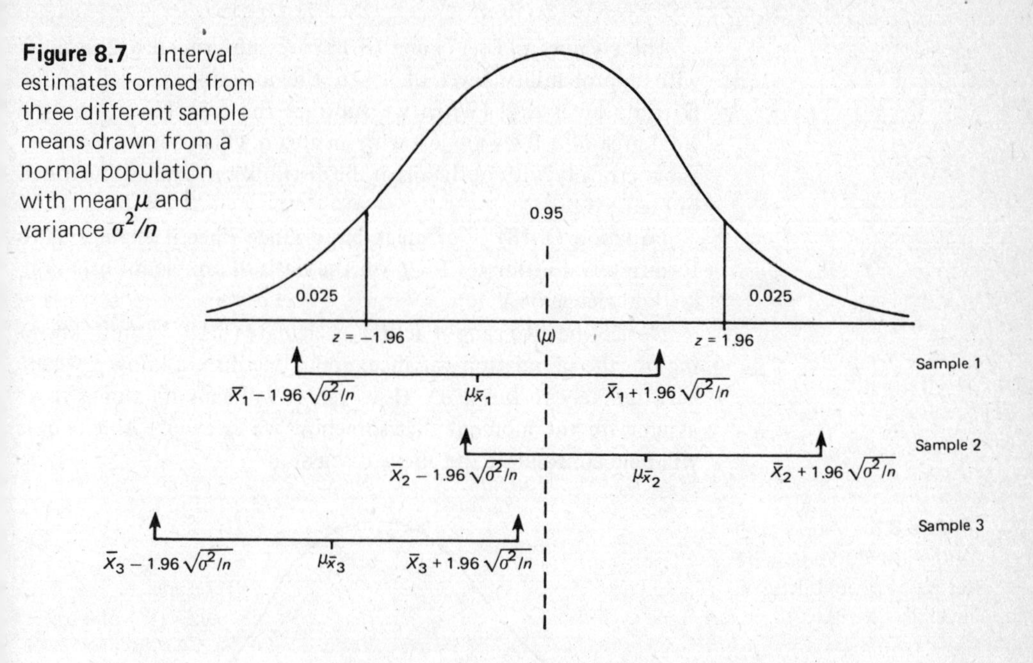

Figure 8.7 Interval estimates formed from three different sample means drawn from a normal population with mean μ and variance σ^2/n

When we want to construct a confidence interval we can choose any level of confidence we like (although 95 per cent is the most often used). The higher the level of confidence, the wider becomes the corresponding interval and the less precise (and therefore the less useful) the interval estimate. In the extreme a 100 per cent confidence interval stretching from $-\infty$ to $+\infty$ would tell us nothing about the value of μ.

Let us now extend the above discussion to cover any level of confidence by rewriting (8.15) and (8.16) in more general terms, i.e.

$$P(\bar{X} - z_{\alpha/2}\sqrt{\sigma^2/n} \leqslant \mu \leqslant \bar{X} + z_{\alpha/2}\sqrt{\sigma^2/n}) = 1 - \alpha \qquad (8.17)$$

and

$$(\bar{X} - z_{\alpha/2}\sqrt{\sigma^2/n} \leqslant \mu \leqslant \bar{X} + z_{\alpha/2}\sqrt{\sigma^2/n}) \qquad (8.18)$$

Expression (8.18) is a $100(1-\alpha)$ per cent confidence interval for μ, and provides us with an interval estimate of μ, and $(1-\alpha)$ is the probability corresponding to the chosen level of confidence, i.e. $100(1-\alpha)$ is the *confidence level*. For example, with 95 per cent $1-\alpha = 0.95$, with 90 per cent $1-\alpha = 0.9$, and so on. Generally a $100(1-\alpha)$ per cent confidence level means that if we took an infinite number of samples of size n from a normal population with a mean of μ and a variance of σ^2, then $100(1-\alpha)$ per cent of the confidence intervals so formed would embrace μ (and 100α per cent would not).

The z values in (8.17) and (8.18) are subscripted $\alpha/2$ because with a probability level of $1 - \alpha$ the area in each tail of the distribution is $\alpha/2$. (When we subtract the area $1 - \alpha$ from the total area of 1.0 we are left with an area α which has to be shared symmetrically with both tails of the distribution. This is illustrated in Figure 8.8.)

Equation (8.18) is of great importance since it enables us to form interval estimates for μ on the basis of sample information, i.e. knowledge of $\bar{X}$.

Notice that we cannot actually calculate (8.18) without knowing σ^2, the population variance, and if we do not know μ we are very unlikely to know σ^2. However, it will simplify things if we assume for the moment that somehow we know σ^2. An example will help consolidate the above discussion.

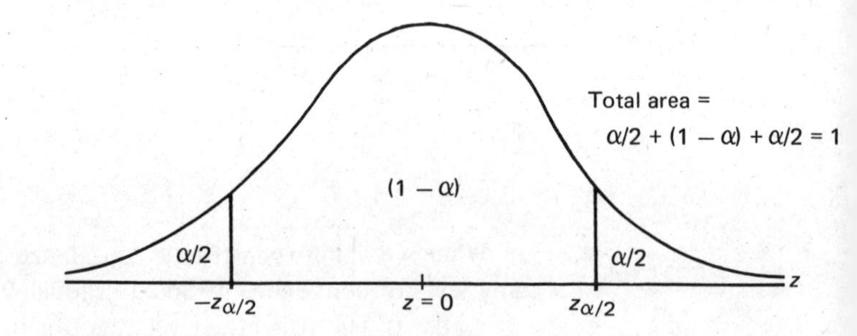

Figure 8.8 Areas under the curve in terms of probability level $1 - \alpha$

Example

Table 4.1 in the 1973 *General Household Survey* reveals that the mean number of persons per household in Great Britain in 1971 was 2.91. This was based on a very large sample of 11 988. Let us calculate the 95 per cent confidence interval for the population mean household size μ.

We shall have to assume that σ^2 is known to equal 0.2. So we have

$$\bar{X} = 2.91$$

$$\sigma^2 = 0.2$$

$$n = 11\,988$$

$$1 - \alpha = 0.95$$

$$z_{\alpha/2} = 1.96$$

Substituting in (8.18) gives

$$(2.91 - 1.96 \times \sqrt{0.2/11\,988} \leqslant \mu \leqslant 2.91 + 1.96 \times \sqrt{0.2/11\,98}$$

which resolves to

$$(2.9020 \leqslant \mu \leqslant 2.9180)$$

which means that we can be 95 per cent confident that the interval from 2.9020 persons per household to 2.9180 persons per household will contain the true population value – which we could write as

$$P(2.9020 \leqslant \mu \leqslant 2.9180) = 0.95$$

Strictly speaking, since the *General Household Survey* sample was not taken with replacement, we should apply the finite population characteristic, but with a population as large as this is (of the order of 25 million) the results would not be noticeably different).

In the preceding discussion on estimation we have assumed that σ^2, the variance of the population, is known, and thus we have been able to form confidence intervals for μ using (8.18). In practice, of course, it is very unlikely that we would know σ^2 and in the next section we consider how we might find a suitable estimator for it.

The ramifications of this discussion are of considerable significance, leading as they do to consideration of two new and important distributions.

8.5 Estimation of population variance σ^2

We have a population which we assume is distributed as

$$X \sim N(\mu, \sigma^2)$$

We have already seen how we can use $\bar{X}$ as a BLU estimator of μ, but the interval estimator assumes knowledge of the value of σ^2, which in general will be unknown.

Since we know that $\bar{X}$ is a BLU estimator of μ an obvious candidate as an estimator of σ^2 would seem to be the sample variance. If we denote the sample variance by $\tilde{\sigma}^2$, then

$$\tilde{\sigma}^2 = \frac{1}{n} \Sigma(X_i - \bar{X})^2 \tag{8.19}$$

Repeated samples will produce different values for $\tilde{\sigma}^2$ and we are interested in determining what the distribution of $\tilde{\sigma}^2$ is and in particular what its average or *expected* value (i.e. its mean) is.

It is not difficult (see, for example, Kane, 1969, p. 185) to demonstrate the following result:

$$E(\tilde{\sigma}^2) = \sigma^2 (n - 1)/n \tag{8.20}$$

Thus the expected or *average* value of the sample variance is *not* equal to the population variance. As we can see from (8.20),

$\tilde{\sigma}^2$ is a biased estimator of σ^2, the value we obtain for $E(\tilde{\sigma}^2)$ is too small, i.e. $\tilde{\sigma}^2$ *underestimates* σ^2.

As it happens, (8.20) suggests a simple way by which we could determine an unbiased estimator of population variance. If we multiply any value of sample variance by $n/(n-1)$, on average we obtain an unbiased estimator of σ^2. If we use s^2 to denote this unbiased estimator, then

$$s^2 = \tilde{\sigma}^2 n/(n-1) \tag{8.21}$$

from which, using (8.19),

$$s^2 = \frac{1}{n-1}\Sigma(X_i - \bar{X})^2 \tag{8.22}$$

and from (8.20) and (8.21) we have

$$E(s^2) = E(\tilde{\sigma}^2)n/(n-1) = [n/(n-1)][(n-1)/n]\sigma^2$$

Therefore

$$E(s^2) = \sigma^2 \tag{8.23}$$

and s^2 is an unbiased estimator of σ^2.[1] Replacing σ^2 by s^2 in (8.17) and (8.18) gives

$$P(\bar{X} - z_{\alpha/2}\sqrt{s^2/n} \leqslant \mu \leqslant \bar{X} + z_{\alpha/2}\sqrt{s^2/n}) = 1 - \alpha \tag{8.24}$$

$$(\bar{X} - z_{\alpha/2}\sqrt{s^2/n} \leqslant \mu \leqslant \bar{X} + z_{\alpha/2}\sqrt{s^2/n}) \tag{8.25}$$

Expression (8.25) provides us with an *interval estimate* of μ. Note, however, that this estimator is now only approximate, since we have replaced the unknown σ^2 with its estimator s^2.

We can illustrate these results by returning to the example considered in section 7.4. This concerned a population consisting of only four members for which $\mu = 23$ and $\sigma^2 = 75$. The values of X for all possible samples of size $n = 2$ were shown in Table 7.6. We shall use this example to determine the sampling distribution of s^2 and $\tilde{\sigma}^2$ for $n = 2$.

To see how we proceed, consider sample number 2, for which

$$X_1 = 12; \quad X_2 = 20; \quad \bar{X} = 16$$

Then, using (8.19),

$$\tilde{\sigma}^2 = \tfrac{1}{2}[(12-16)^2 + (20-16)^2] = 16$$

and, using (8.22),

$$s^2 = \frac{1}{2-1}[(12-16)^2 + (20-16)^2] = 32$$

[1] Hence our reason for using $n - 1$ instead of n in calculating the variance and standard deviation of a sample, first noted on p. 123.

Repeating this for all sixteen different samples in Table 7.6 enables us to construct the sampling distributions shown in Table 8.1.

Table 8.1 Sampling distribution of $\tilde{\sigma}^2$ and s^2 for $n = 2$ from data of Table 7.6

Sample numbers	$\tilde{\sigma}^2$	$P(\tilde{\sigma}^2)$	$\tilde{\sigma}^2 P(\tilde{\sigma}^2)$	s^2	$P(s^2)$	$s^2 P(s^2)$
1,6,11,16	0	$\frac{4}{16}$	0	0	$\frac{4}{16}$	0
7,10	4	$\frac{2}{16}$	0.5	8	$\frac{2}{16}$	1
2,5	16	$\frac{2}{16}$	2	32	$\frac{2}{16}$	4
3,9,12,15	36	$\frac{4}{16}$	9	72	$\frac{4}{16}$	18
8,14	64	$\frac{2}{16}$	8	128	$\frac{2}{16}$	16
4,13	144	$\frac{2}{16}$	18	288	$\frac{2}{16}$	36
Σ		1	37.5		1	75
			$E(\tilde{\sigma}^2) = 37.5$			$E(s^2) = 75$

From Table 8.1 we see that $E(\tilde{\sigma}^2) = 37.5$ and $E(s^2) = 75$. This confirms (8.20), since

$$E(\tilde{\sigma}^2) = 37.5 = \sigma^2(2 - 1)/2 = 75/2 = 37.5$$

and also (8.22), since

$$E(s^2) = 75 = \sigma^2$$

However, we can see that $\tilde{\sigma}^2$ underestimates σ^2 by a considerable amount. If we were to repeat the calculations for larger sample sizes we would find that the degree of underestimation decreases as n increases.

We have shown that s^2 is an unbiased estimator of σ^2: on *average* it will give the correct result. It can also be demonstrated that s^2 and $\bar{X}$ are independent. However, for small samples, any *particular* value of s^2 may well be a considerable distance from the true value σ^2, as inspection of Table 8.1 shows. Here we see that out of the sixteen different possible samples only four have s^2 estimating correctly, while in the other twelve samples (i.e. in 75 per cent of cases) s^2 and σ^2 differ considerably.

This means that when n is large (Common, 1976, p. 169, recommends $n > 50$) we can safely use s^2 as an estimator of σ^2 but for small n we need to know something of the distribution of s^2 around σ^2.

The exact shape of the sampling distribution of s^2 depends on two parameters, population variance σ^2, and the sample size n. The distribution is always markedly right-skewed for small n but becomes increasingly more and more symmetrical as n increases. As we have just seen, the mean of the distribution of s^2 equals

σ^2, and the variance is given by

$$\mathrm{var}(s^2) = 2\sigma^4/(n-1) \tag{8.26}$$

Notice from (8.26) that the variance of s^2 decreases as n increases; taken together with its unbiasedness, this shows s^2 to be a *consistent* estimator of σ^2.

The chi-square distribution

The distribution of s^2 is in fact a special case of what is called the *gamma distribution*, and to form confidence intervals for s^2 would involve finding the appropriate areas under the gamma curve. This curve has a different shape for each pair of values of σ^2 and n, and this makes the task of finding areas under it difficult.

However, recall that the equation of the normal curve is also quite complex and it too depends for its exact appearance on the value of two parameters, in this case μ and σ^2 (although its underlying basic mathematical shape remains constant). We solved that problem by transforming to the unit-normal or z-distribution, for which a table showing the area under the z-curve exists, thus making the problem easy to manage.

We can deal with s^2 in the same way by transforming to another distribution known as the *chi-square distribution*. We define this as

$$\chi^2 = (n-1)\,s^2/\sigma^2 \tag{8.27}$$

Tables showing the area under the χ^2 distribution exist and are given in the back of most statistics or econometrics texts. Such a table is given in the appendix at the end of this book (see Table A.4).

Although the shape of the normal curve is constant, the shape of the distribution of χ^2 changes with sample size. When n is small the curve is skewed to the right but becomes more symmetrical as n increases. It is conventional practice, therefore, because of χ^2's dependence on n, to write (8.27) as

$$\chi^2_{n-1} = (n-1)\,s^2/\sigma^2 \tag{8.28}$$

The subscript $n-1$ on χ^2 is known as the number of *degrees of freedom* and is often abbreviated by d.f. and denoted by the symbol ν. In this context degrees of freedom is just a parameter, equal to 1.0 less than the sample size, which serves to distinguish one particular χ^2-distribution from another, and we need know nothing more about it. However, since it will occur quite often in different contexts later it will be useful to examine the concept here.

Degrees of freedom

Suppose we have two numbers X_1 and X_2 such that

$$\bar{X} = \tfrac{1}{2}(X_1 + X_2) = 5$$

Thus if $X_1 = 7$ then $X_2 = 3$, or if $X_2 = 8$ then $X_1 = 2$, etc. In other words, once we choose the value of one of the numbers then the value of the other is determined by the above relationship (i.e. by the fact that the mean must equal 5). In this example, then, we are *free to choose* one value, and the value of the other number is then fixed.

Suppose now we had three numbers X_1, X_2, and X_3, where

$$\bar{X} = \tfrac{1}{3}(X_1 + X_2 + X_3) = 6$$

Then, if we choose $X_1 = 3$ and $X_2 = 6$, X_3 must equal 9, or if $X_1 = 5$ and $X_3 = 8$ then X_2 must equal 5. In other words, since the mean of the three numbers is 6 then, having chosen the value of two of the numbers, the value of the third number is determined by the above expression.

In the first example we have two numbers and one free choice. In general we may define degrees of freedom as the number of independent values minus the number of restrictions or constraints placed on those values. In the second of our above examples we have three independent numbers and one restriction (that this mean must equal 6) and thus we have $n - 1 = 3 - 1 = 2$ d.f.

In the case of chi-squared, expression (8.28) could be written

$$\chi^2_{n-1} = \Sigma(X_i - \bar{X})^2/\sigma^2 \tag{8.29}$$

by substituting (8.22) into (8.28), and in this context the degrees of freedom relate to the number of independent squared deviation terms in the numerator of (8.29). The total number in this case is not n but $n - 1$ since after calculating any first $(n - 1)$ squared deviation terms the value of the last term is automatically fixed because of the presence of $\bar{X}$, and one of the features of $\bar{X}$ (as we showed in section 4.2) is that

$$\Sigma(X_i - \bar{X}) = 0$$

and this represents one restriction that must be satisfied.

Figure 8.9 shows how the χ^2_{n-1} distribution changes for different values of n, i.e. for different degrees of freedom.

* * *

Returning to our main theme, we can state that the mean and variance of the χ^2-distribution are

$$E(\chi^2_{n-1}) = n - 1$$

$$\text{var}(\chi^2_{n-1}) = 2(n - 1)$$

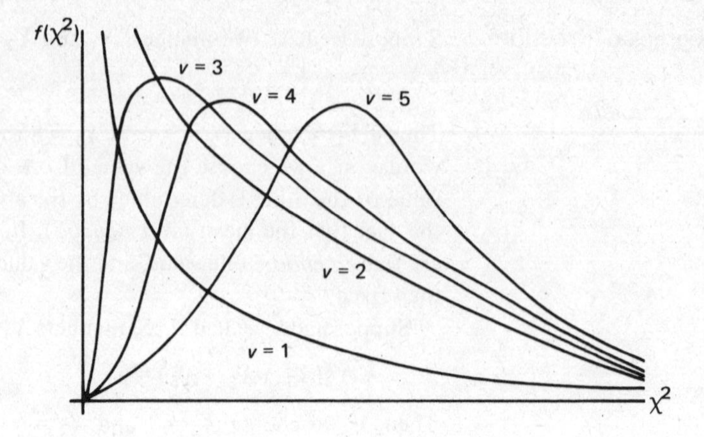

Figure 8.9 The χ^2-distribution for varying d.f.

In other words, what we are saying is that if $X \sim N(\mu, \sigma^2)$ then $(n-1)s^2/\sigma^2 \sim \chi^2_{n-1}[n-1, 2(n-1)]$, though since we are not concerned with the mean and variance of χ^2 we can abbreviate this to

$$\text{if } X \sim N(\mu, \sigma^2), \text{ then } (n-1)s^2/\sigma^2 \sim \chi^2_{n-1} \tag{8.30}$$

We can use (8.30) to make probability statements about s^2 *assuming* that we know σ^2 (which in practice we do not, but we shall deal with that question in a moment).

To see how we can do this first consider Figure 8.10, which shows a χ^2_{n-1}-distribution. Analogous to z_α we now define $\chi^2_{n-1,\alpha}$ as the value for which the area to its right under the χ^2 distribution equals α. This value will (as we have indicated above) depend upon the number of degrees of freedom – hence the double subscript on χ^2. This is shown in Figure 8.10(a). In the same way, $\chi^2_{n-1,\alpha/2}$ is such that the area to its right is $\alpha/2$, while $\chi^2_{n-1,1-\alpha/2}$ is such that the area to its *right* equals $1-\alpha/2$ – see Figure 8.10(b).

We have to make this distinction because the χ^2 distribution, unlike the z-distribution, is not symmetrical.

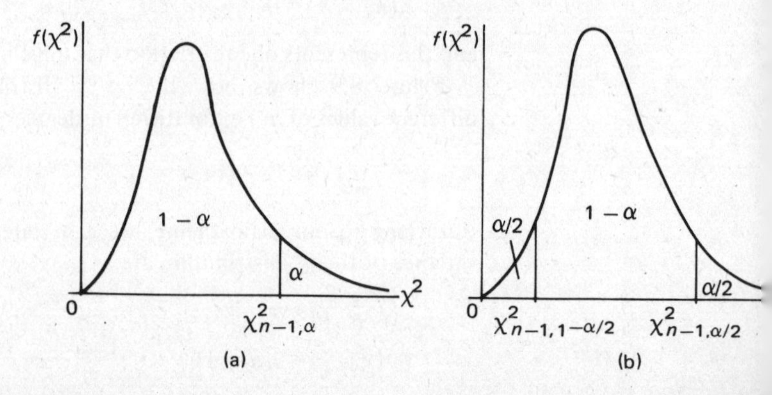

Figure 8.10 χ^2_{n-1} distribution

We adapt (8.30) to a form suitable for making probability statements. Thus

$$P[(n-1)s^2/\sigma^2 \geqslant \chi^2_{n-1,\alpha}] = \alpha \tag{8.31}$$

or

$$P[\chi^2_{n-1,1-\alpha/2} \leqslant (n-1)s^2/\sigma^2 \leqslant \chi^2_{n-1,\alpha/2}] = 1-\alpha \tag{8.32}$$

which we can easily rearrange to give

$$P[s^2 \geqslant \chi^2_{n-1,\alpha}\sigma^2/(n-1)] = \alpha \tag{8.33}$$

and

$$P[\chi^2_{n-1,1-\alpha/2}\sigma^2/(n-1) \leqslant s^2 \leqslant \chi^2_{n-1,\alpha/2}\sigma^2/(n-1)] = 1-\alpha \tag{8.34}$$

For example, suppose we take a sample size of $n = 25$ from a population with a variance of $\sigma^2 = 9$. Then if $\alpha = 0.05$ the appropriate value of $\chi^2_{n-1,\alpha}$ is $\chi^2_{24,0.05}$, which from the table of χ^2 values in the appendix is 36.415. Substituting in (8.33) gives

$$P(s^2 \geqslant 36.415 \times 9/24) = 0.05$$

$$P(s^2 \geqslant 13.66) \qquad = 0.05$$

That is, the probability that the sample variance of a sample taken from a population with $\sigma^2 = 9$ will exceed 13.66 is 0.05.

To use (8.34) the appropriate values of χ^2 are $\chi^2_{24,1-0.05/2}$ and $\chi^2_{24,0.05/2}$, i.e. from the χ^2 table

$$\chi^2_{24,0.975} = 12.401; \quad \chi^2_{24,0.025} = 39.364$$

Substituting into (8.34) gives

$$P(12.401 \times 9/24 \leqslant s^2 \leqslant 39.364 \times 9/24) = 0.95$$

$$P(4.65 \leqslant s^2 \leqslant 14.76) = 0.95$$

Thus there is a probability of 0.95 that the sample variance will be in the interval from 4.65 to 14.76.

As a final point in connection with χ^2 we should note that unlike the normal distribution which not only describes the shape of some important sampling distributions but also of many parent populations occurring in the economic and social sciences. The χ^2-distribution, on the other hand, describes only some special sampling distributions; there are no populations of any interest which are χ^2-distributed.

The problem with (8.34) is that it enables us to make probability statements about s^2 on the assumption that we know the value of the population parameters n and σ^2. Since we do not usually know σ^2, of more use would be an expression enabling us to make inferences about σ^2 on the basis of s^2. It is easy enough to do this by rearranging (8.34) to give

$$P[(n-1)s^2/\chi^2_{n-1,\alpha/2} \leqslant \sigma^2 \leqslant (n-1)s^2/\chi^2_{n-1,1-\alpha/2}] = 1 - \alpha \quad (8.35)$$

and therefore a $100(1-\alpha)$ per cent confidence interval is

$$[(n-1)s^2/\chi^2_{n-1,\alpha/2} \leqslant \sigma^2 \leqslant (n-1)s^2/\chi^2_{n-1,1-\alpha/2}] \quad (8.36)$$

To illustrate the use of (8.36) suppose a sample of size $n = 20$ is taken from a population and the sample variance $s^2 = 2.5$, and we have to construct a 99 per cent confidence interval for σ^2. We have

$$1 - \alpha = 0.99$$

Thus

$$\alpha = 0.01$$

Also

$$\text{d.f.} = 19$$

Then

$$\chi^2_{n-1,1-\alpha/2} = \chi^2_{19,1-0.005} = \chi^2_{19,0.995} = 6.844$$

and

$$\chi^2_{n-1,\alpha/2} = \chi^2_{19,0.005} = 38.582$$

Substituting in (8.36) gives

$$(19 \times 2.5/38.582 \leqslant \sigma^2 \leqslant 19 \times 2.5/6.844)$$

or

$$(1.23 \leqslant \sigma^2 \leqslant 6.94)$$

i.e.

$$P(1.23 \leqslant \sigma^2 \leqslant 6.94) = 0.99$$

In other words, if 100 such samples were taken from this population, then in approximately ninety-nine of the samples the interval from 1.23 to 6.94 would contain σ^2.

8.6 Small sample estimation of population mean μ

In section 8.4 we developed the general expression for an interval estimator of μ, i.e. equation (8.18). We were able to use the

z-distribution in this expression because, as the central limit theorem indicates, whatever the shape of the parent population the shape of the sampling distribution of sample means will be approximately normal (the approximation improving as sample size increases).

The trouble with (8.18) was that it contained the (usually) unknown σ^2. However, as we have seen, σ^2 can be replaced by s^2, an unbiased estimator of σ^2. Substitution of σ^2 by s^2 is acceptable provided sample size is not too small ($n > 50$). When n is small, as we saw in the example accompanying Table 8.1, a particular value of s^2 may be considerably different from σ^2 and introduce an unacceptably high level of error into the estimate. Our object therefore must be to find a method of forming an interval estimate of μ, using s^2 in place of σ^2, which is reliable in situations even where the sample size is small. This is what we propose to do now, but as a preliminary we need to introduce another (and very important) distribution.

The t-distribution

Suppose we have two random variables which are independent of one another. The first random variable, denoted by z, has a unit-normal distribution, the second, denoted by χ^2, has a chi-squared distribution with $n - 1$ degrees of freedom. Then we can define a new random variable, denoted by t:

$$t_{n-1} = z/\sqrt{\chi^2/(n-1)} \qquad (8.37)$$

The random variable t, as defined by (8.37), has $n - 1$ degrees of freedom, and because of the presence of χ^2 in the denominator the actual shape of its distribution is determined by the value of $n - 1$; hence the presence of the subscript attached to t.

The actual calculation of any particular value of t would be tedious but fortunately its values have already been calculated and tabulated in the same way as the z- and χ^2-distributions. However, unlike the χ^2 distribution, the t-distribution is always symmetric around its mean of zero. Its variance is equal to $(n - 1)/(n - 3)$, which is close to unity when n is large. In general the t-distribution is somewhat flatter in shape than the unit-normal distribution, but as n increases the shape of the t-distribution approaches that of the unit-normal distribution; indeed for $n > 30$ the two distributions become difficult to distinguish.

We can use the t-tables to make probability statements about any variable which has a t-distribution; Figure 8.11 illustrates this, and (as can be seen) the situation is analogous to the determination of probability statements about z-distributed variables.

Thus, if a variable T has a distribution with $n - 1$ degrees of

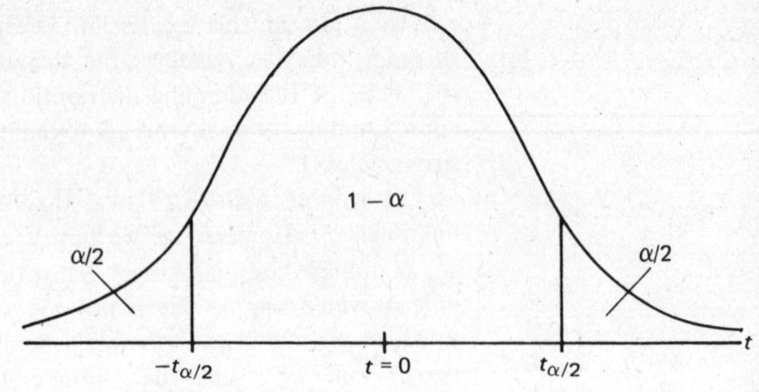

Figure 8.11 Probability statements and the *t*-distribution

freedom at a 100(1 − α) per cent confidence level, then

$$P(T \geqslant t_{n-1,\alpha}) = \alpha \tag{8.38}$$

$$P(T \leqslant -t_{n-1,\alpha}) = \alpha \tag{8.39}$$

$$P(-t_{n-1,\alpha/2} \leqslant T \leqslant t_{n-1,\alpha}) = 1 - \alpha \tag{8.40}$$

As with the χ^2_{n-1}-distribution, the fact that the value of t is a function of its degrees of freedom means that we would need a lot of space and many tables to tabulate the many possible values.

For this reason the usual practice is to provide only one table where, for each value of d.f., only the areas under the curve corresponding to important or commonly used intervals are given. For this reason the t-table is much less detailed than the z-table.

For example, suppose we have a variable T which is t-distributed with 10 d.f. Then from Table A.2 in the appendix and with a 95 per cent confidence level we have

$$1 - \alpha = 0.95$$

$$\alpha = 0.05$$

Thus, substituting in (8.38), (8.39) and (8.40), gives

$$P(T \geqslant +1.812) = 0.05$$

$$P(T \leqslant -1.812) = 0.05$$

$$P(-1.812 \leqslant T \leqslant 1.812) = 0.90$$

In other words, there is a probability of only 0.05 that the value of a t-distributed variable with 10 degrees of freedom would exceed +1.812 or be less than −1.812. Alternatively, there is a probability of 0.90 that such a variable would have a value between −1.812 and +1.812. The situation is illustrated in Figure 8.12.

Let us now return to the problem of forming reliable small

Figure 8.12 Values of t for a 0.95 probability level and 9 degrees of freedom

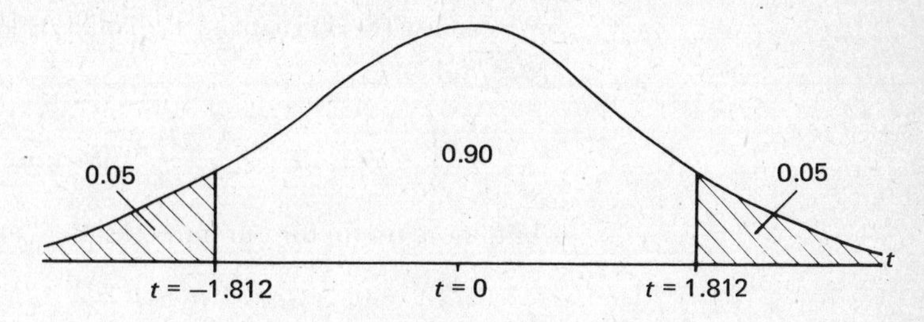

sample confidence intervals for μ. To recapitulate,

$$\text{if} \quad X \sim N(\mu, \sigma^2)$$

$$\text{then} \quad \bar{X} \sim N(\mu, \sigma^2/n)$$

in which case $(\bar{X} - \mu)/\sqrt{\sigma^2/n} \sim N(0, 1)$ from (8.14).

Consider the term $(\bar{X} - \mu)/\sqrt{\sigma^2/n}$, which we can write as

$$\frac{(\bar{X} - \mu)}{s/\sqrt{n}}$$

by replacing σ^2 by its estimator s^2 and taking its square root. If we now divide the numerator and denominator of this expression by σ we have

$$= \frac{(\bar{X} - \mu)}{\sigma} \Big/ \frac{s/\sqrt{n}}{\sigma} = \frac{(\bar{X} - \mu)}{\sigma} \Big/ \frac{s}{\sigma\sqrt{n}} = \frac{(\bar{X} - \mu)}{\sigma} \Big/ \sqrt{\frac{s^2}{\sigma^2 n}}$$

$$= \frac{(\bar{X} - \mu)}{\sigma/\sqrt{n}} \Big/ \sqrt{\frac{(n-1)s^2}{(n-1)\sigma^2}}$$

But as we have just seen above, *if the population is normally distributed*, then the numerator $(\bar{X} - \mu)/(\sigma^2/\sqrt{n})$ is the unit-normal distribution. The denominator is the square root of a χ^2-distributed variable divided by its number of degrees of freedom $n - 1$. In other words,

$$\frac{(\bar{X} - \mu)}{\sqrt{s^2/n}} = \frac{z}{\sqrt{\chi^2_{n-1}/(n-1)}}$$

and as we saw in (8.37) this is just the t-distribution with $n - 1$ degrees of freedom.

We can summarise these results thus:

$$\text{if} \quad X \sim N(\mu, \sigma^2)$$

$$\text{then} \quad \frac{\bar{X} - \mu}{\sqrt{s^2/n}} \sim t_{n-1} \tag{8.41}$$

We can use (8.41) to make probability statements about $(\bar{X} - \mu)/\sqrt{s^2/n}$, i.e.

$$P\left(-t_{n-1,\alpha/2} \leqslant \frac{\bar{X} - \mu}{\sqrt{s^2/n}} \leqslant t_{n-1,\alpha/2}\right) = 1 - \alpha$$

but, more useful for our purposes, this can be rewritten as

$$P(\bar{X} - t_{n-1,\alpha/2}\sqrt{s^2/n} \leqslant \mu \leqslant \bar{X} + t_{n-1,\alpha/2}\sqrt{s^2/n}) = 1 - \alpha \tag{8.42}$$

and the corresponding confidence interval for μ is

$$(\bar{X} - t_{n-1,\alpha/2}\sqrt{s^2/n} \leqslant \mu \leqslant \bar{X} + t_{n-1,\alpha/2}\sqrt{s^2/n}) \tag{8.43}$$

The results expressed in (8.42) and (8.43) are of great importance since they enable us to make inferences about μ without knowing σ^2 and offer reliable results even when n is small.

* * *

This might be a good point at which to highlight the main conclusions in our discussion relating to the estimation of μ.

First of all, if we know σ^2 then we can use (8.17) to make reliable inferences about μ regardless of sample size, i.e.

$$P(\bar{X} - z_{\alpha/2}\sqrt{\sigma^2/n} \leqslant \mu \leqslant \bar{X} + z_{\alpha/2}\sqrt{\sigma^2/n}) = 1 - \alpha$$

Second, if we do not know σ^2 then we can replace it by its unbiased estimator s^2 and use (8.24), i.e.

$$P(\bar{X} - z_{\alpha/2}\sqrt{s^2/n} \leqslant \mu \leqslant \bar{X} + z_{\alpha/2}\sqrt{s^2/n}) = 1 - \alpha$$

But this will only produce reliable estimators of μ if n is large (i.e. $n > 50$).

Finally, if we do not know σ^2 and n is small we can use (8.42), i.e.

$$P(\bar{X} - t_{n-1,\alpha/2}\sqrt{s^2/n} \leqslant \mu \leqslant \bar{X} + t_{n-1,\alpha/2}\sqrt{s^2/n}) = 1 - \alpha$$

As a final point we should be cautious about using (8.42) and (8.43) if the parent population is not normally distributed since these results were established on that premise.

We can illustrate these results with an example. Suppose a sample of size $n = 16$ is taken from a normally distributed population. Assume that $\bar{X} = 20$ and $s^2 = 4$. Let us construct 90 per cent and 95 per cent confidence intervals for population mean μ using (a) the z-distribution, and (b) the t-distribution. We can then compare results.

(a) *z*-distribution

(i) *90 per cent confidence interval.* We have $100(1 - \alpha) = 90$ per cent, i.e. $1 - \alpha = 0.9$ and $\alpha = 0.1$.

Figure 8.13
Appropriate areas under z-curve for $1 - \alpha = 0.90$

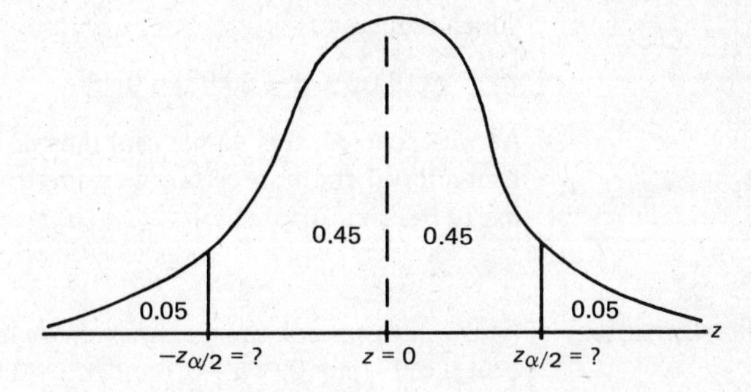

If $1 - \alpha = 0.9$, this means that there is 0.45 of the area in each half of the z-distribution and $\alpha/2 = 0.05$ in each tail. This is shown in Figure 8.13.

Thus we need to look in the z-table to determine the value of $z_{\alpha/2}$ which makes the area between it and $z = 0$ equal to 0.450 (and by symmetry also the area between $-z_{\alpha/2}$ and $z = 0$). This value is found to be $z_{\alpha/2} = \pm 1.64$.

Thus we can substitute into (8.24) to get

$$P(20 - 1.64 \times \sqrt{4/16} \leqslant \mu \leqslant 20 + 1.64\sqrt{4/16}) = 0.90$$

which reduces to

$$P(19.18 \leqslant \mu \leqslant 20.82) = 0.90$$

that is, on *average* the interval from 19.18 to 20.82 would contain μ in 90 per cent of such samples.

(ii) *95 per cent confidence interval*. We have $100(1 - \alpha) = 95$ per cent, i.e. $1 - \alpha = 0.95$ and $\alpha = 0.05$.

Thus we seek the value of $z_{\alpha/2}$ which makes the area between $z = 0$ and it equal to 0.475. This is shown in Figure 8.14.

From the z-table, $z_{\alpha/2} = z_{0.025} = \pm 1.96$. Substituting into (8.24) gives

$$P(20 - 1.96\sqrt{4/16} \leqslant \mu \leqslant 20 + 1.96\sqrt{4/16}) = 0.95$$

Figure 8.14
Appropriate areas under z-curve for $1 - \alpha = 0.95$

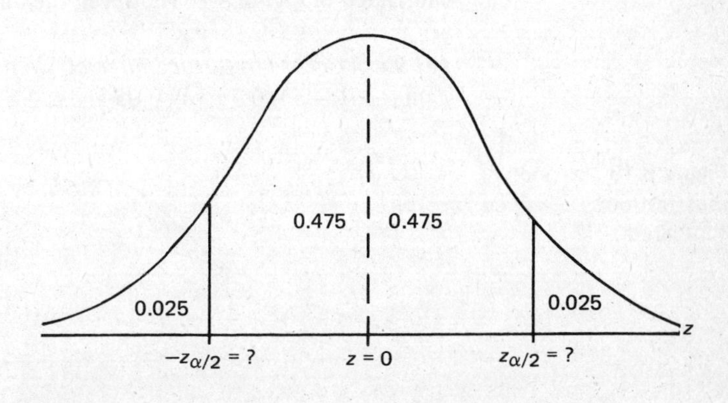

which reduces to

$$P(19.02 \leqslant \mu \leqslant 20.98) = 0.95$$

As we expected, this 95 per cent interval is wider than the 90 per cent interval; the more certain we want to be, the wider the interval has to be.

(b) *t*-distribution

(i) *90 per cent confidence interval.* We have $100(1 - \alpha) = 90$ per cent, i.e. $1 - \alpha = 0.9$, $\alpha = 0.1$ and $\alpha/2 = 0.05$.

Also, t will have $n - 1 = 15$ degrees of freedom, which means we must seek the required value somewhere on the 15th row of the t-table such that 0.05 of the area under the t-distribution lies in each tail. This is shown in Figure 8.15.

Figure 8.15 t-distribution for $1 - \alpha = 0.90$

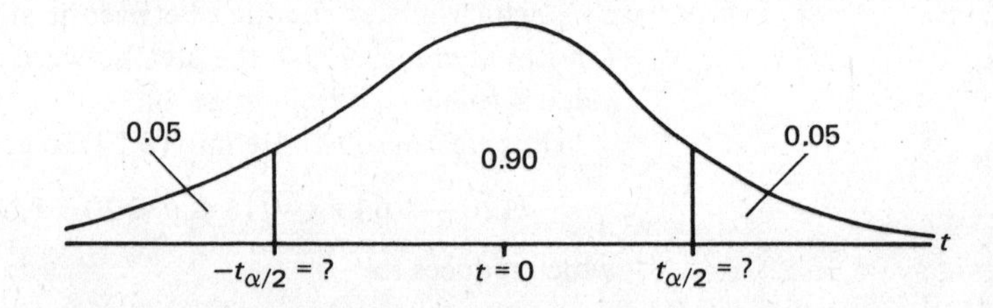

Thus we want the value of $t_{n-1,\alpha/2} = t_{15,0.05} = \pm 1.753$. Substituting in (8.42) gives

$$P(20 - 1.753\sqrt{4/16} \leqslant \mu \leqslant 20 + 1.753\sqrt{4/16}) = 0.90$$

which reduces to

$$P(19.12 \leqslant \mu \leqslant 20.88) = 0.90$$

Notice that this is slightly wider than the corresponding interval constructed using the z-distribution. This is because the shape of the t-distribution is flatter and wider than the z-distribution. Thus to encompass the same area we need a wider base; this concept is illustrated in Figure 8.16 (however, the difference *is* exaggerated).

(ii) *95 per cent confidence interval.* We have $100(1 - \alpha) = 95$ per cent, i.e. $1 - \alpha = 0.95$, $\alpha = 0.05$ and $\alpha/2 = 0.025$.

Figure 8.16 z- and t-distributions compared

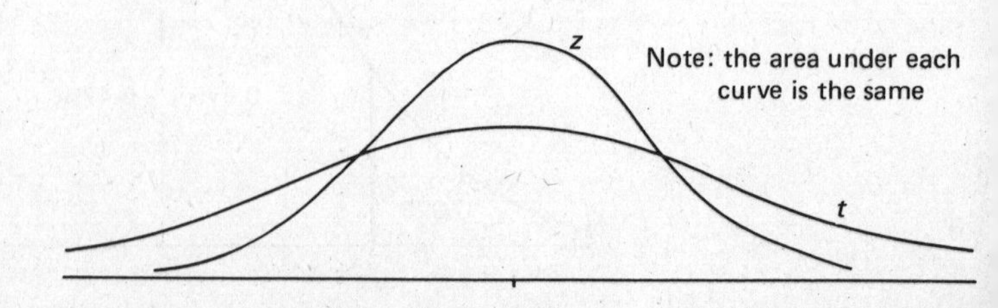

Thus we want the value of $t_{n-1,\alpha/2} = t_{15,0.025} = \pm 2.131$, which will leave 0.025 of the area in each tail. This is illustrated in Figure 8.17.

Substituting in (8.42) gives

$$P(20 - 2.131 \times \sqrt{4/16} \leqslant \mu \leqslant 20 + 2.131\sqrt{4/16}) = 0.95$$

which reduces to

$$P(18.93 \leqslant \mu \leqslant 21.06) = 0.95$$

Again this interval is wider than the z-interval at the same confidence level.

Figure 8.17
t-distribution for
$1 - \alpha = 0.95$

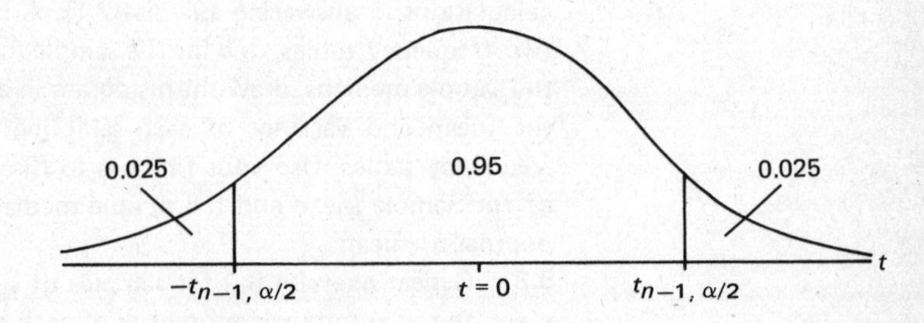

8.7 Summary

In this chapter we started by distinguishing between estimation and hypothesis testing as parallel paths in the inferential process. We then discussed some desirable properties of estimators and showed that $\bar{X}$ is a best linear unbiased (BLU) estimator of μ and coincides in this respect with the maximum likelihood estimator.

A parallel approach to the choice of an estimator is via loss functions, and we noted that both variance and bias have to be considered simultaneously if the quadratic loss function was to be minimised. This led us to the concept of mean square error (m.s.e.).

We then moved on to consider large sample estimators of μ and developed point and interval estimators for μ using s^2 as an unbiased estimator of σ^2. This works for large samples, but the distribution of s^2, which is chi-square, makes the use of s^2 unsatisfactory when n is small.

For this reason we examined the t-distribution, which, provided the parent population is normal, enables us to determine a suitable confidence estimator of μ in terms of s^2 and the t-distribution.

Before we move on to consider hypothesis testing we must briefly discuss estimators for population proportions. This is the subject of the next chapter.

EXERCISES

8.1 The sample mean $\bar{X}$ is to be used as an estimator of the mean of a population. Define the bias and variance of $\bar{X}$ in terms of the expectation operator and show that

$$\text{m.s.e.}(\bar{X}) = \text{var}(\bar{X}) + (\text{bias } \bar{X})^2$$

8.2 Assume that the fifty-five industries listed in the earnings data in exercise 2.1 constitute the population of all industries. Calculate the population mean earnings of male workers. Draw fifty samples each of size 5 from this population (with replacement) and for each sample calculate the sample mean and sample median (you may already have performed some part of these calculations in answering exercise 7.1). Arrange these results into two frequency tables, one for the sample means and the other for the sample medians, draw the histogram in each case and calculate the mean and variance of each sampling distribution from the frequency tables. Use your findings to discuss the relative merits of the sample mean and the sample median as estimators of the population mean.

8.3 Repeat exercise 8.2 for samples of size 10 and hence comment on the asymptotic properties of each estimator.

8.4 The 1978 *Family Expenditure Survey* revealed that, based on a sample of 709 households in the United Kingdom, the mean weekly expenditure on furniture was £1.35, with a standard deviation of £0.138. Construct 90 per cent, 95 per cent and 99 per cent confidence intervals for the population mean weekly household expenditure on furniture.

8.5 A sample of ten women is taken from the occupants of a battered women's refuge and the number of children living each of these women had is shown below:

2 4 5 3 3 5 4 1 2 6

Construct the 95 per cent and 90 per cent confidence intervals for the mean number of children of all the women in the refuge.

9 Estimation of the Population Proportion

OBJECTIVES

After reading this chapter and working through the examples students should understand the meaning of the following term:

error of estimate

Students should be able to:

calculate a point estimator of the population proportion
calculate an interval estimate of the population proportion
calculate the error of the estimate

9.1 Introduction

In this chapter we want to develop estimators for the population proportion Π. In section 7.3 we discussed the sampling distribution of p and established the following results:

$$E(p) = \pi \tag{7.8}$$

$$\mathrm{var}(p) = \pi q/n \tag{7.9}$$

$$\mathrm{s.e.}(p) = \sqrt{\pi q/n} \tag{7.10}$$

In this section we shall examine ways in which we can make inferences about π on the basis of sample evidence on p. The discussion will be briefer than that on estimation of the population mean not only because much of the discussion is necessarily duplicated but also because estimating π is not as difficult as estimating μ.

9.2 A point estimator of π

Equations (7.8) and (7.9) tell us what the sampling distribution of p is in terms of the population proportion π and sample size n. As we pointed out at the time, in practice we do not know π, and hence (7.7) to (7.9) are not of much use.

The problem which usually faces us is to find what value could be placed on Π to ensure that this would be the 'best' guess we could make. 'Best' in this sense means a method which will on average give a result which is closer to the true value than any other method. We have already seen that

$$E(p) = \pi \tag{7.8}$$

(i.e. the mean of the sampling distribution of sample proportions is equal to the population proportion), and intuitively we might feel that since *on average* the sample proportion will equal the population proportion then what better estimate of π, other than p, is there?

Our intuition in this case serves us well and we can define the best estimate of a population proportion to be the sample proportion. Thus

$$p \text{ is a point estimate of } \pi \tag{9.1}$$

This is a very bald statement and tells us nothing about how good a guess our sample value of p is, i.e. how close to Π it is likely to be. Such information would be valuable and is to be found in an interval estimate.

9.3 Interval estimation of π

We have seen already with equation (7.13) that the sampling distribution of the sample proportion is approximately normal, i.e.

$$p \sim \text{approx } N(\pi, \pi q/n)$$

we can standardise p using

$$z = \frac{p - \pi}{\sqrt{\pi q/n}} \tag{9.2}$$

and thus

$$\frac{p - \pi}{\sqrt{\pi q/n}} \sim \text{approx } N(0, 1)$$

and we can use the z-tables to make probability statements about π. We shall show how this is done for a specific level of probability and then generalise the discussion.

We proceed as follows. Recall that for a variable with unit-normal distribution, 95 per cent of all values lie between $z = -1.96$ and $z = +1.96$, i.e. 0.95 of the total area of 1.0 lies between $z = -1.96$ and $z = +1.96$. Remember that we look up $0.95/2 = 0.475$ in the body of the z-tables because of the symmetry of the normal curve around the mean, i.e. 47.5 per cent of all values (or 0.475 of the area) lies between $z = -1.96$ and $z = 0$ and 47.5 per cent (or 0.475 of the area) between $z = 0$ and $z = +1.96$. This is shown in Figure 9.1.

Thus we can write

$$P(-1.96 \leqslant z \leqslant +1.96) = 0.95$$

and, using (9.2), this is

$$P\left(-1.96 \leqslant \frac{p - \pi}{\sqrt{\pi q/n}} \leqslant +1.96\right) = 0.95$$

If we multiply the terms inside the brackets by $\sqrt{\pi q/n}$, subtract p and multiply by -1.0, the above expression becomes

$$P(p - 1.96\sqrt{\pi q/n} \leqslant \pi \leqslant p + 1.96\sqrt{\pi q/n}) = 0.95 \qquad (9.3)$$

The expression means that if we drew an infinite number of samples from the population in question and if for each sample we calculated the interval using (9.3), then approximately 95 per cent of those intervals would contain or embrace the true population parameter π. The value 95 per cent is a measure of our *confidence* that an interval so constructed will contain π. For this reason the interval from $p - 1.96\sqrt{\pi q/n}$ to $p + 1.96\sqrt{\pi q/n}$ is

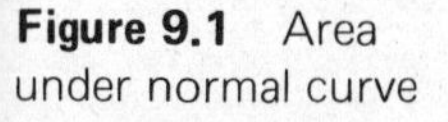

Figure 9.1 Area under normal curve

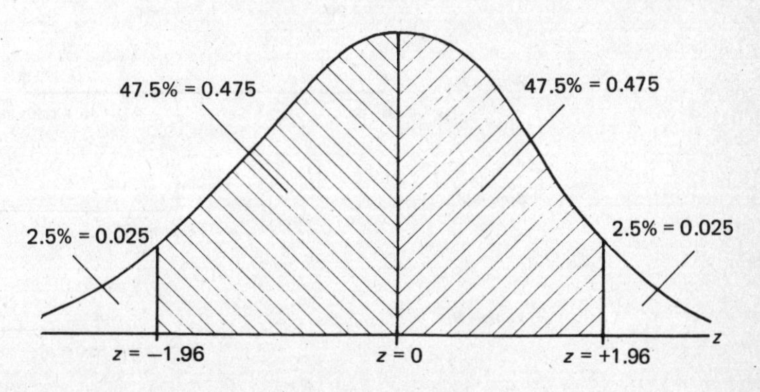

known as the *confidence interval* and it provides us with an *interval estimate* of the population proportion.

It is important to note that we *cannot* say of (9.3) that there is a probability of 0.95 that π will lie in the interval between $p - 1.96\sqrt{\pi q/n}$ and $p + 1.96\sqrt{\pi q/n}$. The population parameter π is not a random variable: it has a *fixed* numerical value for any given population and therefore either does or does not fall in the interval. Unfortunately, since we do not know the value of π, we can not actually say whether it does or not.

So, to repeat, (9.3) indicates that approximately 95 per cent of the time the interval $p - 1.96\sqrt{\pi q/n}$ to $p + 1.96\sqrt{\pi q/n}$ will contain π. This implies that in 5 per cent of samples it will not. We illustrate these possibilities in Figure 9.2. For any given population the interval has a fixed width equal to $3.92\sqrt{\pi q/n}$, but the actual position of the interval will depend on the particular sample value of p. Figure 9.2 shows three different intervals resulting from three different samples from a population with a mean proportion π. Two of these intervals embrace π, samples 1 and 2, but sample 3 does not, as we can see.

When we construct a confidence interval we can choose any level of confidence we like, though 95 per cent is probably the most frequently used by statisticians. Confidence levels of 90 per cent and 99 per cent are also quite common. The higher the chosen level of confidence, the wider becomes the corresponding confidence interval and the less precise (and therefore the less useful) the interval estimate. In the extreme a 100 per cent confidence level would produce a confidence interval stretching from $-\infty$ to $+\infty$ and would effectively tell us nothing about the value of the population parameter we were trying to estimate.

Figure 9.2 Interval estimates of π for three different samples from the same population

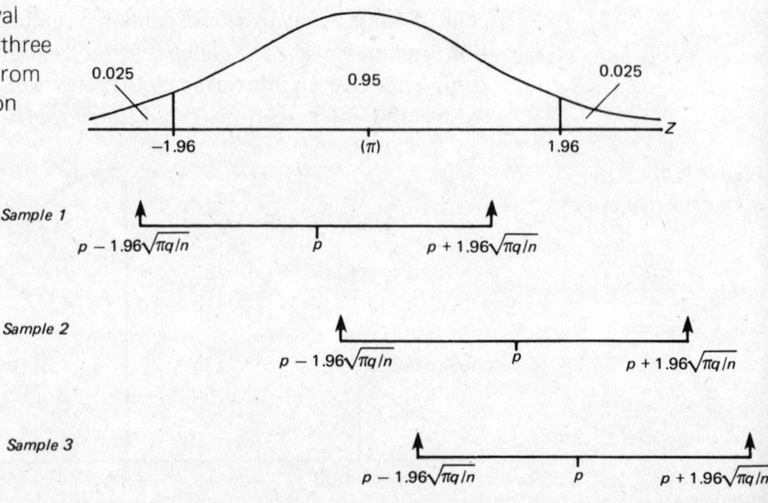

As promised, we now provide a more general treatment of confidence intervals and interval estimates by rewriting (9.3) as follows:

$$P(p - z_{\alpha/2}\sqrt{\pi q/n} \leqslant \pi \leqslant p + z_{\alpha/2}\sqrt{\pi q/n}) = 1 - \alpha \qquad (9.4)$$

where $1 - \alpha$ is the chosen level of probability, i.e. $100(1 - \alpha)$ per cent is the confidence level, and $-z_{\alpha/2}$ and $+z_{\alpha/2}$ are the appropriate z-values.

The z-values are subscripted $\alpha/2$ because with a probability level of $1 - \alpha$, the area in each 'tail' of the distribution is $\alpha/2$, as the following will show:

Total area under curve = 1

Probability area = $1 - \alpha$

Total area left in both tails = Total area − Probability area
$= 1 - (1 - \alpha)$

Area in each tail $= [1 - (1 - \alpha)]/2 = \alpha/2$

Figure 9.3 illustrates these points.

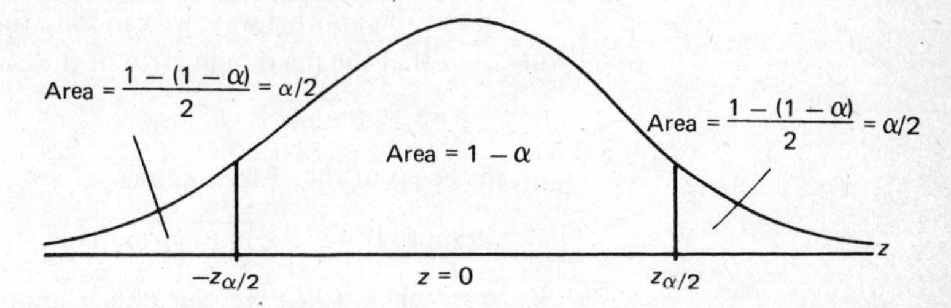

Figure 9.3 Area in each tail of the normal distribution with a probability level of $1 - \alpha$

In the general expression of (9.4) the term $1 - \alpha$ replaces the value of 0.95 in the specific case described by (9.3). Generally, a $100(1 - \alpha)$ per cent confidence level means that if we took an infinite number of samples each of size n from a population with a mean proportion of π, then approximately $100(1 - \alpha)$ per cent of confidence intervals so formed would cover or embrace π (we have to keep inserting the word 'approximately' in these statements since we are approximating the binomial distribution with the normal distribution, and this approximation will be close if the sample size is reasonably large).

Equation (9.4) is thus of an appropriate form for estimating π on the basis of sample information on p. Unfortunately, we still cannot use (9.4) since it contains π (we do not have to know q since this is simply $1 - \pi$). However, we can use p as an estimate of π, i.e. we make the following assumption:

$$\sqrt{\pi q/n} = \sqrt{\pi(1 - \pi)/n} \text{ can be approximated by } \sqrt{p(1 - p)/n}$$

In this case (8.4) becomes

$$P(p - z_{\alpha/2}\sqrt{p(1-p)/n} \leqslant \pi \leqslant p + z_{\alpha/2}\sqrt{p(1-p)/n}) = 1 - \alpha \quad (9.5)$$

Note that this is an *approximate* statement, first of all because we are approximating the binomial with the normal distribution, and second we are approximating π with p.

9.4 Error of the estimate

Finally, we want to consider what is called the *error of the estimate*. This is the size of the error which we may make when we use p as a point estimate of π. We know that the standard err. : of the sampling distribution of sample proportions, first given in equation (7.10), is

$$\text{standard error } \sigma_p = \sqrt{\pi q/n}$$

In other words, there is a probability of $1 - \alpha$ that p will lie within $z_{\alpha/2}$ such standard errors of π. Figure 9.4 illustrates this point.

To put this another way, we can state that there is a probability of $1 - \alpha$ that the maximum error of the estimate is given by

$$e = z_{\alpha/2}\sqrt{\pi q/n}$$

or, in the event that π is unknown,

$$e(\text{approx}) = z_{\alpha/2}\sqrt{p(1-p)/n} \quad (9.6)$$

By rearranging (9.6) we can obtain another useful result which will indicate the sample size required to ensure that there is a probability of $1 - \alpha$ that the error of estimate is no more than some value e. Thus

$$n = p(1-p)(z_{\alpha/2}/e)^2 \quad (9.7)$$

If we have no idea what value to assign to p when using (9.6) or (9.7), or indeed (9.5), then we can make use of the fact that πq (and hence the width of the confidence interval) is a maximum

Figure 9.4 Error of the estimate

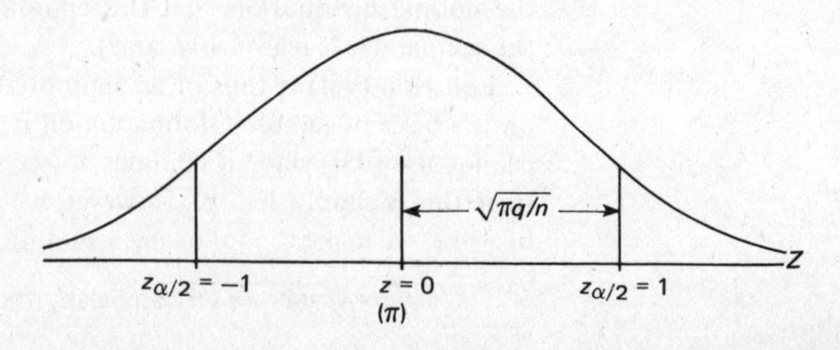

when $\pi = 0.5$. Thus we shall always be on the safe side if we use $p = 0.5$ in the absence of any other information as to its value.

We conclude this chapter by illustrating the above results with an example.

Example

A sample of thirty unemployed workers reveals that twelve of them have been without work for more than three months:

(a) Determine a point estimate of the population proportion of the long-term unemployed.
(b) Calculate the error of estimate of this point estimate at a 95 per cent confidence level.
(c) Calculate a 95 per cent confidence interval for π.
(d) Calculate a 90 per cent confidence interval for π.
(e) Determine the size of sample to reduce the error of estimate to half its original value.
(f) Repeat (b) to (e) above using $p = 0.5$ and compare results.

We proceed as follows:

(a) The point estimate of π is p, from (9.1), i.e.

$$p = 12/30 = 0.4$$

Thus 0.4 is a point estimate of π.

(b) The error of the estimate is found from (9.6). With a 95 per cent confidence level, $z_{\alpha/2} = 1.96$. Therefore, error of estimate

$$e = 1.96\sqrt{0.4(1 - 0.4)/30}$$

i.e.

$$e = 0.1753$$

This means that there is a probability of 0.95 that the sample proportion $p = 0.4$ is no further, at the most, than 0.1753 from π.

(c) To form the 95 per cent confidence interval we use (9.5), i.e.

$$P(0.4 - 1.96\sqrt{0.4(1 - 0.4)/30} \leqslant \pi \leqslant 0.4 + 1.96\sqrt{0.4(1 - 0.4)/30}) = 0.95$$

from which

$$P(0.2247 \leqslant \pi \leqslant 0.5753) = 0.95$$

which means that in approximately 95 per cent of samples (if we were to take repeated samples) the interval 0.2247 to 0.5753 would enclose π.

(d) Again using (9.5), this time $z_{\alpha/2} = 1.65$ and we have

$$P(0.4 - 1.65\sqrt{0.4(1 - 0.4)/30} \leqslant \pi \leqslant 0.4 + 1.65\sqrt{0.4(1 - 0.4/30)}) = 0.90$$

or

$$P(0.2524 \leqslant \pi\ 0.5476) = 0.90$$

which means that in approximately 90 per cent of such samples the interval 0.2524 to 0.5476 would enclose π. As expected, the 90 per cent confidence interval is narrower than the 95 per cent interval.

(e) The error of the estimate calculated in (b) was 0.1753 for a sample size of $n = 30$. What should n be to halve this, i.e. to reduce the error of the estimate to 0.0877. We substitute in (9.7), i.e.

$$n = 0.4(1 - 0.04)(1.96/0.0877)^2$$
$$= 120.0106$$

Effectively, this means a sample of size 121 (since we cannot sample 0.0106 of an unemployed person and we round n upwards to ensure the error condition is satisfied).

In other words, halving the error quadruples the sample size.

(f) We shall now repeat the above calculations in (b) to (e) assuming that we have no knowledge of p and taking $p = 0.5$:

(b) Error of estimate:

$$e = 1.96\sqrt{0.5(1 - 0.5)/30} = 0.1786$$

which differs by about 2 per cent from the value of e calculated using $p = 0.04$.

(c) A 95 per cent confidence interval is

$$P(0.5 - 1.96\sqrt{0.5(1 - 0.5)/30} \leqslant \pi \leqslant 0.5 + 1.96\sqrt{0.5(1 - 0.5)/30}) = 0.95$$

i.e.

$$P(0.3214 \leqslant \pi \leqslant 0.6786) = 0.95$$

As expected, this is wider than the corresponding interval calculated using $p = 0.4$.

(d) A 90 per cent confidence interval is

$$P(0.5 - 1.65\sqrt{0.5(1 - 0.5)/30} \leqslant \pi \leqslant 0.5 + 1.65\sqrt{0.5(1 - 0.5)/30}) = 0.90$$

i.e.

$$P(0.3497 \leqslant \pi \leqslant 0.5503) = 0.90$$

which again is wider than the corresponding interval calculated using $p = 0.4$.

(e) The error of the estimate calculated using $p = 0.5$ was 0.1786. Half of this is 0.0893. Substituting in (9.7) gives

$$n = 0.5(1 - 0.05)(1.96/0.0893)^2$$
$$= 120.0305$$

In other words, we would have to take a sample size of 121.

9.5 Summary

Much of what we have discussed in this chapter parallels the material in the previous chapter on estimation of the population mean. Estimation of population proportions is, perhaps, less frequently encountered than that of means. None the less, the material discussed in this chapter is of some importance and should be treated with the same amount of diligence as that of Chapter 8.

Having discussed estimation of both means and proportions we are now ready to move on to the second branch of statistical inference, that of hypothesis testing.

EXERCISES

9.1 The 1978 *Family Expenditure Survey* reveals that from a sample of 758 households in Wales, 478 owned a car. Construct 90 and 95 per cent confidence intervals for the proportion of all Welsh households owning a car.

9.2 A sample of 100 students in a university revealed that ten were left-handed. Construct a 95 per cent confidence interval for the proportion of all the students in the university who are left-handed.

10 Hypothesis Testing

OBJECTIVES

After reading this chapter and working through the examples students should understand the meaning of the following terms:

null hypothesis	**rejection region**
alternative hypothesis	**acceptance region**
type-I error	**critical value**
type-II error	**decision-rule**
simple hypothesis	**two-tailed test**
significance level	**one-tailed test**

Students should be able to:

set up suitable null and alternative hypotheses

perform a hypothesis test on a population mean

perform a hypothesis test on the difference between two means

perform a hypothesis test on population variance

perform a hypothesis test on a population proportion

10.1 Introduction

At the beginning of Chapter 8 we explained how statistical inference can take two complementary forms, estimation or hypothesis testing. In estimation we seek to determine the value of some unknown population parameter, while in hypothesis testing we seek to confirm or deny some preconceived idea as to the value of a parameter.

For example, suppose we are interested in the annual *per capita* consumption of beer in the United Kingdom and have some evidence indicating (perhaps dating back a couple of years) that the figure is 0.72 gallon per head per year. We could take a new random sample of persons in the United Kingdom and use the sample information resulting from this to *test* whether the *a priori* figure of 0.72 is supported by the sample evidence. Alternatively, suppose we read in a journal article that the mean weekly income (from all sources) of the long-term unemployed in the United Kingdom is £27. If we had reason to suspect this figure was not accurate, we could take a random sample of the long-term unemployed, determine their mean weekly income and use this sample mean to test whether or not the figure quoted appeared to be correct.

In general the assumption which we make about the value of the population parameter in question is known as the *null (or maintained) hypothesis* (it is called the 'null' hypothesis because it is assumed that there is *no* difference between the true value of the population parameter and the hypothesised value). Thus in the first of the two examples above the null hypothesis would be that the population mean $\mu = 0.72$, and in the second example that the population mean $\mu = 27$.

Suppose that the sample evidence is judged *not* to support the value ascribed to μ under the null hypothesis. This means that μ is more likely to have a different value; we describe this different value using the *alternative hypothesis*. Essentially, then, hypothesis testing consists of choosing between two competing hypothesis on the basis of sample evidence. It is important to bear in mind that whatever decision we make regarding the veracity of the null hypothesis we can never be *completely* sure that we have made the *correct* decision. This is because we are using sample evidence as the basis for our decision and it is always possible that chance factors may cause a particular sample to be untypical of the population from which it has been drawn. In this case we may mistakenly either reject the null hypothesis when it is true or accept it when it is in fact false; in the former case we make what is known as a *type-I error*, in the latter case a *type-II error*.

In this chapter we shall discuss hypothesis testing first in the context of population means and variances, and second (and more briefly) with regard to population proportions.

10.2 Testing hypotheses about population means

Let us start by introducing some terminology. The null hypothesis is usually denoted by H_0 and the alternative hypothesis by H_1. Thus if our null hypothesis is that the value of a population mean is equal to μ_0 and our alternative hypothesis is that it is equal to μ_1, then we would write these competing hypotheses as follows:

$$\left.\begin{aligned} H_0&: \mu = \mu_0 \\ H_1&: \mu = \mu_1 \end{aligned}\right\} \quad (10.1)$$

The implication here is that μ is equal to either μ_0 or to μ_1: no other value is considered possible. Hypotheses of this type (i.e. those for which a specific value for μ is postulated) are known as *simple hypotheses*. In practice, however, it would be unusual for us to be able to be so definite about the value of the parameter μ. Far more likely are hypotheses of the type:

$$\left.\begin{aligned} H_0&: \mu = \mu_0 \\ H_1&: \mu \neq \mu_0 \end{aligned}\right\} \quad (10.2)$$

or of the type:

$$\left.\begin{aligned} H_0&: \mu = \mu_0 \\ H_1&: \mu < \mu_0 \end{aligned}\right\} \quad (10.3)$$

or of the type:

$$\left.\begin{aligned} H_0&: \mu = \mu_0 \\ H_1&: \mu > \mu_0 \end{aligned}\right\} \quad (10.4)$$

The alternative hypotheses of (10.2), (10.3) and (10.4) are known as *composite hypotheses*; perhaps (10.2) is the most commonly occurring. Notice that in each of the above cases the null hypothesis is of the simple type, i.e. it represents a very specific statement about the value of μ. This is because it is easier to prove or disprove specific claims than the vague claims represented by composite hypotheses.

Having established the null and alternative hypotheses we then have to devise a procedure for choosing between them, i.e. some method by which (on the basis of the sample information) we can reject or not reject the null hypothesis. In terms of testing hypotheses concerning μ one possibility would be to use the sample mean $\bar{X}$ (which, as we have seen, is an estimator with many desirable properties) and to state that if $\bar{X}$ is 'very different from μ we will reject H_0'. The obvious question is: 'what do we mean by "very different" and "not very different"?'

Alternative hypotheses of the simple type

Let us start by considering competing hypotheses of the type:

$$H_0: \mu = \mu_0$$

$$H_1: \mu = \mu_1 \quad (\text{where } \mu_1 > \mu_0, \text{ say})$$

Although alternative hypotheses of this form are not very common, we can use them to establish some important ideas first and then consider more practical cases later. Let us suppose that we take a sample of size n from a population in which X is distributed $X \sim N(\mu, \sigma^2)$; then we know from section 7.4 that in general the sample mean $\bar{X}$ is distributed $\bar{X} \sim N(\mu, \sigma^2/n)$. In particular,

$$\text{if } H_0 \text{ is true then } \bar{X} \sim N(\mu_0, \sigma^2/n) \qquad \text{Distribution 0}$$

but

$$\text{if } H_1 \text{ is true then } \bar{X} \sim N(\mu_1, \sigma^2/n) \qquad \text{Distribution 1}$$

In other words, we have to decide whether $\bar{X}$ is drawn from the first or the second distribution. The position is illustrated in Figure 10.1.

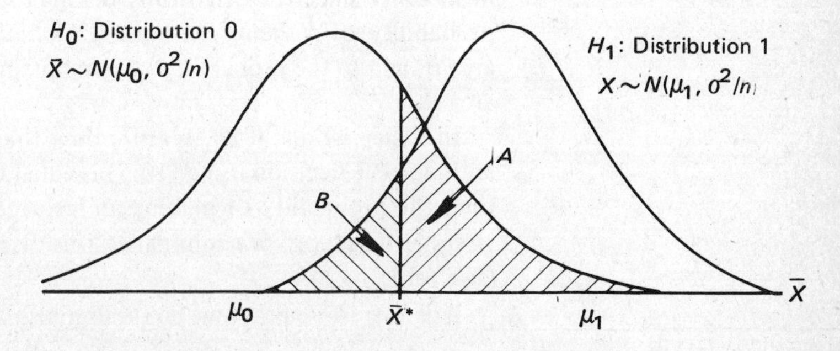

Figure 10.1 Distributions of $\bar{X}$ for H_0 and H_1

Now if the sample value of $\bar{X}$ is less than μ_0, we would be happy to accept H_0, i.e. that $\bar{X}$ is drawn from a population with a mean μ_0. Similarly, if $\bar{X}$ is greater than μ_1, we would accept H_1, i.e. that $\bar{X}$ is drawn from a population whose mean is μ_1. Difficulties arise, however, when $\bar{X}$ lies between μ_0 and μ_1. Then we have to choose some critical value, say $\bar{X}^*$, and use the decision-rule:

$$\left.\begin{array}{l}\text{if } \bar{X} < \bar{X}^* \text{ accept } H_0 \text{ (i.e. reject } H_1) \\ \text{if } \bar{X} > \bar{X}^* \text{ accept } H_1 \text{ (i.e. reject } H_0)\end{array}\right\} \quad (10.5)$$

For obvious reasons the area to the right of $\bar{X}^*$ is known as the *rejection or critical region*, and the area to the left as the *acceptance region.*

We can use these ideas to explore the nature of type-I and type-II errors. Suppose for a moment that H_0 is true: then our decision-rule of (10.5) instructs us to reject H_0 for all those samples whose $\bar{X}$ exceeds (i.e. falls to the right of) $\bar{X}^*$. But we shall be mistakenly rejecting those which lie in the shaded area A since they rightfully belong to Distribution 0. Now the probability of $\bar{X}$ exceeding $\bar{X}^*$ when H_0 is true equals area A, and thus the probability of committing a type-I error is equal to area A.

The probability of making a type-I error is known as the *significance* level of the test. Thus if the probability is 0.05, then the significance level is 5 per cent. One interpretation of a 5 per cent significance level is that if we took an infinite number of samples from a population with mean μ, then in only 5 per cent of the samples would the value of $\bar{X}$ be incorrectly rejected (i.e. fall in the rejection region). Thus when we say $\bar{X}$ is very different from μ we mean so different from μ that in sampling from a population with a mean of μ such values of $\bar{X}$ would only occur in 5 per cent of samples. We normally speak of a test as being *significant* at the α per cent level if the null hypothesis is rejected.

Now suppose that H_1 is true: then the decision-rule tells us to accept H_0 for all those $\bar{X}$s whose values lie to the left of $\bar{X}^*$, but in the case of those lying in area B we shall be accepting them mistakenly since they rightfully belong to Distribution 1. But the probability of $\bar{X}$ being less than $\bar{X}^*$ when H_1 is true is equal to area B, and so the probability of committing a type-II error equals area B.

In other words, if H_0 is true, then the probability of making an incorrect decision using (10.5) is equal to area A. If H_1 is true, then the probability of making an incorrect decision using (10.5) is equal to area B. We summarise this discussion in tabular form (see Table 10.1).

It is easy to see that we could diminish area A and hence make

Table 10.1 Decision-rule and type-I and type-II errors

	H_0 correct	H_1 correct
if $\bar{X} < \bar{X}^*$ accept H_0 (reject H_1)	Correct	Type-II error
if $\bar{X} > \bar{X}^*$ accept H_1 (reject H_0)	Type-I error	Correct

the probability of type-I errors smaller by moving the value of $\bar{X}^*$ to the right. If we did this, we would at the same time be increasing the probability of type-II errors. Thus there is a trade-off between type-I and type-II errors, and when we fixed the value of $\bar{X}^*$ we would bear in mind the cost associated with making each type of error.

An example will help illustrate the above discussion.

Example

An investment analyst will give advice to sell and take profits when an index measuring the rising value of all the shares in his portfolio reaches 200. Although he measures the value of all the shares on a weekly basis, he does not have the resources to do this daily and therefore takes a daily random sample of size $n = 30$ from his portfolio with the intention of issuing selling advice when a sample value is such as to indicate a population value of $\mu = 200$. He knows enough about statistical inference to realise that some values of his sample mean $\bar{X}$ which are below 200 may well have come from a population with a mean of 200 or more. He therefore needs to set a critical value somewhere between the current index value of 190 and the sell value of 200. This value should be chosen in such a way that it takes the costs of errors of types I and II into account. Let us investigate which of these errors is more serious, i.e. more costly.

Suppose a particular sample mean $\bar{X}$ has a value high enough for the analyst to believe the portfolio has risen to 200 when in reality it has not. He issues 'sell' advice, mistakenly believing that the portfolio price will soon fall sharply as profit-takers move in. In hypothesis testing terms he is rejecting $H_0: \mu = 190$ when it is in fact still true, a type-I error. However, the consequences of this error are not serious, since if the market has not yet fallen the investors can buy back in with little loss. Consider now the contrary situation. Suppose the market actually has risen to 200 but the next sample mean $\bar{X}$ has a value low enough to cause the analyst not to issue 'sell' advice. In this case he accepts $H_0: \mu = 190$ when it should have been rejected, a type-II error. The consequences might be serious: by the time he realises the market has actually risen beyond 200, other profit-takers might have come into the market, sold shares and the share index might have fallen.

The sell advice might then come too late and the cost to his clients might be considerable.

Thus the analyst will want to set a critical value of $\bar{X}^*$ which makes the probability of a type-II error much less than a type-I error. Suppose he assesses the cost of a type-II error to be four times as high as a type-I error: thus he will fix his critical value such that area A is four times as large as area B. This is shown in Figure 10.2.

Figure 10.2 The critical value for the investment analyst

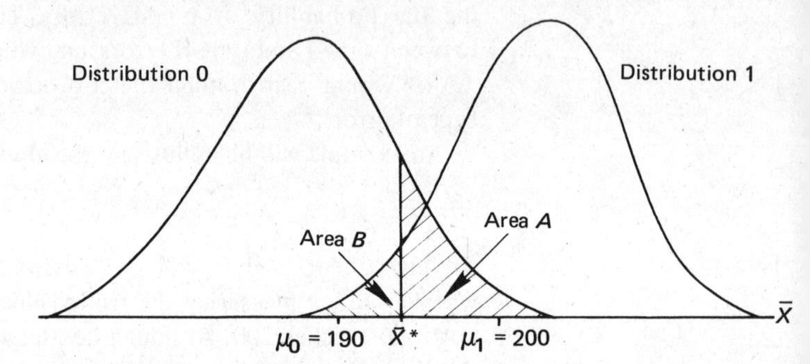

Intuitively we can see that if the analyst is to receive adequate warning of the portfolio's rise to 200, he will want to set his critical value closer to 190. However, he cannot set it too close since this might result in too many false alarms. In other words, we are trying to distinguish between

$$H_0: \mu = 190$$

and

$$H_1: \mu = 200$$

Let us set a trial value of $\bar{X}^* = 192$. Then

$$P(\text{type-I error}) = \text{area } A = P(\bar{X} > 192 \mid H_0: \mu_0 = 190)$$
$$= P(\bar{X} > 192 \mid \bar{X} \sim N(190, \sigma_0^2/n)$$

Let us assume that the analyst knows from previous experience that $\sigma_0^2 = 1470$; then we have

$$P(\text{type-I error}) = P[\bar{X} > 192 \mid \bar{X} \sim N(190, 1470/30)]$$
$$= P\left[z > \frac{192 - 190}{\sqrt{1470/30}} \mid z \sim N(0, 1)\right]$$
$$= P(z > 0.286)$$
$$= 0.388$$

using the z-tables. In other words, the significance level of this test is 0.388, or 38.8 per cent.

Similarly,

$$P(\text{type-II error}) = \text{area } B = P(\bar{X} < 192 \mid H_1: \mu_1 = 200)$$
$$= P[(\bar{X} < 192 \mid \bar{X} \sim N(200, \sigma_1^2/n)]$$

Assume that the analyst knows that $\sigma_1^2 = 1470$; then

$$P(\text{type-II error}) = P[\bar{X} < 192 \mid \bar{X} \sim N(200, 1470/30)]$$
$$= P\left[z < \frac{192 - 200}{\sqrt{1470/30}} \mid z \sim N(0, 1)\right]$$
$$= P(z < -1.143)$$
$$= 0.126$$

Area A, although bigger than area B, is not four times as large. Clearly the value of $\bar{X}^*$ must be lower. A value of $\bar{X}^* = 191$ gives $P(\text{type-I error}) = 0.443$ and $P(\text{type-II error}) = 0.090$. Thus the correct value of $\bar{X}^*$ will be between 191 and 192 and closer to 191.

The point of this example has been to illustrate the trade-off between the two types of error. Moving the critical value in one direction will increase one while it decreases the other; moving it in the opposite direction has the reverse effect. Intuitively we may feel that if we are using consistent estimators (as $\bar{X}$ is), then the 'performance' of the hypothesis test should improve as the variance of the sampling distributions decreases. In other words, by increasing n we should be able to effect a reduction in the probability of *both* types of error. This is illustrated in Figure 10.3, which shows a reduction in both areas A and B (and hence in both types of error) when n is increased (for a given critical value).

The power of a test

We want to consider briefly the *power* of a test. Consider the two sets of competing hypotheses:

(a) $H_0: \mu = \mu_0$

$H_1: \mu = \mu_1$

(b) $H_0: \mu = \mu_0$

$H_1: \mu = \mu_2$ (where $\mu_2 > \mu_1$)

These are illustrated in Figure 10.4.

We can see that, for a given level of significance, the probability of a type-II error (area B) is greater when μ_0 and μ_1 are close (Figure 10.4(a)) than when the alternative hypothesis value μ_2 is further away (Figure 10.4(b)). Thus, if the null hypothesis is

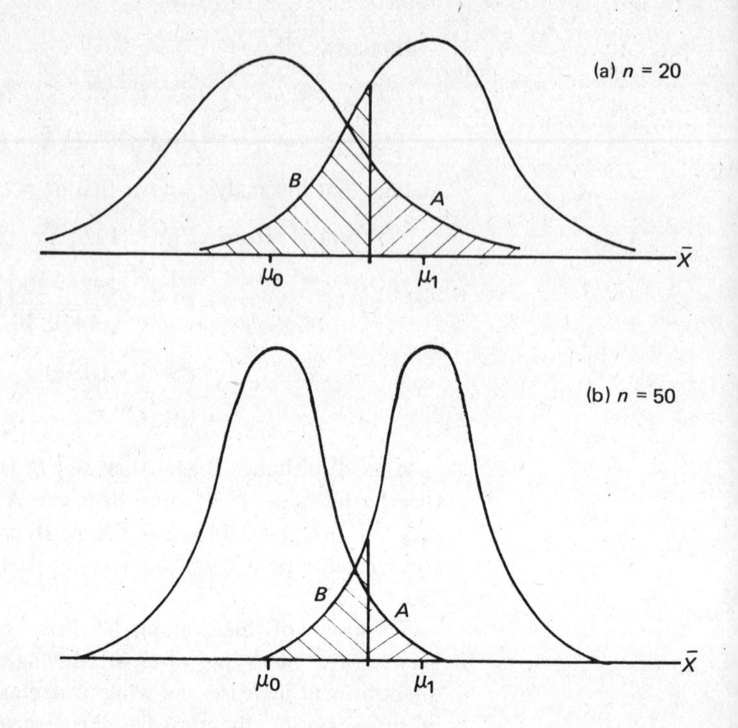

Figure 10.3 Effect of increasing sample size on the value of the two types of error

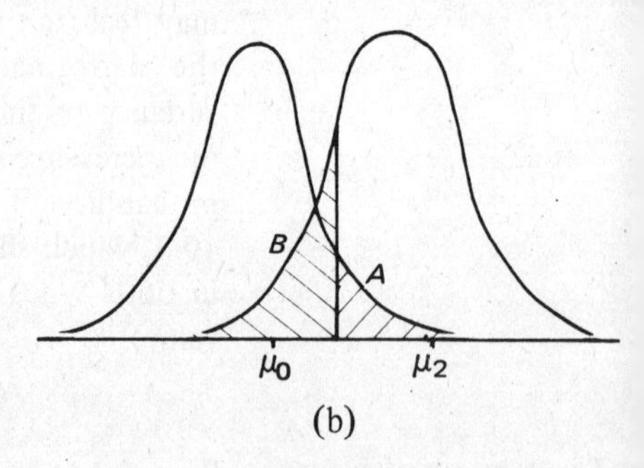

(a) (b)

Figure 10.4 The power of a test

H_0: $\mu = \mu_0$, then the probability of *not* rejecting H_0 is clearly greater if the true mean is μ_1 than it is if the true mean is μ_2. The smaller this probability is, the better is the test at discriminating between true and false hypotheses.

The test discriminates better in Figure 10.4(b) than it does in Figure 10.4(a). We say that the lower the probability of *not* rejecting H_0 when it is false (i.e. the lower the probability of committing a type-II error), the more *powerful* is the test.

In other words, the *power* of a test is the probability of rejecting H_0 when it is false, which is 1.0 minus the probability of *not* rejecting it; therefore

$$\text{The power of a test} = 1 - P(\text{type-II error}) \qquad (10.6)$$

Alternative hypotheses of the composite type

So far we have only considered alternative hypotheses of the simple type. In practice we are unlikely to know so much about μ that we can ascribe specific numeric values to it. We are much more likely to have to choose between competing hypotheses of the type given in (10.2), i.e.

$$\left.\begin{aligned} H_0&: \mu = \mu_0 \\ H_1&: \mu \neq \mu_0 \end{aligned}\right\} \quad (10.2)$$

In this form of the alternative hypothesis we say only that μ is *not* equal to μ_0 but make no statement as to any alternative value. In principle the basis of hypothesis testing with a *composite* alternative hypothesis is the same as with the simple type. We take a sample and if the sample mean $\bar{X}$ is 'very different' from μ_0 we reject H_0 in favour of H_1. Since we are no longer dealing with a specific value for μ under H_1, 'very different' from μ can mean that $\bar{X}$ is 'very much larger' or 'very much smaller' than μ. In a statistical sense we mean that if we are sampling from a population with $X \sim N(\mu, \sigma^2)$ then 'very much larger' and 'very much smaller' values of $\bar{X}$ taken from such a population would occur by chance only very rarely (perhaps only in one sample in a hundred, or one in fifty, or whatever).

Obviously, since we no longer have a specific value for the alternative hypothesis, we can no longer draw Distribution 1 as we did in Figure 10.1, and thus we cannot calculate the probability of a type-II error. This means that we cannot choose a critical value which will achieve some desired balance between type-I and type-II errors. Instead we choose a critical value such that some desired level of significance is achieved. We can illustrate this approach using Figure 10.5.

Let us denote the value of $\bar{X}$ (which we would consider to be very much greater than μ_0) by $\bar{X}_H^*$ and the value which we would consider very much smaller by $\bar{X}_L^*$. Thus if a particular sample value of $\bar{X}$ is less than $\bar{X}_L^*$ or more than $\bar{X}_H^*$ we shall consider it to be too different from μ_0 to have been taken from a population

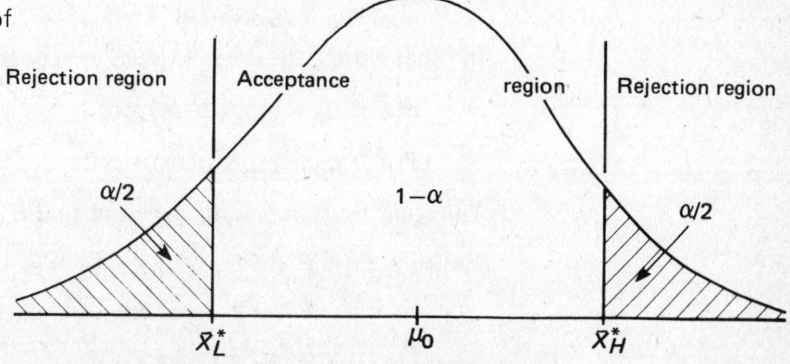

Figure 10.5 Choice of critical region when $H_1: \mu \neq \mu_0$ with a significance level of α

with $\mu = \mu_0$, and accordingly we would *reject* H_0. Thus the regions to the left of $\bar{X}_L^*$ and to the right of $\bar{X}_H^*$ are together termed the *rejection regions*. However, if the value of $\bar{X}$ lies *between* $\bar{X}_L^*$ and $\bar{X}_H^*$, then we do not consider this to be so different from μ_0 to enable us to reject H_0, i.e. we would *accept* H_0. Thus the region between $\bar{X}_L^*$ and $\bar{X}_H^*$ is termed the *acceptance region*. The percentage of the *total* area under the sampling distribution which lies in the *two* rejection regions is the *significance level* of the test, denoted by α, and we say that a test is *significant* at the α level if $\bar{X}$ falls in the rejection region.

Now the rejection region and hence the significance level has to be shared equally between the two tails of the sampling distribution of $\bar{X}$ in Figure 10.5. If the over-all significant level is α, then this means that the area in each tail is $\alpha/2$. Thus we choose the two critical values such that

$$\left.\begin{aligned} P(\bar{X} < \bar{X}_L^*) &= \alpha/2 \\ P(\bar{X} > \bar{X}_H^*) &= \alpha/2 \end{aligned}\right\} \quad (10.7)$$

Unfortunately in general we do not know which values of $\bar{X}_L^*$ and $\bar{X}_H^*$ will leave $\alpha/2$ of the area under the normal sampling distribution in each tail. But, as we have seen before, it is easy enough to find the corresponding value for the unit-normal distribution in the z-tables. For a sample taken from $X \sim N(\mu_0, \sigma^2)$ the sampling distribution of $\bar{X}$ will be such that the appropriate values of z will be

$$\left.\begin{aligned} -z_{\alpha/2} &= \frac{\bar{X}_L^* - \mu_0}{\sqrt{\sigma^2/n}} \\ &\text{and} \\ z_{\alpha/2} &= \frac{\bar{X}_H^* - \mu_0}{\sqrt{\sigma^2/n}} \end{aligned}\right\} \quad (10.8)$$

We can rearrange (10.8) to give

$$\left.\begin{aligned} \bar{X}_L^* &= \mu_0 - z_{\alpha/2}\sqrt{\sigma^2/n} \\ \bar{X}_H^* &= \mu_0 + z_{\alpha/2}\sqrt{\sigma^2/n} \end{aligned}\right\} \quad (10.9)$$

In other words, the critical values are chosen such that

$$\left.\begin{aligned} P(\bar{X} < \mu_0 - z_{\alpha/2}\sqrt{\sigma^2/n}) &= \alpha/2 \\ P(\bar{X} > \mu_0 + z_{\alpha/2}\sqrt{\sigma^2/n}) &= \alpha/2 \end{aligned}\right\} \quad (10.11)$$

Thus our decision-rule, in the event that $H_1: \mu \neq \mu_0$, is

$$\left.\begin{aligned} &\text{reject } H_0 \text{ if: } \bar{X} < \bar{X}_L^* \quad (\text{where } \bar{X}_L^* = \mu_0 - z_{\alpha/2}\sqrt{\sigma^2/n}) \\ &\qquad\quad \text{or if: } \bar{X} > \bar{X}_L^* \quad (\text{where } \bar{X}_H^* = \mu_0 + z_{\alpha/2}\sqrt{\sigma^2/n}) \\ &\text{do not reject } H_0 \text{ if: } \bar{X}_L^* \leqslant \bar{X} \leqslant \bar{X}_H^* \end{aligned}\right\} \quad (10.11)$$

For obvious reasons this test is known as a *two-tailed* test.

Suppose we decided that we considered $\bar{X}$ to be significantly different from μ_0 if such a value of $\bar{X}$ occurred only in one sample in twenty. This is a significance level of 5 per cent, i.e. $\alpha = 0.05$, which must be shared equally between the two tails of the sampling distribution. Thus $\alpha/2 = 0.025$, and in this case our decision-rule would be

$$\text{reject } H_0 \text{ if: } \bar{X} < \mu_0 - z_{0.025}\sqrt{\sigma^2/n}$$

$$\text{or if: } \bar{X} > \mu_0 + z_{0.025}\sqrt{\sigma^2/n}$$

$$\text{do not reject } H_0 \text{ if: } \mu_0 - z_{0.025}\sqrt{\sigma^2/n} \leqslant \bar{X} \leqslant \mu_0 + z_{0.025}\sqrt{\sigma^2/n}$$

We shall now work through an example to illustrate these results.

Example

Suppose that employees in a certain company work an average 8.5 hours of overtime per week. After a change in production conditions the company takes a sample of fifty employees' work records and finds that the sample mean number of hours of overtime is $\bar{X} = 7.5$. Apply a hypothesis test at a 5 per cent significance level to determine whether the sample mean is indicative of a significant change in the pattern of overtime working.

Let us first assume that the population variance is known (highly unlikely) and is $\sigma^2 = 100$. Thus we have

$$H_0: \mu = 8.5$$

$$H_1: \mu \neq 8.5$$

$$\alpha = 0.05$$

From the z-tables we see that the value of z that leaves $\alpha/2 = 0.025$ of the area in each tail is $z_{0.025} = \pm 1.96$ (remember that we look up $0.5 - 0.025 = 0.475$ in the body of the table). We can now substitute into (10.11) with $\mu_0 = 8.5$. This gives

$$\bar{X}_L^* = \mu_0 - z_{0.025}\sqrt{\sigma^2/n} = 8.5 - 1.96\sqrt{100/50}$$

i.e.

$$\bar{X}_L^* = 5.73$$

and

$$\bar{X}_H^* = \mu_0 + z_{0.025}\sqrt{\sigma^2/n} = 8.5 + 1.96\sqrt{100/50}$$

i.e.

$$\bar{X}_H^* = 11.27$$

These two critical values define the rejection and acceptance regions, as shown in Figure 10.6.

Figure 10.6
Distribution of mean weekly overtime hours showing critical values

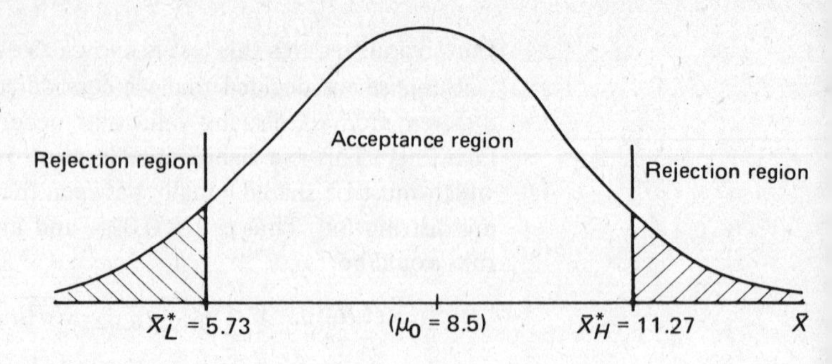

Now

$$\bar{X}_L^* \leqslant \bar{X} \leqslant \bar{X}_H^*$$

i.e.

$$5.73 \leqslant 7.5 \leqslant 11.27$$

Thus the sample mean falls in the acceptance region for H_0: $\mu = 8.5$. We thus conclude that the sample mean value of $\bar{X} = 7.5$ is not sufficiently different from our *a priori* value of $\mu = 8.5$ to lead us to think that it is likely to have come from a population with a different mean. The mean number of overtime hours per week does not appear to have changed in any statistically significant sense.

* * *

It is very unlikely that we would know σ^2 but we can use the sample variance s^2 as an unbiased estimator of σ^2 (see section 8.5) provided that the sample size is large enough ($n > 50$). In this case we can proceed as above except for the substitution of s^2 for σ^2 in (10.11). The decision-rule then becomes

$$\left.\begin{aligned} &\text{reject } H_0 \text{ if: } \bar{X} < \bar{X}_L^* \text{ where } (\bar{X}_L^* = \mu_0 - z_{\alpha/2}\sqrt{s^2/n}) \\ &\qquad\text{or if: } \bar{X} > \bar{X}_H^* \text{ where } (\bar{X}_H^* = \mu_0 + z_{\alpha/2}\sqrt{s^2/n}) \\ &\text{do not reject } H_0 \text{ if: } \bar{X}_L^* \leqslant \bar{X} \leqslant \bar{X}_H^* \end{aligned}\right\} \quad (10.12)$$

We now want to discuss the situation where the alternative hypotheses are of the form given in (10.3) and (10.4). We start by considering (10.3), i.e.

$$\left.\begin{aligned} H_0&: \mu = \mu_0 \\ H_1&: \mu < \mu_0 \end{aligned}\right\} \quad (10.13)$$

The alternative hypothesis in this case states merely that μ is less than μ_0. Quite often we shall have sound reasons for believing that if μ is not equal to μ_0 it is most likely to be larger, or anyway that is the hypothesis we wish to consider. In this case it is a question of $\bar{X}$ being close enough to μ_0 for us to accept H_0 or of it being so much less than μ_0 that we must reject H_0 (the question of $\bar{X}$ being much greater than μ_0 does not arise with this form of the alternative hypothesis).

The single critical value $\bar{X}^*$ in this situation is chosen so that the probability of any sample mean attaining such a low value is equal to the desired level of significance α. Thus if we want such a sample mean to occur by chance only in one in a hundred samples we would choose a significance of 1 per cent, i.e. $\alpha = 0.01$. The position is illustrated in Figure 10.7, where the whole of α is shown in the left-hand tail.

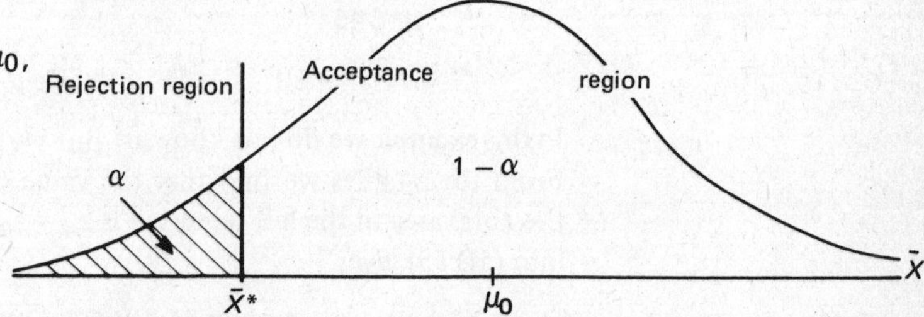

Figure 10.7 Critical region when $H_1: \mu < \mu_0$, significance level α

As we see, the critical value is chosen such that

$$P(\bar{X} < \bar{X}^*) = \alpha \tag{10.13}$$

and we can standardise this to determine the corresponding value in the unit-normal distribution, i.e.

$$-z_\alpha = \frac{\bar{X} - \mu_0}{\sqrt{\sigma^2/n}} \tag{10.14}$$

where the minus sign attached to z_α indicates that it is the *left-hand* tail of the distribution with which we are concerned. Rearranging gives

$$\bar{X}^* = \mu_0 - z_\alpha\sqrt{\sigma^2/n} \tag{10.15}$$

so that the critical value is chosen such that

$$P(\bar{X} < \mu_0 - z_\alpha\sqrt{\sigma^2/n}) = \alpha$$

Thus the decision-rule in the event that the alternative hypothesis is $H_1: \mu < \mu_0$ is

$$\left.\begin{array}{ll} \text{reject } H_0 \text{ if:} & \bar{X} < \bar{X}^* \\ \text{do not reject } H_0 \text{ if:} & \bar{X} > \bar{X}^* \\ \text{where} & \bar{X}^* = \mu_0 - z_\alpha\sqrt{\sigma^2/n} \end{array}\right\} \tag{10.16}$$

In the event that σ^2 is not known, then, provided n is large enough, we can replace it by its estimator s^2, in which case $\bar{X} = \mu_0 - z_\alpha\sqrt{s^2/n}$. This is a *one-tailed test.*

Example

A claim is made by the government that the mean weekly income of single-parent families is £46. A charitable organisation thinks it is less than this and takes a random sample of fifty single-parent families. The sample mean weekly income $\bar{X}$ = £41 and the sample variance $s^2 = 20$. Does the sample evidence support the government's contention or not? (Test at a 10 per cent significance level.)

We have

$$H_0:\ \mu = 46$$

$$H_1:\ \mu < 46$$

$$\alpha = 0.1;\ s^2 = 20;\ \bar{X} = 41;\ n = 50;\ \mu_0 = 46$$

In this example we do not know σ^2 but we may use s^2 in its place. From the z-tables we find that the value of z which leaves 0.1 of the total area in the left-hand tail is $z_\alpha = z_{0.1} = 1.28$. Substituting into (10.15) gives

$$\bar{X}^* = \mu_0 - z_\alpha\sqrt{s^2/n} = 46 - 1.28\sqrt{20/50} = 45.19$$

This critical value defines the 10 per cent rejection and acceptance regions in Figure 10.8.

Since $\bar{X} = 41$ is *less* than $\bar{X}^* = 45.19$, i.e. $\bar{X}$ falls in the rejection region, then we reject H_0 in favour of H_1. The sample evidence does not support the government's figure.

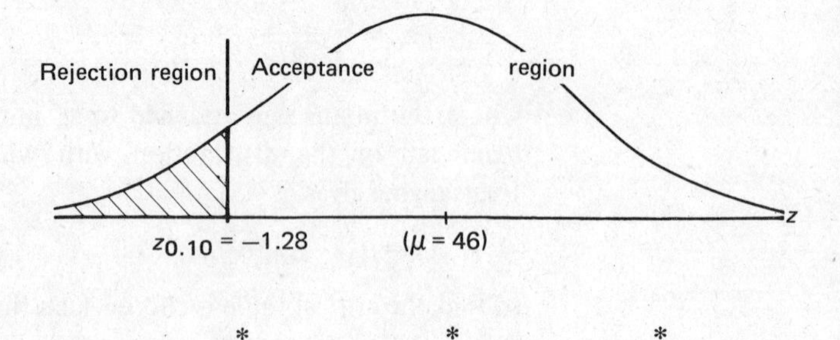

Figure 10.8 Unit-normal distribution of single-parent family income showing $\alpha = 0.1$ critical value

*   *   *

Finally we want to consider alternative hypotheses of the type indicated by (10.4), i.e.

$$H_0:\ \mu = \mu_0$$

$$H_1:\ \mu > \mu_0$$

Now we are concerned with values of $\bar{X}$ that are so much bigger than μ_0 that we ask whether they can possibly have been drawn from a population with a mean of μ_0.

If the level of significance is α, then the critical value $\bar{X}^*$ will be chosen to leave α of the area in the *right-hand* tail of the sampling distribution, as shown in Figure 10.9, such that

$$P(\bar{X} > \bar{X}^*) = \alpha$$

The decision-rule is thus

$$\left.\begin{array}{ll} \text{reject } H_0 \text{ if:} & \bar{X} > \bar{X}^* \\ \text{do not reject } H_0 \text{ if:} & \bar{X} < \bar{X}^* \\ \text{where } \bar{X}^* = \mu_0 + z_\alpha\sqrt{\sigma^2/n} & \end{array}\right\} \quad (10.17)$$

or where $\bar{X}^* = \mu_0 + z_\alpha\sqrt{s^2/n}$, if we are using s^2 as an estimator of $\sigma^2 (n > 50)$.

This is also a *one-tailed* test.

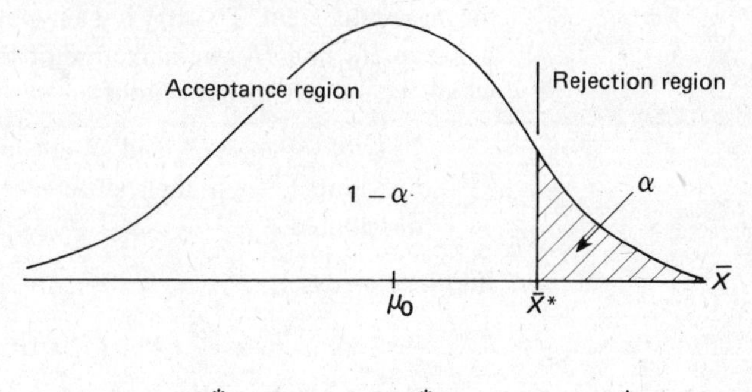

Figure 10.9 Critical regions when $H_1: \mu > \mu_0$

* * *

This concludes our discussion of hypothesis testing of population means when n is large enough to allow the substitution of σ^2 by s^2. We shall shortly consider the more realistic position when n is *not* large enough to allow us to use s^2 as an unbiased estimator of σ^2 without introducing an unwelcome degree of error. Before we do that, however, there are a couple of other topics we want to examine briefly.

To summarise the above discussion, when H_1 is of the form $\mu \neq \mu_0$ we use a two-tailed test, i.e. $\alpha/2$ in each tail. When H_1 is of the form $\mu < \mu_0$ or $\mu > \mu_0$ then we used a one-tailed test with α in either the left- or right-hand tail.

10.3 Testing the difference between two means

We now want briefly to consider the application of hypothesis testing to the difference between the means of two independent normal populations. Suppose these two populations are distributed as follows:

$$\text{first population:} \qquad X \sim N(\mu_1, \sigma_1^2)$$

$$\text{second population:} \qquad X \sim N(\mu_2, \sigma_2^2)$$

and we have to decide, on the basis of a sample mean from each population, $\bar{X}_1$ and $\bar{X}_2$, whether the two populations have different means. The competing hypotheses are

$$\left.\begin{aligned} H_0&: \mu_1 = \mu_2 \\ H_1&: \mu_1 \neq \mu_2 \end{aligned}\right\} \qquad (10.18)$$

and the form of H_1 tells us immediately that this is a two-tailed test. If the null hypothesis is true, then $\mu_1 - \mu_2 = 0$, and the two sample means should show little difference between them, i.e. $\bar{X}_1 - \bar{X}_2$ should be close to zero. If $\bar{X}_1 - \bar{X}_2$ is very different from zero, we can take this as evidence that the two population means *are* different. To carry out a hypothesis test to differentiate between H_0 and H_1 we make use of the following relationship, which we present without proof:

> If two variables X and Y are independent and normally distributed, then their difference $X - Y$ is also normally distributed.

In other words,

$$\left.\begin{aligned} &\text{if} \quad X \sim N(\mu_1, \sigma_1^2) \text{ and } Y \sim N(\mu_2, \sigma_2^2) \\ &\text{then} \quad (X - Y) \sim N[(\mu_1 - \mu_2), (\sigma_1^2 + \sigma_2^2)] \end{aligned}\right\} \qquad (10.19)$$

In our case the two independent normal distributions are the sampling distributions of the sample means $\bar{X}$ and $\bar{Y}$, and the distribution of $\bar{X} - \bar{Y}$, using (10.19), is therefore

$$(\bar{X} - \bar{Y}) \sim N[\mu_1 - \mu_2, (\sigma_1^2/n_1 + \sigma_2^2/n_2)] \qquad (10.20)$$

where n_1 and n_2 are the sizes of the two samples.

We know from the form of the alternative hypothesis that this is a two-tailed test and analogous to the form of (10.9). We can develop the two critical values thus:

$$\text{Lower critical value} = (\mu_1 - \mu_2) - z_{\alpha/2, \bar{X}_1 - \bar{X}_2}\sqrt{\sigma_1^2/n_1 + \sigma_2^2}$$

$$\text{Upper critical value} = (\mu_1 - \mu_2) + z_{\alpha/2, \bar{X}_1 - \bar{X}_2}\sqrt{\sigma_1^2/n_1 + \sigma_2^2/}$$

However, under the null hypothesis, $\mu_1 = \mu_2$, i.e. $\mu_1 - \mu_2 = 0$, and in these circumstances the two critical values become

$$\left.\begin{aligned} \bar{X}_L^* &= -z_{\alpha/2, \bar{X}_1 - \bar{X}_2}\sqrt{\sigma_1^2/n_1 + \sigma_2^2/n_2} \\ \bar{X}_H^* &= z_{\alpha/2, \bar{X}_1 - \bar{X}_2}\sqrt{\sigma_1^2/n_1 + \sigma_2^2/n_2} \end{aligned}\right\} \qquad (10.21)$$

and the decision-rule is

$$\left.\begin{array}{ll}\text{reject } H_0 \text{ if:} & (\bar{X}_1 - \bar{X}_2) < \bar{X}_L^* \\ \text{or if:} & (\bar{X}_1 - \bar{X}_2) > \bar{X}_H^* \\ \text{do not reject } H_0 \text{ if:} & \bar{X}_L^* \leqslant (\bar{X}_1 - \bar{X}_2) \leqslant \bar{X}_H^* \end{array}\right\} \quad (10.22)$$

In the event that σ^2 is unknown and $n > 50$ we can substitute s^2 in place of σ^2 in (10.21).

Example

Suppose a claim is made that male and female employees in the United Kingdom work the same number of hours. A sample of size $n = 100$ is taken of male and female workers and $\bar{X}_M = 46.3$ and $\bar{X}_F = 38.4$. If $\sigma_M^2 = 25$ and $\sigma_F^2 = 20$, test at the 5 per cent significance level whether the claim is true.

The value of $z_{\alpha/2}$ which leaves 2.5 per cent (i.e. 0.025) of the area under the normal curve in each tail is 1.96. Substituting in (10.21) gives

$$\bar{X}_L^* = -1.96\sqrt{25/100 + 20/100} = -1.31$$

$$\bar{X}_H^* = 1.96\sqrt{25/100 + 20/100} = +1.31$$

Now $\bar{X}_1 - \bar{X}_2 = 46.3 - 38.4 = 7.9$. Thus $(\bar{X}_1 - \bar{X}_2) > \bar{X}_H^*$, and we reject H_0 in favour of H_1, i.e. there does seem to be a *significant* difference between the means of the two populations: males and females do not work the same number of weekly hours.

10.4 A test of population variance

Let us now briefly consider a test concerning σ^2. Suppose we wish to choose between the following hypotheses relating to the value of σ^2:

$$\left.\begin{array}{l} H_0: \sigma^2 = \sigma_0^2 \\ H_1: \sigma^2 \neq \sigma_0^2 \end{array}\right\} \quad (10.23)$$

We can use the standard deviation from a sample of size n to perform the hypothesis test, i.e. from (8.22) we have

$$s^2 = \frac{1}{n-1}\left[\Sigma(X_i - \bar{X})^2\right]$$

and we can define the acceptance and rejection regions for the distribution of s^2 using the result originally given in equation (8.30), i.e.

$$(n-1)s^2/\sigma^2 \sim \chi_{n-1}^2$$

Thus we choose values Q_L^* and Q_H^* such that for a level of significance of α

$$\left.\begin{aligned} P[(n-1)s^2/\sigma_0^2 < Q_L^*] &= \alpha/2 \\ P[(n-1)s^2/\sigma_0^2 > Q_H^*] &= \alpha/2 \end{aligned}\right\} \quad (10.24)$$

where, as can be seen from Figure 8.10(b),

$$Q_L^* = \chi^2_{1-n-1,\alpha/2}$$
$$Q_H^* = \chi^2_{n-1,\alpha/2}$$

where the subscripts $1-\alpha/2$ and $\alpha/2$ refer to the area to the *right* of the particular critical value of χ^2. Thus the decision-rule is

reject H_0 if: $(n-1)s^2/\sigma_0^2 < \chi^2_{1-\alpha/2,n-1}$

or if: $(n-1)s^2/\sigma_0^2 > \chi^2_{\alpha/2,n-1}$

do not reject H_0 if: $\chi^2_{1-\alpha/2,n-1} \leqslant (n-1)s^2/\sigma_0^2 \leqslant \chi^2_{\alpha/2,n-1}$

Example

It is suggested that the variance in the distribution of women's earnings (£ per week) in a certain industry is the same as that of men, whose variance is known to be 25. A sample of twenty women yields a sample variance $s^2 = 29$. Perform a hypothesis test at a 1 per cent level of significance to determine whether the sample evidence upholds the suggestion.

We have

$$H_0:\ \sigma^2 = 25$$
$$H_1:\ \sigma^2 \neq 25$$
$$s^2 = 29;\ \alpha = 0.01;\ \alpha/2 = 0.005;\ 1-\alpha/2 = 0.995$$

From the χ^2 tables with 19 d.f. we have

$$\chi^2_{19,0.005} = 38.582$$

which means that this value of χ^2 would only be exceeded with a probability of 0.005:

$$\chi^2_{19,0.995} = 6.844$$

which means that this value of χ^2 would be exceeded with a probability of 0.995.

The test-statistic in this example takes the value

$$(n-1)s^2/\sigma_0^2 = (20-1)29/25 = 22.04$$

and since

$$6.844 \leqslant 22.04 \leqslant 38.582$$

we do not reject H_0. In other words, the variance of women's weekly earnings in the population cannot be statistically distinguished, at $\alpha = 0.01$, from that of men's earnings. The position is illustrated in Figure 10.10.

It is not difficult to establish the test criteria for alternative hypotheses of the form H_1: $\sigma^2 < \sigma_0^2$ and H_1: $\sigma^2 > \sigma_2^2$, but we will leave this task to the reader.

Figure 10.10
Hypothesis test of σ^2

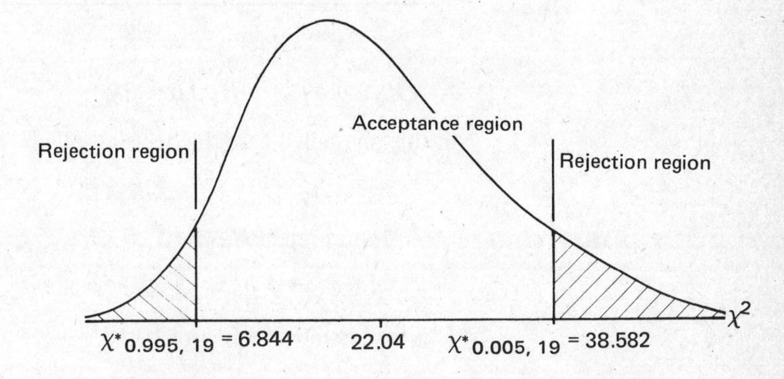

10.5 Small sample hypothesis testing of population means

In all of the hypothesis tests of μ so far we have assumed that the sample size is large enough to enable us to use s^2 as an estimator of the usually unknown σ^2. More often than not in economics our sample size will not be so large as to enable us to make the substitution without introducing potentially serious error into the results. We now want to consider the more realistic situation where n is small. Recall that in section 8.5 when n was small we made use of the t-distribution to enable us to use s^2 in place of σ^2 when estimating μ. We do exactly the same thing in hypothesis testing. We want to consider each of the three types of alternative hypotheses in turn, but since the development of the decision-rules and the test-statistic exactly parallels that for large sample hypothesis testing – equations (10.7) through to (10.17) – except that t is used instead of z, we need only present the decision-rules in the three cases.

1. H_0: $\mu = \mu_0$; H_1: $\mu \neq \mu_0$

In this two-tailed case the decision-rule is

$$\left.\begin{aligned} &\text{reject } H_0 \text{ if: } \bar{X} < \bar{X}_L^* \text{ (where } \bar{X}_L^* = \mu_0 - t_{n-1,\alpha/2}\sqrt{s^2/n}) \\ &\qquad\text{or if: } \bar{X} > \bar{X}_H^* \text{ (where } \bar{X}_H^* = \mu_0 + t_{n-1,\alpha/2}\sqrt{s^2/n}) \\ &\text{do not reject } H_0 \text{ if: } \bar{X}_L^* \leqslant \bar{X} \leqslant \bar{X}_H^* \end{aligned}\right\} \quad (10.25)$$

This is illustrated in Figure 10.11.

Figure 10.11 Critical values when $H_1: \mu \neq \mu_0$

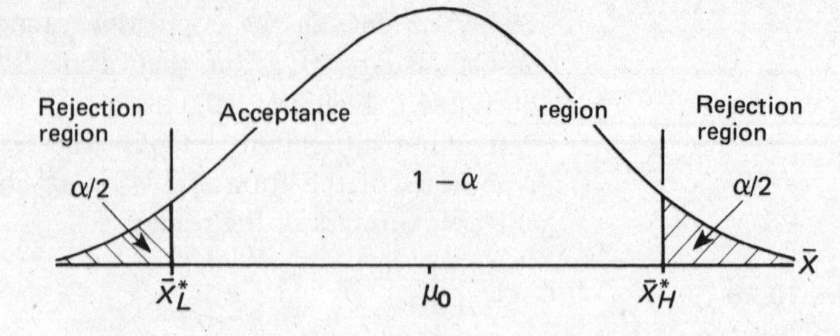

2. H_0: $\mu = \mu_0$; H_1: $\mu < \mu_0$

In this one-tailed case the decision-rule is

$$\left.\begin{array}{ll} \text{reject } H_0 \text{ if:} & \bar{X} < \bar{X}^* \\ \text{do not reject } H_0 \text{ if:} & \bar{X} > \bar{X}^* \\ \text{where } \bar{X}^* = \mu_0 - t_{n-1,\alpha}\sqrt{s^2/n} & \end{array}\right\} \quad (10.26)$$

This is illustrated in Figure 10.12.

Figure 10.12 Critical value when $H_1: \mu < \mu_0$

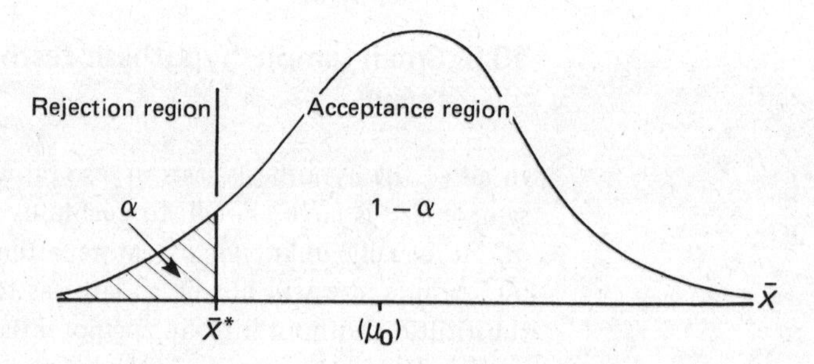

3. H_0: $\mu = \mu_0$; H_1: $\mu > \mu_0$

In this one-tailed case the decision-rule is

$$\left.\begin{array}{ll} \text{reject } H_0 \text{ if:} & \bar{X} > \bar{X}^* \\ \text{do not reject } H_0 \text{ if:} & \bar{X} < \bar{X}^* \\ \text{where } \bar{X}^* = \mu_0 + t_{n-1,\alpha}\sqrt{s^2/n} & \end{array}\right\} \quad (10.27)$$

This is illustrated in Figure 10.13.

Figure 10.13 Critical value when $H_1: \mu > \mu_0$

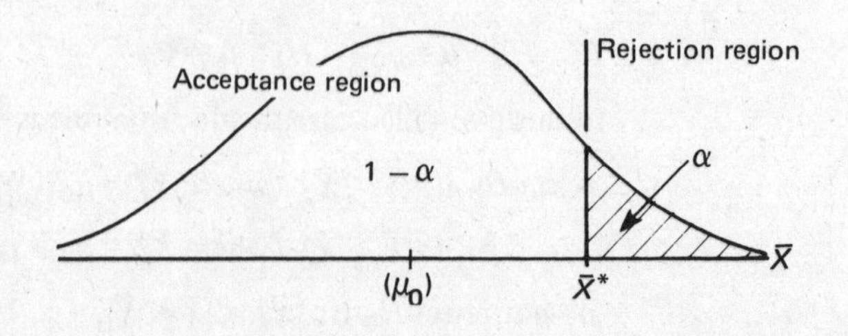

* * *

We now illustrate the use of (10.25), (10.26) and (10.27) with three examples.

Example 1

It is suggested that the mean rate of return on capital invested in public-sector enterprises is the same as that for large private-sector companies. If the latter value is $\mu = 14.2$ and a sample of nine public-sector enterprises yields a mean of $\bar{X} = 12.9$ and an $s^2 = 9$, test whether this contention is true using a 5 per cent significance level.

We have

$$H_0: \mu = 14.2$$

$$H_1: \mu \neq 14.2$$

$$\alpha = 0.05; \quad s^2 = 9; \quad \bar{X} = 12.9; \quad n = 9; \quad \mu_0 = 14.2$$

The value of t which leaves $\alpha/2 = 0.025$ in each tail of the distribution is $t_{8,0.025} = \pm 2.306$. Therefore, from (10.25),

$$\bar{X}_L^* = \mu_0 - t_{8,0.025}\sqrt{s^2/n} = 14.2 - 2.306\sqrt{9/9} = 11.894$$

$$\bar{X}_H^* = \mu_0 + t_{8,0.025}\sqrt{s^2/n} = 14.2 + 2.306\sqrt{9/9} = 16.506$$

Now $\bar{X} = 12.9$; thus

$$11.894 \leqslant 12.9 \leqslant 16.506$$

Thus, applying (10.25), we accept H_0: there does not appear to be a significant difference between the rates of return.

Example 2

A psychology lecturer believes that programmed learning will produce a better examination performance by his students. The mean examination mark prior to the introduction of programmed learning was 54.2 per cent. In the first year after the introduction the mean mark of fifteen students was 57.1 per cent, with $s^2 = 8.8$. Test whether this is a significant improvement using a 5 per cent significance level.

We have

$$H_0: \mu = 54.2$$

$$H_1: \mu > 54.2$$

$$\alpha = 0.05; \quad s^2 = 8.8; \quad \bar{X} = 57.1; \quad n = 15; \quad \mu_0 = 54.2$$

The value of t which leaves 0.05 of the area in the right-hand tail is $t_{n-1,\alpha} = t_{0.05,14} = 1.761$. Therefore, from (10.27),

$$\bar{X}^* = \mu_0 + t_{n-1,\alpha}\sqrt{s^2/n} = 54.2 + 1.761\sqrt{8.8/15} = 55.55$$

But $\bar{X} = 57.1$; thus

$$57.1 > 55.55$$

Thus, applying (10.27), we reject H_0; the sample evidence supports the belief in an increased examination performance. There is a probability of less than 0.05 that a sample with a mean of 57.1 could have been drawn from a population with a mean of 54.2.

Example 3

A large mining company has a mean level of absenteeism of ninety-eight workers per 1000 at any one time. It introduces a new attendance bonus to improve this figure. A subsequent sample of nine of the company's mines yields a mean of ninety absent workers per 1000, with $s^2 = 16$. Test whether absenteeism has declined using a 5 per cent level of significance.

We have

$$H_0: \mu = 98$$

$$H_1: \mu < 98$$

$$\bar{X} = 90; \quad s^2 = 16; \quad \alpha = 0.05; \quad n = 9; \quad \mu_0 = 98$$

The value of t which leaves 0.05 of the area in the left-hand tail is $t_{n-1,\alpha} = t_{0.05,8} = -1.86$. Therefore, from (10.26), we have

$$\bar{X}^* = \mu_0 - t_{n-1,\alpha}\sqrt{s^2/n} = 98 - 1.86\sqrt{16/9} = 95.52$$

but $\bar{X} = 90$; thus

$$90 < 95.52$$

i.e.

$$\bar{X} < \bar{X}_L^*$$

Hence we reject H_0 in favour of H_1; we conclude that absenteeism has probably declined.

* * *

We want to conclude this section by presenting an algorithm for small sample testing of population means. This is shown in Figure 10.14.

A small sample test of the difference between two means

In section 10.3 we established some results relating to the difference between the means of two populations in the event that the sample sizes were large enough to justify the substitution of s^2 for σ^2. This section presents a parallel result for the small sample case except that with small n we use the t- rather than the z-distribution.

We know from (10.19) that if $\bar{X}_1 \sim N(\mu_1, \sigma_1^2/n_1)$ and $\bar{X}_2 \sim N(\mu_2, \sigma_2^2/n_2)$, and if $\bar{X}_1$ and $\bar{X}_2$ are independent, then

$$(\bar{X}_1 - \bar{X}_2) \sim N[(\mu_1 - \mu_2), (\sigma_1^2/n_1 - \sigma_2^2/n_2)]$$

Figure 10.14 An algorithm for small sample hypothesis tests of μ

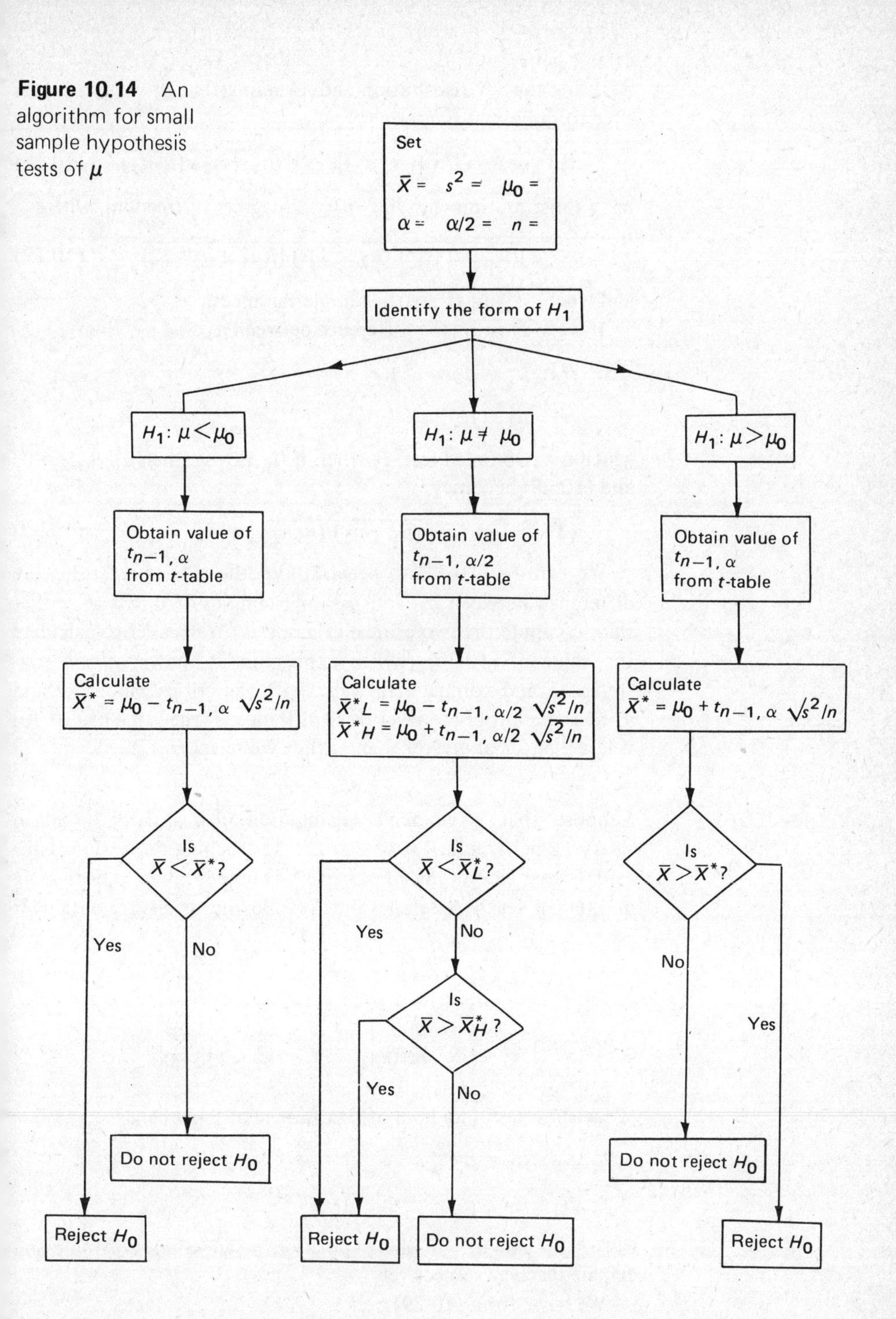

By using the χ^2 distribution and assuming that $\sigma_1^2 = \sigma_2^2$ it is possible to show that the term

$$[(\bar{X}_1 - \bar{X}_2) - (\mu_1 - \mu_2)]/\{s\sqrt{(n_1 + n_2)/(n_1 n_2)} \qquad (10.28)$$

has a t-distribution with $n_1 + n_2 - 2$ degrees of freedom, where

$$s = \sqrt{[(n_1 - 1)s_1^2 + (n_2 - 1)s_2^2]/(n_1 + n_2 - 2)} \qquad (10.29)$$

and where s_1^2 and s_2^2 are the sample variances.

If we are testing the difference between μ_1 and μ_2, then

H_0: $\mu_1 = \mu_2$

H_1: $\mu_1 \neq \mu_2$

which is a two-tailed test. Further, if H_0 is true, then $\mu_1 - \mu_2 = 0$, and (10.28) becomes

$$(\bar{X}_1 - \bar{X}_2)/\{s\sqrt{(n_1 + n_2)/(n_1 n_2)}\} \sim t_{(n_1 + n_2 - 2)} \qquad (10.30)$$

We can use (10.30) to perform hypothesis tests regarding the difference between μ_1 and μ_2. In examples of this type, rather than compute the two critical values, it is often easier to calculate the value of (10.30) (which expression is known as the *test-statistic*), and compare this value with the critical values $-t$ and $+t$ at the appropriate level of significance. If the value of (10.30) is less than $-t$ or greater than $+t$ then we reject H_0.

Example

Suppose that a women's organisation claims that the mean salary (£ per annum) of women in the teaching profession is different from that of male teachers. A sample is taken of both male and female teachers' salaries and the following results are obtained:

Male	*Female*
$n_1 = 18$	$n_2 = 22$
$\bar{X}_1 = 5900$	$\bar{X}_2 = 5500$
$s_1^2 = 160\,000$	$s_2^2 = 115\,000$

We wish to test, at a level of significance of 5 per cent,

H_0: $\mu_1 = \mu_2$

H_1: $\mu_1 \neq \mu_2$

where μ_1 and μ_2 are the population mean salaries of male and female teachers respectively.

We have, from (10.29),

$$s = \sqrt{[(18 - 1)160{,}000 + (22 - 1)115{,}000]/(18 + 22 - 2)}$$

$$s = \sqrt{135\,131.58} = 367.6025$$

Also, the value of t which leaves 0.025 in each tail is (by interpolation)

$$t_{(n_1+n_2-2),\alpha/2} = t_{38,0.025} = \pm 1.96$$

Now $\overline{X}_1 - \overline{X}_2 = 5900 - 5500 = 400$, and $(n_1 + n_2)/(n_1 n_2) = (18 + 22)/(18 \times 22) = 0.1010$. Thus the value of the test-statistic as given by (10.30) is equal to

$$400/(367.6025\sqrt{0.1010}) = 3.42$$

Since this value exceeds $t = 1.96$ we are able to reject H_0. On the basis of the sample evidence there does appear to be a significant difference between male and female mean salaries.

10.6 Testing hypotheses about population proportions

In this section we shall consider the testing of hypotheses relating to a population proportion π. Much of this discussion parallels that for tests of population means, and for that reason (and because in economics we are more frequently concerned with population means) we shall be brief. We want to consider hypothesis testing first of a single population proportion and then of the difference between two population proportions.

We will examine the situation which arises for each of the three different alternative hypotheses.

1. H_0: $\pi = \pi_0$; H_1: $\pi \neq \pi_0$

The sign in the alternative hypothesis indicates that this is a two-tailed test situation. Recall that if a population has a mean of π then the sampling distribution of the sample proportion p is approximately normal, that is, from (7.13),

$$p \sim \text{approx } N(\pi, \pi q/n)$$

and we can make use of this result to establish the values of the two critical values p_L^* and p_H^* in each tail. Analogous with (10.9) we have, therefore,

$$\left.\begin{aligned} p_L^* &= \pi_0 - z_{\alpha/2}\sqrt{\pi_0 q_0/n} \\ p_H^* &= \pi_0 + z_{\alpha/2}\sqrt{\pi_0 q_0/n} \end{aligned}\right\} \quad (10.31)$$

In other words,

$$P(p < \pi_0 - z_{\alpha/2}\sqrt{\pi_0 q_0/n}) = \alpha/2 \quad (10.32)$$

$$P(p > \pi_0 + z_{\alpha/2}\sqrt{\pi_0 q_0/n}) = \alpha/2 \quad (10.33)$$

This provides us with the decision-rule

$$\left.\begin{array}{ll} \text{reject } H_0 \text{ if:} & p < p_L^* \\ \quad\text{or if:} & p > p_H^* \\ \text{do not reject } H_0 \text{ if:} & p_L^* \leqslant p \leqslant p_H^* \end{array}\right\} \quad (10.34)$$

where p_L^* and p_H^* are as given in (10.31).

The situation is illustrated in Figure 10.15.

Figure 10.15 Critical values when $H_1: \pi \neq \pi_0$

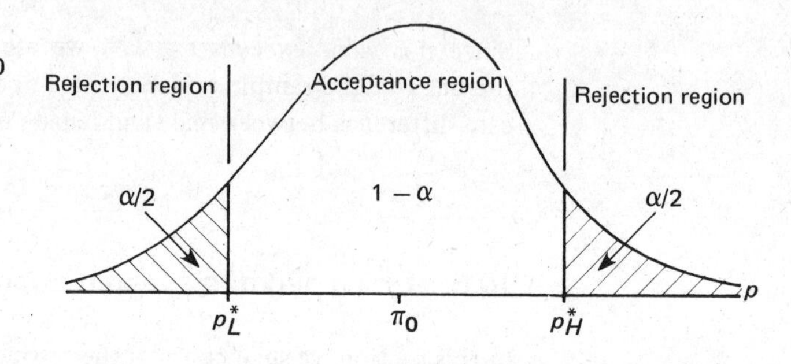

Example

Suppose it is believed that the proportion of pre-1914 housing is 0.3071. A sample of 100 houses provides a sample proportion of $p = 0.280$. Test whether this sample supports the null hypothesis using a 1 per cent significance level.

The competing hypotheses are

$$H_0: \pi = 0.3071$$

$$H_1: \pi \neq 0.3071$$

We have

$$\pi_0 = 0.3071; \quad p = 0.280; \quad n = 100; \quad \alpha = 0.01$$

The value of $z_{\alpha/2}$ which leaves 0.005 of the area in each tail is $z_{0.005} = \pm 2.58$ (remember that we look up $0.5 - 0.005 = 0.495$ in the body of the z-table). Therefore, substituting in (10.31) will give us the critical values, i.e.

$$p_L^* = \pi_0 - z_{\alpha/2}\sqrt{\pi_0 q_0/n} = 0.3071 - 2.58\sqrt{0.3071(1 - 0.3071)/100}$$

$$p_H^* = \pi_0 + z_{\alpha/2}\sqrt{\pi_0 q_0/n} = 0.3071 + 2.58\sqrt{0.3071(1 - 0.3071)/100}$$

from which

$$p_L^* = 0.188$$

$$p_H^* = 0.426$$

Therefore,

$$0.188 \leqslant 0.280 \leqslant 0.426$$

i.e.

$$p_L^* \leqslant p \leqslant p_H^*$$

and, using (10.34), we do not reject H_0: the sample evidence supports the *a priori* value of 0.3071.

2. H_0: $\pi = \pi_0$; H_1: $\pi < \pi_0$

This is a one-tailed test; thus there will be one critical value p^* located such as to leave an area α in the left-hand tail of the sampling distribution of p. Therefore, analogous to (10.15), we can define the value of p^* as being

$$p^* = \pi_0 - z_\alpha \sqrt{\pi_0 q_0 / n} \tag{10.35}$$

where p^* is chosen such that

$$p(p < p^*) = \alpha$$

This provides the decision-rule

$$\left. \begin{array}{ll} \text{reject } H_0 \text{ if:} & p < p^* \\ \text{do not reject } H_0 \text{ if:} & p > p^* \\ \text{where } p^* = \pi_0 - z_\alpha \sqrt{\pi_0 q_0 / n} & \end{array} \right\} \tag{10.36}$$

This is illustrated in Figure 10.16.

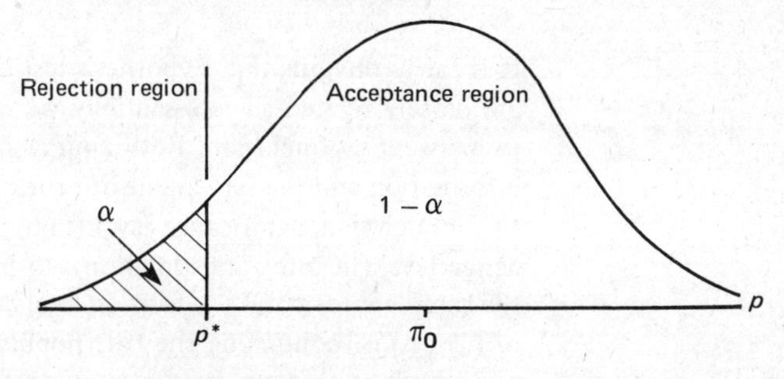

Figure 10.16 Critical value when $H_1: \pi < \pi_0$

3. H_0: $\pi = \pi_0$; H_1: $\pi > \pi_0$

This is a one-tailed test with a critical value p^* located such as to leave an area α in the right-hand tail of the sampling distribution of p. The critical value is therefore given by

$$p^* = \pi_0 + z_\alpha \sqrt{\pi_0 q_0 / n} \tag{10.37}$$

such that

$$P(p > p^*) = \alpha$$

This provides the decision-rule

$$\left.\begin{array}{ll}\text{reject } H_0 \text{ if:} & p > p^* \\ \text{do not reject } H_0 \text{ if:} & p < p^* \\ \text{where } p^* = \pi_0 + z_\alpha\sqrt{\pi_0 q_0/n} & \end{array}\right\} \quad (10.38)$$

This is illustrated in Figure 10.17.

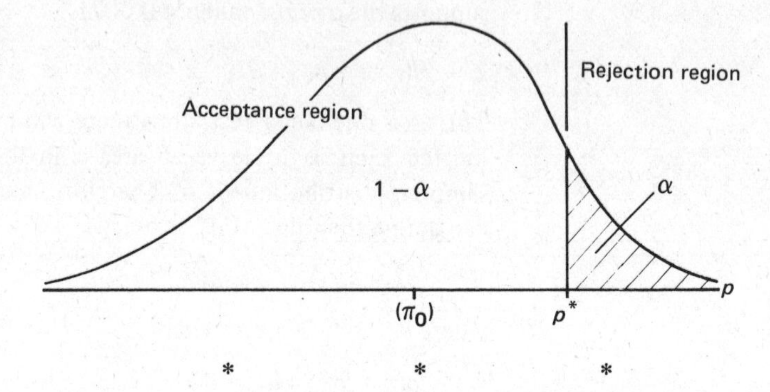

Figure 10.17 Critical value when $H_1: \pi > \pi_0$

* * *

This ends our discussion of hypothesis testing. Before we conclude this chapter we want to examine the relationship between interval estimation and hypothesis testing.

10.7 Interval estimation and hypothesis testing compared

It is fairly obvious that hypothesis testing and interval estimation are closely related since essentially we use the same information to arrive at a conclusion. Both approaches use the same sampling information and the same value of t (or z, if n is large) appropriate to some chosen significance level (in hypothesis testing) or confidence level (in interval estimation). In interval estimation we use the sample information to form an interval which has a probability of $(1 - \alpha)$ of containing the true population parameter value. In hypothesis testing we use the sample information to decide (with a probability of $1 - \alpha$ of being correct) whether the true population parameter value has some *a priori* value. Clearly if a hypothesis test results in a rejection of some *a priori* value of μ (or π), then we would not expect the confidence interval to contain that same *a priori* value. The acceptance region in the hypothesis test is exactly the interval estimate in the interval estimation case (provided of course that $\mu_0 = \bar{X}$) and

$$\text{confidence level} = 100 - \text{significance level}$$

Let us illustrate these points with an example.

Example

Suppose that X, defined as the mean age of the economically active population of the United Kingdom, is believed to be distributed $X \sim N(62,90)$. A sample of thirty of the economically active yields values of $\bar{X} = 55$ and $s^2 = 60$.

(a) Construct an interval estimate for μ using a confidence level of 95 per cent.

(b) Test whether the sample evidence supports the belief that $\mu = 62$, using a significance level of 5 per cent.

(a) With a confidence level of 95 per cent we have

$$100(1 - \alpha) = 95$$

Therefore, $\alpha = 0.05$ and $\alpha/2 = 0.025$. From (8.43) we know that a $100(1 - \alpha)$ per cent confidence interval for μ is

$$(\bar{X} - t_{n-1,\alpha/2}\sqrt{s^2/n}, \bar{X} + t_{n-1,\alpha/2}\sqrt{s^2/n})$$

Since $t_{29,0.025} = \pm 2.045$ we have

$$(55 - 2.045\sqrt{60/30}, 55 + 2.045\sqrt{60/30})$$

i.e.

$$(52.1089, 57.892)$$

i.e. from (8.42)

$$P(52.108 \leqslant \mu \leqslant 57.892) = 0.95$$

In other words, there is a probability of 0.95 that the interval from 52.108 to 57.892 will contain μ.

(b) The hypothesis test must differentiate between $H_0: \mu = 62$ and $H_1: \mu \neq 62$, and this is a two-tailed test. Since the interval estimate was from 52.108 to 57.892, and since $\mu_0 = 62$ does not fall in this interval, then we would expect H_0 to be rejected.

From (10.25) we have

$$\bar{X}_L^* = \mu_0 - t_{n-1,\alpha/2}\sqrt{s^2/n} = 62 - 2.045\sqrt{60/30} = 59.108$$

$$\bar{X}_H^* = \mu_0 + t_{n-1,\alpha/2}\sqrt{s^2/n} = 62 + 2.045\sqrt{60/30} = 64.892$$

Since

$$55 < 59.108$$

i.e.

$$\bar{X} < \bar{X}_L^*$$

we reject H_0. This sample was probably *not* drawn from a population with a mean of 62. Thus the results from the hypothesis test and from the interval estimate support each other.

10.8 Summary

In this chapter we have discussed the second of the two parallel branches of statistical inference and we should now know how the two approaches are equivalent. We examined the nature of type-I and type-II errors and saw how in general the probability of a type-II error remains indeterminate. We have distinguished between one- and two-tailed tests and have performed such tests on population means and variances and on population proportions. In the former regard, although we discussed both small and large sample testing, in practice we shall normally find ourselves using the *t*-distribution because of the prevalence of small samples in the economic and social sciences.

Although the material we have discussed so far is important and enables us to indulge in a wide variety of statistical analyses, we have yet to discuss one of the most important and powerful techniques available to statisticians. What we have done so far has laid the necessary foundations for what is to come in the remaining two chapters of this book.

EXERCISES

10.1 Discuss the relationship between the significance level of a hypothesis test and the probability of a type-I error.

10.2 The net profit per person employed by a large supermarket chain is £10 000 per annum. Management changes are introduced to try to improve this figure, and a subsequent sample of sixty-four outlets reveals a sample mean net profit per employee of £10 500 with a sample standard deviation of £2000. Test whether this is indicative of a significant increase in the profit level. Test at $\alpha = 0.10$ and $\alpha = 0.05$.

10.3 The government believes that the average stay on the unemployment register is ten weeks, but because it is uncertain about this figure it takes a sample of 100 unemployed persons. This yields a sample mean of eleven weeks with a standard deviation of four weeks. Test whether this result confirms the governments belief or not. Use $\alpha = 0.05$ and $\alpha = 0.01$.

10.4 A similar study to that in the previous question but using a sample of twenty-five unemployed persons from an inner-city area yields a mean of twelve weeks and a standard deviation of 5 weeks. Does this result support the government's contention?

10.5 A relief organisation in a certain country knows from previous studies that the average distance which each family has to walk to fetch water is 5.6 miles. A small capital investment programme in new wells is initiated in order to reduce this figure. A subsequent sample of thirty families reveals that the sample mean distance is five miles with a standard deviation of 1.4 miles. Test whether this represents a significant improvement. Use $\alpha = 0.05$.

10.6 A transport undertaking knows that the average number of passenger-miles per employee carried on its vehicles is 60.5 per journey. It reduces fares in an attempt to increase this figure, and a subsequent sample of twenty-five vehicle journeys yields a sample mean of sixty-seven miles with a standard deviation of ten miles. Does this represent a significant improvement? use $\alpha = 0.10$.

10.7 Take a random sample of twenty male and twenty female earnings figures from the data in exercise 2.1 and test whether there is a significant difference between male and female earnings. Use $\alpha = 0.05$ and $\alpha = 0.01$.

10.8 A researcher believes that there is a significant difference in the suicide rates between city and rural communities. A survey yields the following results in terms of the suicide rate per 1000 persons:

City	$n = 12$	$\bar{X} = 0.08$	$s = 0.02$
Rural	$n = 10$	$\bar{X} = 0.11$	$s = 0.03$

What do you conclude from these results?

10.9 An advertising company believes that 50 per cent of households in its area own an automatic washing-machine. A sample of 100 households reveals that forty-four own these machines. Does this result support the company's belief?

10.10 A television company believes that more people watch its programmes than those of a rival channel. A sample of thirty-six people shows that twenty watch the company's programmes in preference to those of its rival. Does this support the television company's contention?

PART IV

REGRESSION ANALYSIS

11 Simple Regression

OBJECTIVES

After reading this chapter and working through the examples students should understand the meaning of the following terms

- bivariate statistics
- multivariate statistics
- mathematical model
- function
- dependent variable
- independent variable
- explanatory variable
- causality
- linear relationship
- constant or intersect parameter

slope parameter
deterministic model
stochastic model
correlation coefficient
spurious correlation
population regression equation
population regression line
random disturbance term
sample regression equation
sample regression line
estimated regression line
error or residual term
classical normal linear regression model
basic assumptions
ordinary least squares
normal equations
variance of an estimator
standard error of an estimator
standard error of the regression
goodness of fit
coefficient of determination R^2
variation of Y
total sum of squares
regression or explained sum of squares
error sum of squares
significance of an estimator
confidence interval of an estimator
point prediction
mean value prediction
particular value prediction
standard error of a predictor
confidence interval of a predictor

Students should be able to:

draw a scatter diagram
calculate the correlation coefficient
calculate the values of the OLS estimators for a simple model
test the significance of the parameters
construct confidence intervals of the parameters
calculate the coefficient of determination R^2
use a model for prediction

11.1 Introduction

Apart from a brief reference to joint distribution and covariance in Chapter 6 (p. 114) all of the discussion so far has been concerned with *univariate statistics*, i.e. with statistical measures on a single variable. Although in the economic and social sciences we are often interested in describing the principal statistical features of a single variable, more often we are concerned with the relationship between two or more variables. Obvious examples include the relationship between the price of a good and the quantity sold, between money supply and inflation, between trade-union membership and wage claims, between profit and capital investment, and so on. All these examples refer to the relationship between two variables (the study of which we might refer to as *bivariate statistics*), but the study of the relationships between three or more variables (which we call *multivariate statistics*) is just as common – for example, the relationship between the price of some commodity, the quantity sold, the prices of complementary and substitute commodities, incomes, tastes, and so on.

Such relationships are usually expressed in the form of *models*, which may be defined as (more or less) simplified representations of reality, and will frequently take mathematical forms. For example, if we believe that the number of new motor-cars sold (denoted by a variable Y and measured in thousands) is related in some way to the average price of a new motor-car (denoted by the variable X and measured in £000) we might express this in the following way:

$$Y = f(x) \tag{11.1}$$

which is read as 'Y is a function of X'. Note that in (11.1) we have not specified the *exact* form of the relationship, simply that a relationship of some sort does exist. However, (11.1) does imply that the value of Y, i.e. the number of cars sold, is dependent on X, the price of a new car, and *not* that the price of a car is dependent on the number sold (which may well be true but is not stated by this model). The value of X in (11.1) is considered to be independent of the value of Y. For this reason, when we write down an economic relationship in the form of a model of the type given in (11.1) the variable on the left-hand side of such a model (Y in this case) is referred to as the *dependent variable*. The variable on the right-hand side (X in this case) is known as the *independent* or *explanatory variable*. Thus a model of the form of (11.1) contains within it an implication of *causality*: Y 'is caused by' X. (Strictly speaking, the causality springs from the theory underlying the model.)

By observing the behaviour of Y and X, i.e. by taking a sufficient number of observations on each, it will usually be possible to be more explicit about the nature of the relationship between them. One of the simplest possibilities is a *linear* relationship, expressed by a model of the form:

$$Y = a + bX \tag{11.2}$$

In such a model a and b are referred to as the *parameters* of the model: a is referred to as the *constant* or *intersect* parameter, b as the *slope* parameter. The difference between variables and parameters is that whereas the former may assume any one of a range of values, the latter are numeric *constants* (whose values may or may not be known) in any given model. For example, suppose $a = 1000$ and $b = -100$; then

$$Y = 1000 - 100X \tag{11.3}$$

Thus at one time if $X = 5$, then $Y = 1000 - (100 \times 5) = 500$; whereas at another time if $X = 6$, then $Y = 1000 - (100 \times 6) =$

400. Thus the values of X and Y may change in any given model but the parameters a and b are fixed numbers.

Notice that once we set the value of X in (11.2) the value of Y is completely and *exactly* determined: there is no uncertainty. We call such a model *deterministic*. Most economic relationships in the real world are not of this exact nature, as usually Y may take any one of a number of values which together constitute the probability distribution of Y. For example, when $X = 5$, Y might take the value 450 with a probability of 0.25, the value 500 with a probability of 0.5, or the value 550 with a probability of 0.25. What causes Y to take on one particular value rather than some other will depend not only on the value of X but also on a host of other influences, many of which cannot be identified, never mind measured. The presence of these other factors lends a *random* or *stochastic* element to the value taken by Y. To take these into account (11.2) would have to be written:

$$Y = a + bX + (\text{a stochastic term}) \tag{11.4}$$

Models such as this are termed *stochastic* and form the basis for all of our subsequent work. If X and Y are stochastically related, then any observations we make of Y for a given value of X have to be considered as a *sample*. How does all this fit in with the work we have done so far? Essentially, we need the techniques and ideas we have discussed to enable us to investigate economic relations and analyse economic models.

This chapter's major concern is the statistical analysis of economic models. The techniques we examine (known collectively as *regression analysis*) constitute a powerful armoury not only for the investigation of economic relationships but also for those in many other of the social and behavioural sciences.

We start by considering an extension of the idea of covariance which will enable us to investigate the strength of bivariate relationships.

11.2 Correlation

In section 6.6 of Chapter 6 we introduced the concept of *covariance* as a measure of the strength of the relationship between two variables. In equation (6.14) we defined the covariance between two variables X and Y as

$$\text{cov}(X, Y) = E[X_i - E(X)]\,[Y_i - E(Y)]$$

Analogously, if we have n observations on two variables X and Y, then the sample covariance between X and Y is given by

$$\text{cov}(X, Y) = \frac{1}{n}\sum_{i=1}^{n} (X_i - \bar{X})(Y_i - \bar{Y}) \tag{11.5}$$

The rationale behind (11.5) is as follows. For those values of X_i above its mean all the terms $(X_i - \bar{X})$ will be positive and so $\Sigma(X_i - \bar{X})$ will be positive. For all those values of X_i below its mean all the terms $(\bar{X}_i - \bar{X})$ will be negative and $\Sigma(X_i - \bar{X})$ will be negative. The same argument applies to Y. When high values of X (i.e. values above its mean) are associated with high values of Y, then both $\Sigma(X_i - \bar{X})$ and $\Sigma(Y_i - \bar{Y})$ will be positive and their product in (11.5) will be positive. When low values of X are associated with low values of Y, both $\Sigma(X_i - \bar{X})$ and $\Sigma(Y_i - \bar{Y})$ will be negative and thus their product in (11.5) will be positive. Thus when high values of X are associated with high values of Y and low values of X are associated with low values of Y then their covariance, cov(X, Y), will be positive.

In the same way, if high values of X, producing a positive $\Sigma(X_i - \bar{X})$, are associated with low values of Y, producing a negative $\Sigma(Y_i - \bar{Y})$, then their product in (11.5) will be negative, i.e. cov(X, Y) will be negative. Similarly, low values of X, producing a negative $\Sigma(X_i - \bar{X})$, associated with high values of Y, producing a positive $\Sigma(Y_i - \bar{Y})$, will result in a negative value for their product, and again cov(X, Y) will be negative.

In other words, when X and Y are *positively* associated, i.e. high values of X with high values of Y and low values of X with low values of Y, then their covariance will be *positive*. But when X and Y are *negatively* or inversely related, i.e. high values of one variable with low values of the other, then their covariance will also be *negative*.

The scatter diagrams of Figure 11.1 are another way of looking at these ideas. Points in any of the top left-hand quadrants will have Y_is bigger than $\bar{Y}$, and so $(Y_i - \bar{Y})$ will be positive, but the X_is will be smaller than $\bar{X}$, and so $\Sigma(X_i - \bar{X})$ will be negative. Points in the top right-hand quadrant will have X_is bigger than $\bar{X}$ and Y_is bigger than $\bar{Y}$, so that both $\Sigma(X_i - \bar{X})$ and $\Sigma(Y_i - \bar{Y})$ will be positive. Points in the bottom right-hand quadrant will have X_is bigger than $\bar{X}$, so $\Sigma(X_i - \bar{X})$ will be positive, but the Y_is will be smaller than $\bar{Y}$, so $\Sigma(Y_i - \bar{Y})$ will be negative. Finally, points in the bottom left-hand quadrant will have the X_is less than $\bar{X}$ and the Y_is less than $\bar{Y}$, and so both $\Sigma(X_i - \bar{X})$ and $\Sigma(Y_i - \bar{Y})$ will be negative.

Now consider Figure 11.1(a). High values of X are associated with high values of Y and low values of X with low values of Y; thus the variables are positively related. And the points in the top

Figure 11.1 Illustrating covariance

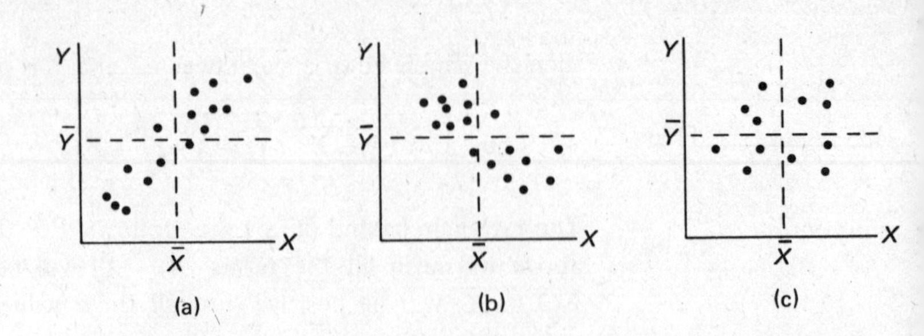

right-hand quadrant and in the bottom left-hand quadrant will both produce a positive $\Sigma(X_i - \bar{X})(Y_i - \bar{Y})$, i.e. a positive covariance.

In Figure 11.1(b) high values of X are associated with low values of Y and low values of X with high values of Y; therefore X and Y are negatively related. And the points in the top left-hand quadrant and the bottom right-hand quadrant will both produce a negative $\Sigma(X_i - \bar{X})(Y_i - \bar{Y})$, i.e. a negative covariance.

In Figure 11.1(c) there does not appear to be any definite relationship between X and Y. Some of the points will produce a positive $\Sigma(X_i - \bar{X})(Y_i - \bar{Y})$ and some a negative value and the cancelling-out effect will produce a zero or near-zero covariance.

The sum total of all this is that if we are investigating the relationship between two variables X and Y, then a scatter diagram will give us an indication of whether the relationship between them is positive, negative or uncertain.

Clearly we can measure the strength of the association by calculating the covariance, as we did in the example in Chapter 6. However, the problem with covariance is that it is sensitive to the units of measurement of the variables X and Y. For example, X measured in, say, pence will produce a different result for cov(X, Y) than if it is measured in pounds. Clearly this is unsatisfactory and we would like a measure of the association to be *invariable* (i.e. not sensitive) to the units in which X and Y are measured. Such a measure, known as the coefficient of linear correlation (or just the *correlation coefficient* and denoted by r), is available and is obtained by dividing the covariance of X and Y by the square root of their respective variances, i.e.

$$r = \frac{\text{cov}(X, Y)}{\sqrt{\text{var}(X)\,\text{var}(Y)}} \tag{11.6}$$

Expression (11.6) is more conveniently written

$$r = \frac{\Sigma(X_i - \bar{X})(Y_i - \bar{Y})}{\sqrt{\Sigma(X_i - \bar{X})^2\ \Sigma(Y_i - \bar{Y})^2}} = \frac{\Sigma x_i y_i}{\sqrt{\Sigma x_i^2\ \Sigma y_i^2}} \tag{11.7}$$

where, of course, x_i and y_i represent deviations from the mean. r

can vary from −1 (implying a perfect negative relationship between X and Y) through 0 (implying no relationship) to +1 (implying a perfect positive relationship), i.e.

$$-1 \leqslant r \leqslant 1 \tag{11.8}$$

Example

The data in Table 11.1 relate to the number of trade-union members in the United Kingdom (in millions) X and to real wages and salaries (in £ billion) Y, between 1966 and 1975. Notice that it does not make any difference which variable we use to denote which quantity since causality is not involved. However, it will be convenient for later work to use X and Y as indicated here.

Table 11.1 Real wages and salaries, and trade-union membership between 1966 and 1975

Year	1966	1967	1968	1969	1970	1971	1972	1973	1974	1975
X (millions)	10.3	10.2	10.2	10.5	11.2	11.1	11.4	11.4	11.8	12.0
Y (£ billion)	20.4	20.6	21.0	21.4	22.4	22.5	23.5	22.4	25.7	26.7

Many economists, for example Hines (1964), have argued that trade-union militancy (for which membership is a possible proxy) has caused increases in wages. In other words, we might expect a positive relationship between the two. A scatter diagram is shown in Figure 11.2.

The fact that most of the points lie either in the top right- or the bottom left-hand quadrant suggests a positive association

Figure 11.2 Scatter diagram of real wages and salaries against union membership

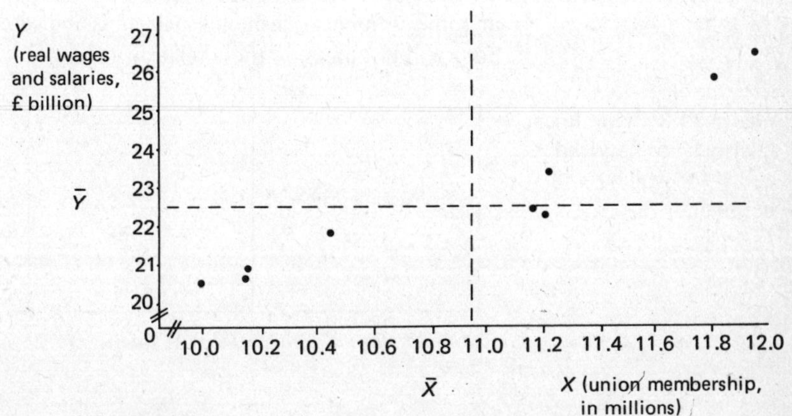

between X and Y. In Table 11.2 we show the calculations required to obtain the summary measures necessary to calculate r.

Table 11.2 Calculation of summary measures for r

X_i (millions of members)	Y_i (£ billion)	$x_i = (X_i - \bar{X})$	$y_i = (Y_i - \bar{Y})$	$x_i y_i$	x_i^2	y_i^2
10.3	20.4	−0.71	−2.26	1.6046	0.5041	5.1076
10.2	20.6	−0.81	−2.06	1.6686	0.6561	4.2436
10.2	21.0	−0.81	−1.66	1.3446	0.6561	2.7556
10.5	21.4	−0.51	−1.26	0.6426	0.2601	1.5876
11.2	22.4	0.19	−0.26	−0.0494	0.0361	0.0676
11.1	22.5	0.09	−0.16	−0.0144	0.0081	0.0256
11.4	23.5	0.39	0.84	0.3276	0.1521	0.7056
11.4	22.4	0.39	−0.26	−0.1014	0.1521	0.0676
11.8	25.7	0.79	3.04	2.4016	0.6241	9.2416
12.0	26.7	0.99	4.04	3.9996	0.9801	16.3216
Σ 110.1	226.6			11.8240	4.0290	40.1240
$\bar{X}$ = 11.01	$\bar{Y}$ = 22.66					

From Table 11.2 we have

$$\Sigma x_i y_i = 11.8240; \quad \Sigma x_i^2 = 4.0290; \quad \Sigma y_i^2 = 40.1240$$

Therefore, substituting into (11.7) gives

$$r = \frac{11.8240}{\sqrt{40.1240 \times 4.0290}} = 0.93$$

The correlation, positive as expected, indicates a strong association between the variables, since the value of r is close to 1.0. We should emphasise that r is a measure only of the strength of *linear* relationships; two variables X and Y may well be strongly related in some non-linear fashion, such as is indicated in Figure 11.3, but r will not measure these relationships.

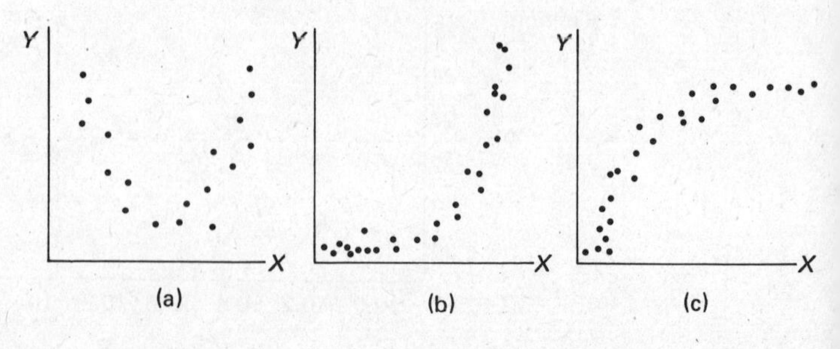

Figure 11.3 Non-linear relationships between X and Y for which r will be small or zero

* * *

There are three other points concerning the correlation coefficient which we must consider. First, there is no implication of causality between the variables in question. When we calculate a value of r for two variables X and Y we make no assumptions of the sort 'X depends on Y' or 'Y depends on X'. Second, r has no predictive power; knowledge of the value of the correlation coefficient does not enable us to predict the value of one variable given the value of the other. For example, knowing $r = 0.10$ and $X = 21.0$ does not enable us to determine a value for Y. Finally, we should be very cautious about what are termed *spurious* correlations; there may well be a strong positive correlation between the number of students in higher education and the incidence of crime but nobody would seriously suggest that there is any real association between the two variables. We must be careful to establish a possible relationship between two variables on theoretical grounds before we use correlation as a way of examining the strength or otherwise of this relationship.

There is one final aspect to r which we must mention. In the above example the data given for X and Y constitute a sample from all possible values of X and Y in all possible years. This means that r is in fact an *estimator* of the true population correlation coefficient ρ (Greek rho). Thus if we wished to say of some particular value of r 'Yes, this indicates a significant relationship between X and Y' we would have to distinguish between the hypotheses $H_0: \rho = 0$ and $H_1: \rho \neq 0$. Unfortunately, the sampling distribution of r is somewhat complicated and depends not only on n, the sample size, but also on ρ, being symmetric when ρ is 0 but skewed (and therefore making r a biased estimator of ρ) to the right for negative ρ and to the left for positive ρ. It is possible to develop a formal test for r (see, for example, Koutsoyiannis, 1977, p. 95; or Games and Klare, 1967, pp. 366–7) but we cannot consider it here. In any case, even if r is judged to be significant it does not mean that X and Y will have sufficient 'in common' to be of interest. For example, with $n = 100$ and $\alpha = 0.05$ a value of $r = 0.2$ is significant but (as we shall see shortly) this implies that X and Y have only 4 per cent common variance; this might not be sufficient, even though r is significant, to interest us. That the values of significant r depend upon n can be seen if we know that, with $\alpha = 0.05$, for r to be considered significantly different from zero it would have to be $r = 0.71$ when $n = 8$, $r = 0.5$ when $n = 16$ and $r = 0.35$ when $n = 32$.

We shall develop a test relating to the significance of r later, so we leave the subject until then.

In terms of investigating and analysing the economic models discussed in the introduction to this chapter the correlation coefficient is of very limited value. All it can do is give an indication of the strength of the relationship; it cannot contribute to an understanding of *how* the variables may be related, nor does it enable us to express the relationship as a causal function, or enable us to predict Y knowing X. For these reasons we now turn to *regression analysis*.

11.3 The simple regression model

We shall discuss regression analysis by investigating one of the commonest of economic relationships, i.e. that between household income and household expenditure. Economic theory indicates that as household income increases, so does household expenditure. We are going to postulate that the relationship between them is linear.

Imagine that we have a ficticious economic community which contains nine households whose weekly income is £100. Clearly we would not expect all nine households to have exactly the same level of expenditure in some given week. However, we could calculate the mean level of expenditure of these households easily enough. For example, let us suppose that one of the households spends £91 per week, two spend £93 per week, three of them spend £95 per week, two spend £97 per week and one spends £99 per week. Let us denote weekly income by X and weekly expenditure by Y, and let X_1 represent an income of £100 per week and Y_1 the expenditures of those households with X_1 income. The frequency distribution of Y_1 is given in Table 11.3.

We can write the frequency of a variable as its probability distribution, and if we do this for the frequency distribution of Table 11.3 we obtain Table 11.4, the first two columns of which show the probability distribution of Y_1 *given* X_1.

Thus if we picked one family at random there would be a probability of 0.11 that this family would have an expenditure

Table 11.3 Frequency distribution of Y_1

Household expenditure Y_1	Frequency (no. of households) f	Relative frequency
91	1	0.11
93	2	0.22
95	3	0.33
97	2	0.22
99	1	0.11
Σ	9	1.00

Table 11.4 Probability distribution of $Y_1 \mid X_1$ from Table 11.3

Household expenditure Y_1	$P(Y_1)$	$Y_1P(Y_1)$	$[Y_1 - E(Y_1)]^2\ P(Y_1)$
91	0.11	10.11	1.78
93	0.22	20.66	0.89
95	0.33	31.66	0.00
97	0.22	21.55	0.89
99	0.11	11.00	1.78
Σ	1.00	95.00	5.33
		$E(Y_1 \mid X_1) = 95.00$	$\text{var}(Y_1 \mid X_1) = 5.33$

of £91 per week; and so on. We can calculate the mean level of expenditure as well as the variance for all these families with incomes of X_1 = £100 per week following the method described in section 6.5 of Chapter 6 and illustrated in Table 6.8; the calculations for this are also shown in Table 11.4 (after allowing for rounding errors). These results indicate that the mean level of expenditure of Y_1 households, i.e. those in receipt of incomes of X_1 = £100 per week, is £95 per week, which we can write as

$$E(Y_1 \mid X_1) = 95$$

and the dispersion in the expenditure of these households is given by

$$\text{var}(Y_1 \mid X_1) = 5.33$$

Suppose our economic community also contains nine households each with an income of £120 per week. Assume that we find that one of these households spends £109, two spend £111, three spend £113, two spend £115 and one spends £117. We can develop the probability distribution of Y_2 given X_2 = £120 in the same way as for X_1 = £100; this is shown in the first two columns of Table 11.5, while the remaining columns show the calculation of the mean and variance of Y_2.

Therefore we have

$$E(Y_2 \mid X_2) = 113.00$$

$$\text{var}(Y_2 \mid X_2) = 5.33$$

Notice that the variance of $Y_2 \mid X_2$ is the same as the variance of $Y_1 \mid X_1$. This is no accident since we deliberately chose the values of Y_1 and Y_2 to give this result. (One of the assumptions which we want to make is that the variance of household expenditure is

Table 11.5 Probability distribution of $Y_2 \mid X_2$

Household expenditure Y_2	$P(Y_2)$	$Y_1P(Y_2)$	$[Y_2 - E(Y_2)]^2\ P(Y_2)$
109	0.11	12.11	1.78
111	0.22	24.66	0.89
113	0.33	·37.66	0.00
115	0.22	25.55	0.89
117	0.11	13.00	1.78
Σ	1.00	113.00	5.33
		$E(Y_2 \mid X_2) = 113.00$	$\text{var}(Y_2 \mid X_2) = 5.33$

a constant. We shall examine this assumption again shortly.) We have also chosen the values of expenditure at each income level so that the probability distribution of $Y_1 \mid X_1$ and $Y_2 \mid X_2$ have the same shape. (In fact a second assumption we want to make is that the distributions of Y_1 and Y_2 are normal. We shall also examine this assumption again shortly.) The distributions of Y_1 and Y_2 are shown in the probability histograms of Figure 11.4.

Thus we have seen that the probability distribution of household expenditure for the two levels of household income X_1 and X_2 differ *only* in their mean or expected value. The mean level of

Figure 11.4 Probability distribution of Y_1 and Y_2

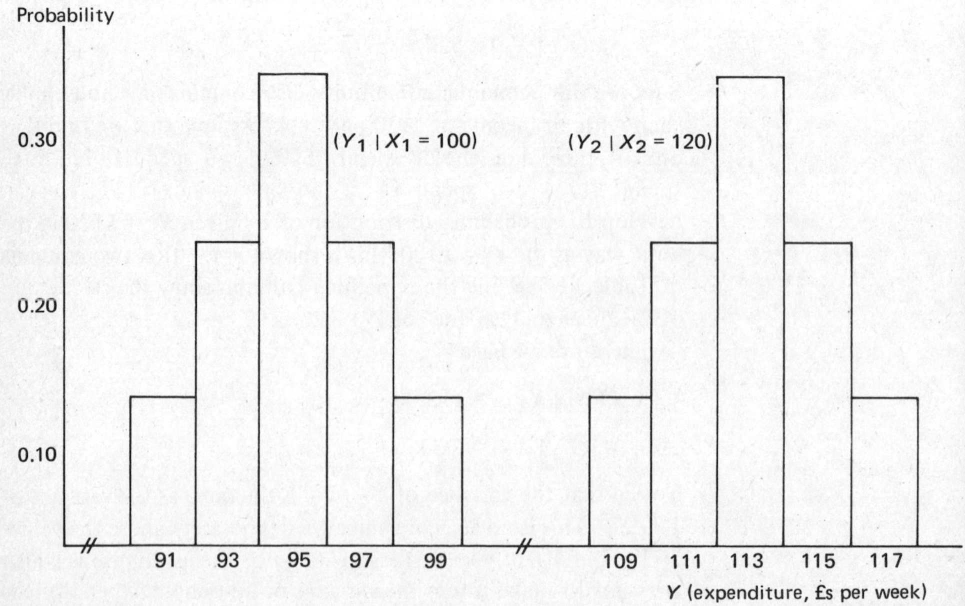

expenditure for households with incomes of £100 per week is £95 per week, the mean level of expenditure of households with incomes of £120 per week is £113 per week. Of course, individual households will not in general have exactly these mean levels of expenditure – some will spend more than the mean, some less, i.e. each household will have an expenditure which *deviates* from the mean by a lesser or greater amount.

Let us now introduce a new variable u which, for each household, measures these deviations. Thus u_i measures the difference above or below the mean of households with an income X_i. In general, therefore, we have

$$u_i \mid x_i = Y_i - E(Y_i \mid X_i) \tag{11.9}$$

($u_i \mid x_i$ means u_i given x_i, i.e. u_i conditional upon x_i. This explicit conditioning is, for simplicity, dropped in subsequent expressions.) Thus when $X_i = X_1 = 100$, as we have seen $E(Y \mid X = 100) = 95$; therefore (11.9) becomes

$$u_1 = Y_1 - 95 \tag{11.10}$$

Thus, for the household which spends £91 per week, we have

$$u_1 = 91 - 95 = -4$$

and, for the two households which spend £93 per week,

$$u_1 = 93 - 95 = -2$$

We could carry on and determine all the u_1 values for $X = 100$ and these values together with their respective probabilities (which will necessarily be the same as the probabilities of Y_1 in Table 11.4) are given in Table 11.6, which also shows the calculation of the mean and variance of u_1. We can draw the probability histogram of u_1 from Table 11.6, and this is shown in Figure 11.5. From Table 11.6 we see that the mean of u_1 is zero, a result we might have anticipated from the definition of u given in (11.9),

Table 11.6 Probability distribution of u_1

Y_1	u_1	$P(u_1)$	$u_1 P(u_1)$	$[u_1 - E(u_1)]^2 P(u_1)$
91	−4	0.11	−0.44	1.78
93	−2	0.22	−0.44	0.89
95	0	0.33	0.00	0.00
97	2	0.22	0.44	0.89
99	4	0.11	0.44	1.78
Σ			0.00	5.33
			mean $E(u_1 \mid X_1) = 0$	$\text{var}(u_1 \mid X_1) = 5.33$

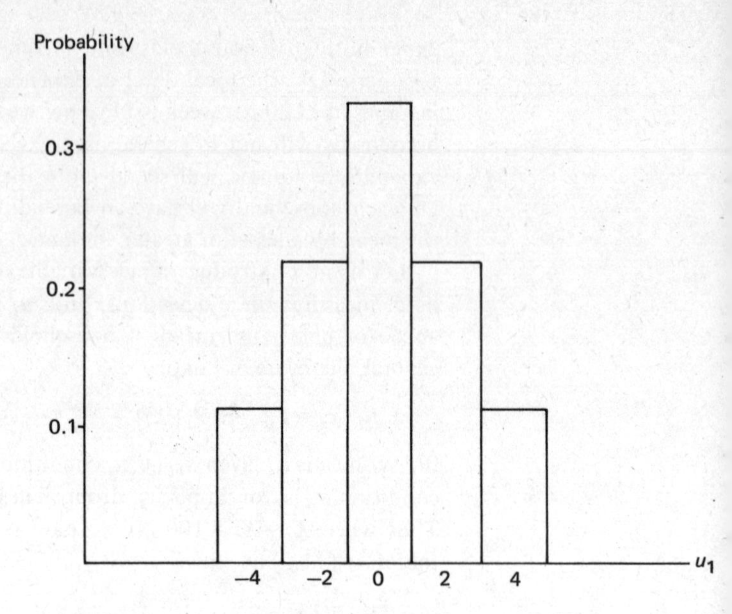

Figure 11.5
Probability histogram of u_1

i.e. if we take expected values of (11.10) we have

$$E(u_1) = E(Y_1) - 95 = 95 - 95 = 0$$

Further, the variance of u_1 is the same as the variance of Y_1, another result which we might have anticipated since the values of u_1 are simply the values of Y_1 minus a constant (in this example, 95). As we saw in exercise 5.3, the variance of two sets of numbers which differ only by the addition or subtraction of a constant is the same. We are not surprised, therefore, to find that the shape of the distribution of u_1 in Figure 11.5 is the same as the shape of the distribution of Y_1.

We can establish similar results for the deviations u_2 of those nine households with incomes $X_2 = 120$, i.e.

$$E(u_2) = 0$$

$$\text{var}(u_2) = 5.33$$

This is a convenient point to illustrate the relationship between household income and household expenditure in the form of a scatter diagram, shown in Figure 11.6. Each dot in the diagram represents a certain level of expenditure, and the number of households at each level of expenditure is indicated by the number of concentric circles around each dot.

Thus we have seen that as household income X increases from $X_1 = 100$ to $X_2 = 120$, mean household expenditure increases from $E(Y_1) = 95$ to $E(Y_2) = 113$. We postulated at the beginning of this section that economic theory indicates that as household

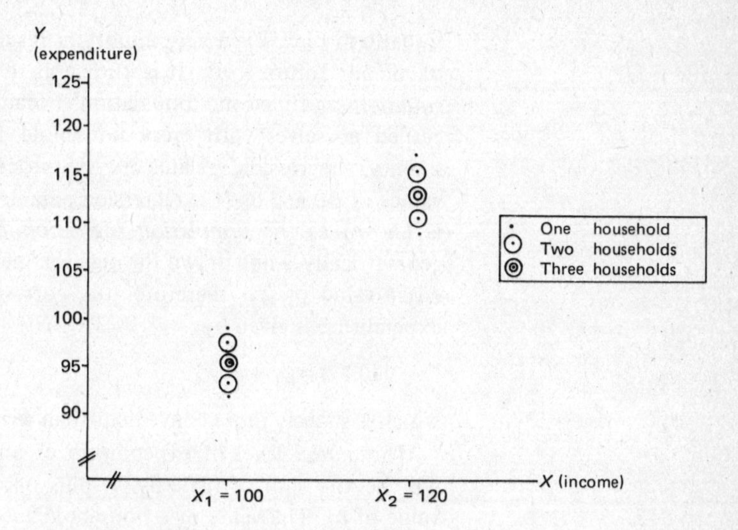

Figure 11.6 Distribution of expenditure for two levels of income

income increases so does household expenditure. We took this a step further by suggesting that the relationship between the two variables might be considered to be linear, e.g. of the form given by (11.2). We have just demonstrated that the mean value of Y does increase as X increases, and by adding the linear assumption we can write

$$E(Y_i \mid X_i) = \beta_1 + \beta_2 X_i \tag{11.11}$$

It is not difficult to calculate the value of the parameters β_1 and β_2 since when $X_i = 100$, $E(Y_i) = 95$, and when $X_i = 120$, $E(Y_i) = 113$. Thus we have

$$95 = \beta_1 + 100\beta_2$$

$$113 = \beta_1 + 120\beta_2$$

from which $\beta_2 = 0.9$ and $\beta_1 = 5$; (11.11) can be written

$$E(Y_i \mid X_i) = 5 + 0.9X_i \tag{11.12}$$

In the context of our income–expenditure model, the slope parameter $\beta_2 = 0.9$ is the marginal propensity to consume, since for every £1 increase in X_i, the mean level of expenditure will increase by £0.9, and the intersect parameter $\beta_1 = £5$ represents autonomous expenditure since it represents the expenditure which takes place regardless of the level of, or changes in, income.

The population regression equation

Let us now substitute (11.11) into (11.9) and obtain:

$$u_i = Y_i - (\beta_1 + \beta_2 X_i)$$

i.e.

$$Y_i = \beta_1 + \beta_2 X_i + u_i \tag{11.13}$$

Equation (11.13) is a very important result and is the foundation of all our future work. It is known as the *population regression equation* or function: 'population' because so far we have concerned ourselves with every household in our economic community; 'regression' relates to the process of determining the values of β_1 and β_2 (the regression parameters). Equation (11.11) is known as the *population regression line* since it represents geometrically a line drawn through the mean values of Y. For any *given* value of X_i, therefore, the corresponding *mean* level of expenditure is given by

$$E(Y_i) = \beta_1 + \beta_2 X_i \qquad (11.14)$$

which is slightly more convenient than writing (11.11).

The *actual* level of expenditure of any particular household will be the sum of $(\beta_1 + \beta_2 X_i)$ plus or minus that household's value of u_i. The value of a household's u_i term will depend upon many factors, some measurable (such as family size), some not (such as taste), and some whose value can neither be anticipated nor measured, for example a household might have visitors in a particular week which causes their expenditure in that week to increase, or bad weather might restrict their shopping expeditions and so reduce their expenditure in that week, and so on. The variable u is therefore essentially a random or stochastic variable which *disturbs* the value of Y_i from simply being equal to $\beta_1 + \beta_2 X_i$. For these reasons u is known as the *random disturbance term*.

It will be helpful to draw the linear relationship expressed by (11.11), i.e. the population regression line, on to the scatter diagram, and the result is shown in Figure 11.7. It is possible to visualise Figure 11.7 in three dimensions, with probability density as the vertical axis (see Figure 11.8).

If our population also contained households with other levels of income, we could determine their corresponding $E(Y_i \mid X_i)$ values. The population regression line would pass through all these mean Y values and the population regression equation of (11.13) would represent the linear relationship between Y and X for all values of X.

The sample regression equation

So far we have only been concerned with *population* values of income and expenditure, i.e. our population consists of only eighteen households, nine with an income of £100 per week and nine with an income of £120 per week, and we were able to measure the exact expenditure of each (recorded in the first column of Tables 11.3 and 11.5) and hence determine the value of u_i for each household, using (11.9).

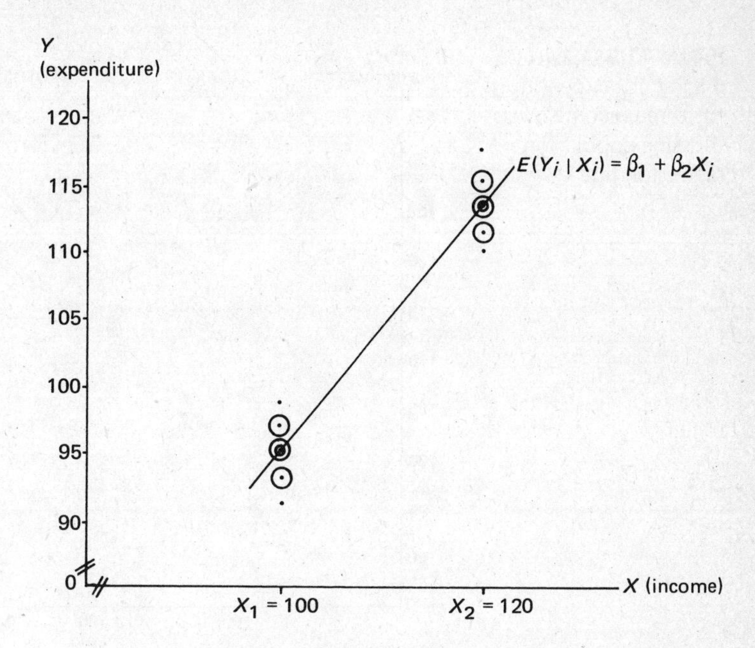

Figure 11.7 The line $E(Y_i \mid X_i) = \beta_1 + \beta_2 X_i$

In practice, however, we would not be able to measure the income and expenditure of every household in the population and we would have to make our calculations on the basis of a single random sample.

For the purpose of exposition let us assume that our population also contains nine households, each with a weekly income of £140, and of these one spends £127, two spend £129, three spend £131, two spend £133, and one spends £135. If we plot all twenty-seven income–expenditure values on a scatter diagram, we obtain Figure 11.9.

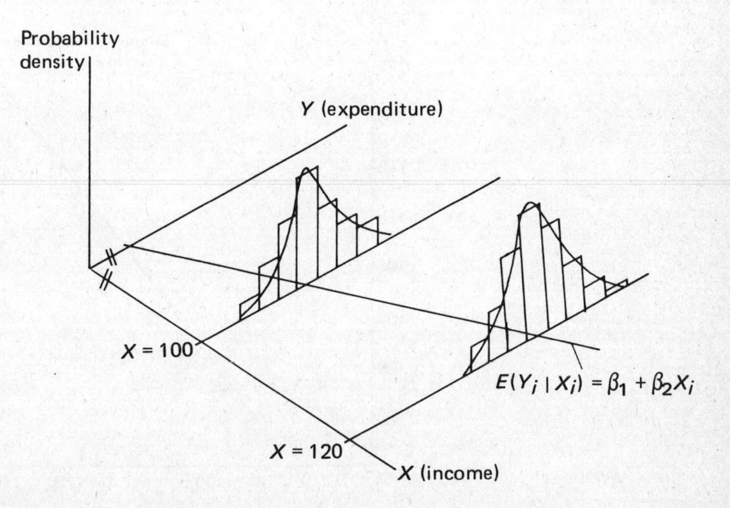

Figure 11.8 Probability distribution of expenditure: the line $E(Y_i \mid X_i) = \beta_1 + \beta_2 X_i$

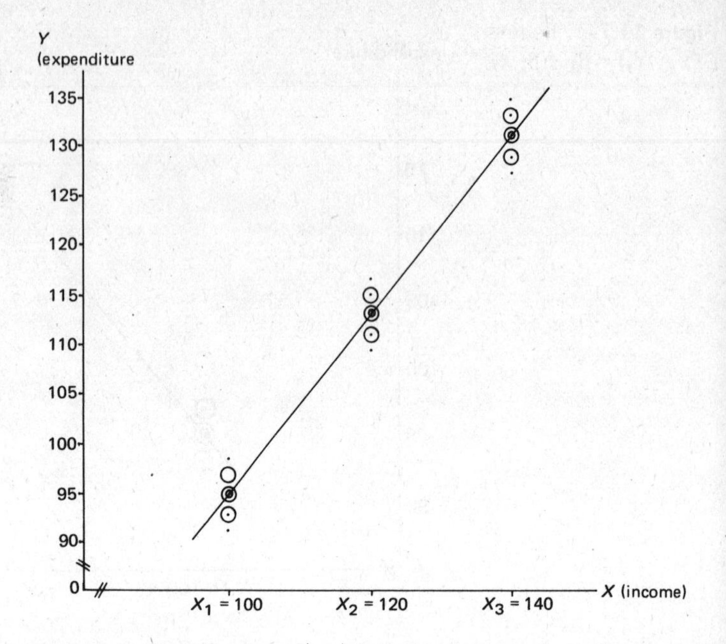

Figure 11.9 Scatter diagram of expenditure for three income levels showing population regression line

Let us suppose that we take a sample of three households in such a way that one household is drawn from each income level. Assume that these three households have expenditures of £93, £115 and £131 respectively. We do not of course know anything of the expenditure levels of the other twenty-four households and all we observe are the three points in Figure 11.10. There is no

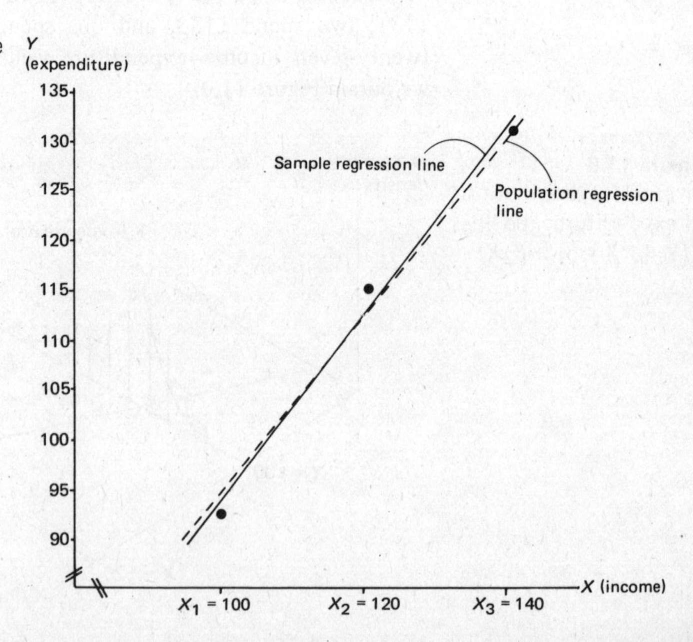

Figure 11.10 A sample regression line and the true line

reason to expect the sample points to lie on a straight line; if we want to use the three observed values to determine the values of β_1 and β_2 we can do so by drawing the 'best' straight line we can through the three points in Figure 11.10.

In this context (for the moment anyway) 'best' means a line drawn by eye which passes as close to all three points as possible. Since we do not know the true population regression line, then we do not know the true values of β_1 and β_2. However, the 'best' straight line through the three points in Figure 11.10 will provide us with an *estimate* of β_1 and β_2, and hence an estimate of the true line. We call this best line fitted to the sample observations the *sample regression line* and the estimated values of β_1 and β_2 we denote by $\hat{\beta}_1$ and $\hat{\beta}_2$ respectively. Thus the *sample regression line* is

$$\hat{Y}_i = \hat{\beta}_1 + \hat{\beta}_2 X_i \tag{11.15}$$

where $\hat{Y}_i$ is an estimator of $E(Y_i \mid X_i)$.

For obvious reasons the sample regression line is also known as the *estimated regression line*. We have drawn a sample regression line on Figure 11.10 as well as the true (but we must emphasise usually *unknown*) regression line, for comparison.

A glance at Figure 11.10 reveals that the sample regression line does not pass through all of the observed points (inevitable, since in general the points will not lie on a straight line) and the vertical difference between each observed value Y_i and the sample regression line is known as the *error* or *residual* term and is denoted by $\hat{u}_i$ (or alternatively by e_i). Thus analogous to the population regression equation there is an equivalent *sample regression equation* given by

$$Y_i = \hat{\beta}_1 + \hat{\beta}_2 X_i + \hat{u}_i \tag{11.16}$$

We must emphasise the difference between the variable u and the variable $\hat{u}$. The former is the difference between each observed value of Y and the true line, while the latter is the difference between the observed value of Y and the estimated line. In a sense we can think of $\hat{u}$ as an estimator of u. We can illustrate this difference by expanding that part of Figure 11.10 which corresponds to the observed value of Y of £93 per week; this is shown in Figure 11.11. In other words

$$u_i = Y_i - E(Y_i \mid X_i) \tag{11.17}$$

$$\hat{u}_i = Y_i - \hat{Y}_i \tag{11.18}$$

* * *

To sum up, we have an unknown population regression equation

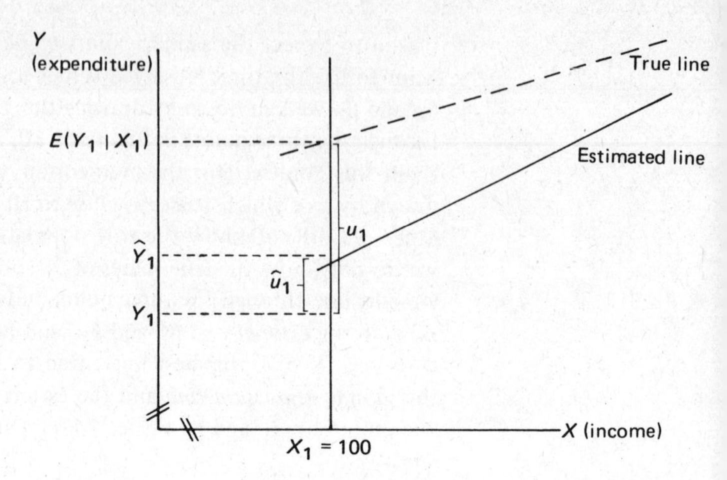

Figure 11.11
Sample and population regression line

which expresses the true relationship between Y and X of the form

$$Y_i = \beta_1 + \beta_2 X_i + u_i \tag{11.13}$$

and our object in regression analysis is to use sample information on X and Y to estimate this true line using the sample regression equation

$$Y_i = \hat{\beta}_1 + \hat{\beta}_2 X_i + \hat{u}_i \tag{11.16}$$

Essentially, as we shall see, this amounts to determining the values of the estimators $\hat{\beta}_1$ and $\hat{\beta}_2$ in such a way that the sample regression line

$$\hat{Y}_i = \hat{\beta}_1 + \hat{\beta}_2 X_i \tag{11.15}$$

is the 'best' fit to the observed values of X and Y. The crucial question is: What do we mean by the 'best' line and how do we find the values of $\hat{\beta}_1$ and $\hat{\beta}_2$ which identify this line? We can now see how all the work we have done so far on statistical inference fits in. We are in the position of trying to *estimate* the value of some unknown population parameters (in this case β_1 and β_2) by inferring from sample information on the estimators $\hat{\beta}_1$ and $\hat{\beta}_2$ (themselves derived from observations on X and Y). Having estimated β_1 and β_2 we can then use the ideas of *hypothesis testing* to test the significance or otherwise of our estimators, and we can also examine our estimators to see whether they have any of the usual *desirable properties*.

Before we consider how to find the best values of $\hat{\beta}_1$ and $\hat{\beta}_2$ we want to emphasise the point that any single sample we take, and any values of $\hat{\beta}_1$ and $\hat{\beta}_2$ we obtain on the basis of that sample, is *only one* of many different samples we could (in theory, if not in practice) take. The stochastic nature of the regression model implies that for every value of X there is a whole probability distribution of values of Y. The uncertainty of the Y values is caused by the stochastic nature of the disturbance term u, which imparts its randomness to Y; without a disturbance element all the observations would lie *exactly* on the true regression line, which could therefore be identified without any difficulty. Thus the probability distribution of Y is completely determined by the values of X (the deterministic part of the relationship) and by the probability distribution of u, a random variable. Thus Y is a random variable, as must be the estimators $\hat{\beta}_1$ and $\hat{\beta}_2$.

We want to draw together some of the assumptions made so far about the disturbance term and add to them other crucial assumptions which form the basis of the regression model. Having done this we can then proceed directly to consideration of determining the best straight line, i.e. estimating β_1 and β_2.

11.4 Basic assumptions of the classical normal linear regression model

The method for determining the 'best' straight line depends crucially on the following set of assumptions:

$$Y_i = \beta_1 + \beta_2 X_i + u_i \quad \text{(i.e. the model is linear)} \tag{A.1}$$

u_i is normally distributed (A.2)

the mean of u_i is zero, i.e. $E(u_i) = 0$ (A.3)

the variance of u_i is a constant, i.e. $E(u_i^2) = \sigma^2$ (A.4)

the u_is are independent, i.e. $E(u_i u_j) = 0 \quad i \neq j$ (A.5)

X_i is a non-stochastic variable with fixed values (A.6)

Taken together these six assumptions constitute what is known as the classical normal linear regression model (CNLRM); they are of such crucial importance that we want to consider each in turn.

Assumption (A.1) states that the dependent variable Y is assumed to be a linear function of the explanatory variable X of the form given by (11.13).

Assumption (A.2) states that the random disturbance term u is normally distributed, i.e. u ranges in value from $-\infty$ to $+\infty$, is symmetrically distributed around its mean and is fully described by two parameters, its mean and its variance.

Assumption (A.3) states that the mean of the distribution of u is zero, i.e. $E(Y_i) = \beta_1 + \beta_2 X_i$.

Assumption (A.4) states that the variance of u is a constant, and we denote this constant by σ^2. We now want to show that we can write the variance of a random variable as the expected value of its square. Using equation (6.9) we can write the variance of u_i as

$$\begin{aligned}\text{var}(u_i) &= E[u_i - E(u_i)]^2 \\ &= E\{u_i^2 - 2u_iE(u_i) + [E(u_i)]^2\} \\ &= E(u_i^2) - 2E(u_i)E(u_i) + [E(u_i)]^2\end{aligned}$$

but $\quad E(u_i) = 0.$

Therefore,

$$\text{var}(u_i) = E(u_i^2)$$

Assumption (A.5) states that the value of the disturbance term corresponding to the first observation on Y and X in no way influences or is influenced by the value of u for the second observation or indeed by any other value of u. In other words, all the u values resulting from a sample will be uncorrelated each from another. However, for normally distributed random variables, uncorrelatedness also implies independence. In other words, the disturbance terms are all mutually independent. It follows that their covariance is zero, i.e. $\text{cov}(u_iu_j) = 0$ (see equation (6.16)). We shall now show that the covariance between two random variables can be written in terms of the expected value of their product. From (6.14) we can write:

$$\text{cov}(u_iu_j) = E[u_i - E(u_i)][u_j - E(u_j)]$$

but from (A.3) $E(u_i) = E(u_j) = 0$; therefore

$$\begin{aligned}\text{cov}(u_iu_j) &= E[u_i - 0][u_j - 0] \\ &= E(u_iu_j)\end{aligned}$$

As we have seen, under the assumption of independence,

$$E(u_iu_j) = 0 \text{ (for all } i \neq j)$$

We have to exclude the case where $i = j$ since $E(u_iu_j)$ would then become $E(u_i^2)$ or $E(u_j^2)$ and thus relates not to covariance but to variance, as we have seen above.

Assumption (A.6) implies that the values of X can be chosen, i.e. fixed, before sampling of Y takes place. In other words, if we were investigating the relationship between household income X and household expenditure Y, then we could pre-select values of X, i.e. choose certain values of household income, and then sample Y corresponding to these chosen values of X_1, and if necessary always using the same fixed values of X for repeated samples.

This assumption implies that u and X are independent since if

X is chosen to have certain fixed values and u is a randomly determined variable then they must be independent.

In practice, we cannot usually fix the values of X but have to sample X and Y together, treating them both as random variables. For the time being, however, we are going to assume that all the assumptions of the classical normal linear regression model are fulfilled (we shall examine below the consequences of each of these assumptions not being met).

The model described by (A.1) through to (A.6) has four parameters: β_1, β_2, the mean of u and the variance of u. However, the mean of u is assumed to be zero, and the model thus has three unknown parameters which have to be estimated: β_1, β_2 and σ^2.

As a final point we can use some of these assumptions to describe the probability distribution of Y. Taking expected values of (A.1) confirms our result in (11.14), that the mean of Y is $\beta_1 + \beta_2 X_i$. Using (6.9) the variance of Y can be written as

$$\begin{aligned} \text{var}(Y_i) &= E[Y_i - E(Y_i)]^2 \\ &= E[(\beta_1 + \beta_2 X_i + u_i) - (\beta_1 + \beta_2 X_i)]^2 \\ &= E(u_i^2) \\ &= \sigma^2 \end{aligned}$$

(using (A.5)).

Finally, since from (A.1) Y_i is a linear function of u_i and from (A.2) u_i is normally distributed, then Y_i is also normally distributed. In other words,

$$Y_i \sim N(\beta_1 + \beta_2 X_i, \sigma^2) \qquad (11.19)$$

11.5 Ordinary least squares estimation

We now want to describe a method for obtaining values of the estimators $\hat{\beta}_1$ and $\hat{\beta}_2$. The method to be discussed is certainly the most widely used approach and is known as ordinary least squares (OLS). Essentially the idea behind OLS lies in minimising the sum of the squared-error terms. To see what we mean consider the three observed values of Y in the scatter diagram of Figure 11.12, which also shows the three error terms $\hat{u}_1, \hat{u}_2$ and $\hat{u}_3$.

Using OLS we seek to minimise the sum

$$(\hat{u}_1^2 + \hat{u}_2^2 + \hat{u}_3^2)$$

i.e.

$$\min \sum_{i=1}^{n} \hat{u}_i^2 \qquad (11.20)$$

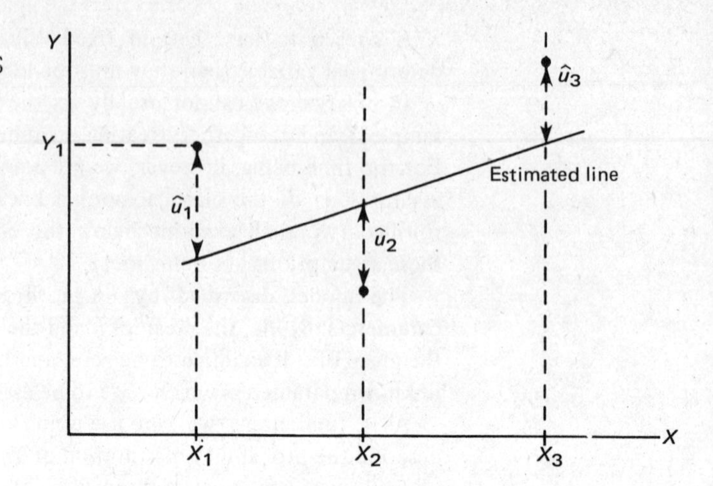

Figure 11.12
Illustration of the OLS principle

The term $\Sigma\hat{u}_i^2$ is usually referred to as the *error sum of squares*. We shall show that using the criterion embodied in (11.20) leads to estimators which, given the assumptions of the classical normal linear regression model, have all the desirable properties. Indeed, estimators calculated in this way turn out to be BLU estimators. We start by developing a method for estimating β_1 and β_2 given a set of observations on X and Y (estimating β_1 and β_2 is of course the same as estimating the mean of Y, from (11.19)).

Let:

$$S = \Sigma\hat{u}_i^2$$

i.e.

$$S = \Sigma(Y_i - \hat{Y}_i)^2$$
$$= \Sigma(Y - \hat{\beta}_1 - \hat{\beta}_2 X_i)^2$$

Using calculus we have

$$\frac{\partial S}{\partial \hat{\beta}_1} = -2\Sigma(Y_i - \hat{\beta}_1 - \hat{\beta}_2 X_i)$$

and

$$\frac{\partial S}{\partial \hat{\beta}_2} = -2\Sigma X_i(Y_i - \hat{\beta}_1 - \hat{\beta}_2 X_i)$$

Equating each of these expressions to zero gives

$$-2\Sigma(Y_i - \hat{\beta}_1 - \hat{\beta}_2 X_i) = 0$$

$$-2\Sigma X_i(Y_i - \hat{\beta}_1 - \hat{\beta}_2 X_i) = 0$$

from which

$$\Sigma Y_i = n\hat{\beta}_1 + \hat{\beta}_2 \Sigma X_i \tag{11.21}$$

$$\Sigma X_i Y_i = \hat{\beta}_1 \Sigma X_i + \hat{\beta}_2 \Sigma X_i^2 \tag{11.22}$$

(using 4.4). Or alternatively,

$$\Sigma \hat{u}_i = 0 \tag{11.21a}$$

$$\Sigma X_i \hat{u}_i = 0 \tag{11.21b}$$

since $\hat{u}_i = Y_i - \hat{\beta}_1 - \hat{\beta}_2 X_i$.

These two pairs of equations are known as the *normal equations*. Equations (11.21) and (11.22) can be used to solve for $\hat{\beta}_1$ and $\hat{\beta}_2$ by substituting the calculated values of ΣY_i, ΣX_i, $\Sigma X_i Y_i$ and ΣX^2 together with n. However, it is usually more convenient to rearrange these equations so that $\hat{\beta}_1$ and $\hat{\beta}_2$ are given directly, i.e.

$$\hat{\beta}_2 = \Sigma x_i y_i / \Sigma x_i^2 \tag{11.23}$$

$$\hat{\beta}_1 = \bar{Y} - \beta \bar{X} \tag{11.24}$$

Before we go on to examine the properties of these estimators let us illustrate their use by returning to the example on trade-union membership and real wages and salaries considered earlier in the context of correlation.

From the data in Table 11.2 we have calculated that

$$\bar{X} = 11.01; \quad \bar{Y} = 22.66; \quad \Sigma x_i y_i = 11.8240; \quad \Sigma x_i^2 = 4.0290$$

Substituting in (11.23) gives, as an estimator of β_2,

$$\hat{\beta}_2 = 11.8240/4.0290 = 2.9347$$

and in (11.24) gives, as an estimator of β_1,

$$\hat{\beta}_1 = 22.66 - (2.9347 \times 11.01) = -9.6510$$

Thus from (11.15) the sample regression line is

$$\hat{Y}_i = -9.6510 + 2.9347 X_i \tag{11.25}$$

Thus the estimated relationship between real wages and salaries and union membership implies that a change in X of one unit (one million members) brings about a corresponding change in real wages of £2.9347 billion. The positive sign of $\hat{\beta}_2$ in (11.25) indicates that an *increase* in X brings about an *increase* in Y. It is often difficult to ascribe any real economic meaning to the intercept term β_1 and this example is no exception. The value of $\hat{\beta}_1$ gives the value of $\hat{Y}$ when $X_i = 0$, i.e. the level of real wages and salaries when union membership is zero. The value of $\hat{\beta}_1 = -9.6510$ implies that real wages and salaries will be –£9.6510 billion when union membership is zero, clearly an unrealistic result! However, $\hat{\beta}_1$ is *arithmetically* important since it is needed to determine the correct value of $\hat{Y}$ when the value of X is given. For example, when $X_i = 11$ million, then $\hat{Y}$ is found by substituting into (11.25), i.e.

$$\hat{Y}_i = -9.6510 + (2.9347 \times 11) = £22.6307 \text{ billion}$$

(We should note that it is usual practice to give the value of the regression estimates to four decimal places.)

In Figure 11.13 we have drawn the estimated regression line through the scatter observations. We note immediately that the line passes through the point of means $(\bar{X}, \bar{Y})$ and it follows that the mean value of the actual Y is equal to the mean value of the estimated Y, i.e. $\bar{Y} = \bar{\hat{Y}}$. If the line were extended to the left, it would cut the Y-axis at the point $\hat{\beta}_1$, i.e. at the value -9.6510 (note that the zeros have been suppressed on both axes); it is for this reason that β_1 is often referred to as the intersect parameter and the slope of the regression line β_2 as the slope parameter.

The OLS regression line has been derived in such a way as to minimise the error sum of squares, i.e. $\Sigma\hat{u}_i^2$. The value of the error sum of squares for the regression line calculated in this example can be found using (11.18):

$$\hat{u}_i = Y_i - \hat{Y}_i = Y_i - \hat{\beta}_1 - \hat{\beta}_2 X_i$$

Thus

$$\hat{u}_1 = Y_1 - \hat{Y}_1 = Y_1 - \hat{\beta}_1 - \hat{\beta}_2 X_1 = 20.4 - (-9.6510) - (2.9347 \times 10.3) = -0.1764$$

$$\hat{u}_2 = Y_2 - \hat{Y}_2 = Y_2 - \hat{\beta}_1 - \hat{\beta}_2 X_2 = 20.6 - (-9.6510) - (2.9347 \times 10.2) = 0.3171$$

Figure 11.13
Estimated regression line drawn through scatter of observations

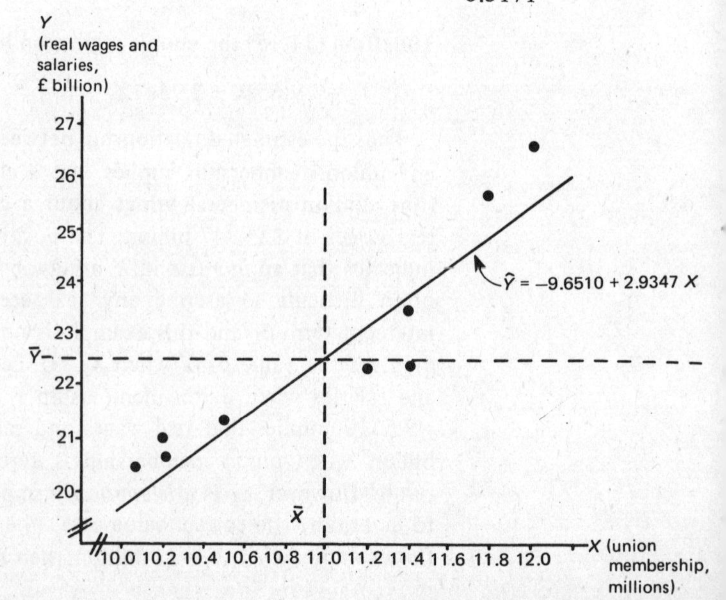

and so on up to $\hat{u}_{10}$. Thus, by squaring each $\hat{u}_i$ term and summing we have

$$\Sigma\hat{u}_i^2 = 5.3880$$

and no other line drawn using any other criteria would produce a smaller error sum of squares.

Properties of the OLS estimators

In this section we shall discuss the properties of the OLS estimators. We consider the linearity property first and then the properties of unbiasedness and minimum variance. Dealing with $\hat{\beta}_2$ first, (11.23) states that

$$\hat{\beta}_2 = \Sigma x_i y_i / \Sigma x_i^2$$

(note that this result was established on the basis of assumption (A.1)).

If we let

$$w_i = x_i / \Sigma x_i^2 \tag{11.26}$$

then

$$\hat{\beta}_2 = \Sigma w_i y_i \tag{11.27}$$

i.e.

$$\hat{\beta}_2 = w_1 y_1 + w_2 y_2 + \ldots + w_n y_n$$

Since the w_i are fixed (being combinations of the X_i fixed by assumption (A.6)) they are effectively constants, and (11.27) implies that $\hat{\beta}_2$ is a *linear function* of the observations on the dependent variable Y. A similar result can be demonstrated for $\hat{\beta}_1$.

Now let us see whether the OLS estimators are unbiased. We start by considering $\hat{\beta}_2$. From (11.13)

$$Y_i = \beta_1 + \beta_2 X_i + u_i \tag{11.13}$$

Expressing this in mean deviation terms gives

$$y_i + \bar{Y} = \beta_1 + \beta_2(x_i + \bar{X}) + u_i = \beta_1 + \beta_2 x_i + \beta_2 \bar{X} + u_i \tag{11.28}$$

but taking the summation of (11.13) and dividing by n gives

$$\Sigma Y_i/n = \beta_1 + \beta_2 \Sigma X_i/n + \Sigma u_i/n$$

i.e.

$$\bar{Y} = \beta_1 + \beta_2 \bar{X} + \bar{u} \tag{11.29}$$

where $\bar{u}$ is the mean value of the u_i in the particular sample.

Substituting (11.29) into (11.28):

$$y_i + \beta_1 + \beta_2 \bar{X} + \bar{u} = \beta_1 + \beta_2 x_i + \beta_2 \bar{X} + u_i$$

Therefore,

$$y_i = \beta_2 x_i + (u_i - \bar{u}) \tag{11.30}$$

Thus expressing the population equation (11.13) in deviation terms produces equation (11.30) (which we may note is Assumption (A.1) in deviation form).

Substituting (11.30) into (11.27) gives

$$\begin{aligned}\hat{\beta}_2 &= \Sigma w_i(\beta_2 x_i + u_i)\\ &= \beta_2 \Sigma w_i x_i + \Sigma w_i u_i \end{aligned} \tag{11.31}$$

Taking expected values we have

$$\begin{aligned}E(\hat{\beta}_2) &= \beta_2 E(\Sigma w_i x_i) + E\Sigma(w_i u_i)\\ &= \beta_2 \Sigma w_i x_i + E\Sigma(w_i u_i)\end{aligned}$$

Since w_i, β and x_i are all either constants or non-stochastic variables with fixed values, the expectation operator leaves them unchanged.

Now $E\Sigma\, w_i u_i$; therefore,

$$E(\hat{\beta}_2) = \beta_2 \Sigma w_i x_i$$

but if we multiply both sides of (11.26) by Σx_i we have

$$\begin{aligned}\Sigma w_i x_i &= \Sigma\,[(x_i/\Sigma x_i^2)x_i]\\ &= \Sigma(x_i^2/\Sigma x_i^2) = \Sigma x_i^2/\Sigma x_i^2\\ &= 1\end{aligned}$$

Therefore,

$$E(\hat{\beta}_2) = \beta_2 \tag{11.32}$$

In other words, the OLS estimator $\hat{\beta}_2$ is an unbiased estimator of β_2. A similar result can be established for $\hat{\beta}_1$. Thus the OLS estimators are unbiased.

Note that to establish this result we used assumptions (A.1), (A.3) and (A.6).

We now turn to the minimum variance property. We shall derive an expression for the *variance of the estimators* and then show that this variance is a minimum.

From (6.9):

$$\text{var}(\hat{\beta}_2) = E\,[\hat{\beta}_2 - E(\hat{\beta}_2)]^2 = E(\hat{\beta}_2 - \beta_2)]^2 \tag{11.33}$$

From (11.31)

$$\beta_2 - \beta_2\ \Sigma\, w_i x_i = \Sigma\, w_i u_i$$

or $\beta_2 - \beta_2 = \Sigma\, w_i u_i$

since $\Sigma\, w_i x_i = 1$

Substituting this last result in (11.33) gives

$$\begin{aligned}\text{var}(\hat{\beta}_2) &= E(\Sigma w_i u_i)^2 \\ &= E(w_1 u_1 + w_2 u_2 + \ldots + w_n u_n)^2 \\ &= E(w_1^2 u_1^2 + w_2^2 u_2^2 + \ldots + w_n^2 u_n^2 + 2w_1 w_2 u_1 u_2 \\ &\quad + 2w_1 w_3 u_1 u_3 + \ldots + 2w_{n-1} w_n u_{n-1} u_n)\end{aligned}$$

This expression contains terms in u_i^2 and $u_i u_j$, but by assumption (A.4) $E(u_i u_j) = 0$ for all $i \neq j$, and therefore all the terms to the right of the term $w_n^2 u_n^2$ drop out, leaving

$$\text{var}(\hat{\beta}_2) = E(w_1^2 u_1^2 + w_2^2 u_2^2 + \ldots + w_n^2 u_n^2)$$

But by assumption (A.5) $E(u_i^2) = \sigma^2$ for every i; therefore;

$$\begin{aligned}\text{var}(\hat{\beta}_2) &= \sigma^2 E(w_1^2 + w_2^2 + \ldots + w_n) \\ &= \sigma^2 E(\Sigma w_i)^2 \\ &= \sigma^2 \Sigma w_i^2\end{aligned}$$

Since by assumption (A.6) the w_i, being combinations of the x_i, are fixed and taking expectations leaves them unchanged. Therefore,

$$\text{var}(\hat{\beta}_2) = \sigma^2 \Sigma(x_i/\Sigma x_i^2)^2 = \sigma^2 \Sigma x_i^2/(\Sigma x_i^2)^2$$

giving the result

$$\text{var}(\hat{\beta}_2) = \sigma^2/\Sigma x_i^2 \qquad (11.34)$$

and the *standard error of the estimator* $\hat{\beta}_2$ is therefore given by

$$s_{\hat{\beta}_2} = \sqrt{\sigma^2/\Sigma x_i^2} \qquad (11.35)$$

Following a similar argument to that above, we can derive expressions for the variance and standard error of $\hat{\beta}_1$, i.e.

$$\text{var}(\hat{\beta}_1) = \sigma^2 \Sigma X_i^2/n\Sigma x_i^2 \qquad (11.36)$$

$$s_{\hat{\beta}_1} = \sqrt{\sigma^2 \Sigma X_i^2/n\Sigma x_i^2} \qquad (11.37)$$

Notice that to establish these expressions for the variance of the OLS estimators we used assumptions (A.1), (A.4), (A.5) and (A.6).

We now want to demonstrate that the OLS estimators have the smallest variance of all linear unbiased estimators.

Let us start by considering some other linear estimator of β_2, say $\tilde{\beta}_2$, which is thus assumed to be a linear function of the Y values, using assumption (A.1), i.e.

$$\tilde{\beta}_2 = a_1 Y_1 + a_2 Y_2 + \ldots + a_n Y_n = \Sigma a_i Y_i \tag{11.38}$$

where the a_i are constants, i.e.

$$\tilde{\beta}_2 = \Sigma a_i(\beta_1 + \beta_2 X_i + u_i) = \beta_1 \Sigma a_i + \beta_2 \Sigma a_i X_i + \Sigma a_i u_i \tag{11.39}$$

Taking expected values:

$$E(\tilde{\beta}_2) = E(\beta_1 \Sigma a_i + \beta_2 \Sigma a_i X_i + \Sigma a_i u_i)$$

Since a_i are defined as constants and X_i are fixed by assumption (A.6), taking expected values leaves them unchanged. Therefore,

$$E(\tilde{\beta}_2) = \beta_1 \Sigma a_i + \beta_2 \Sigma a_i X_i$$

since $E(u_i) = 0$ by Assumption (A.3). Thus, for $\tilde{\beta}_2$ to be an unbiased estimator of β_2, we require that

$$\Sigma a_i = 0$$

$$\Sigma a_i X_i = 1$$

Now if these conditions hold, then

$$\begin{aligned} \Sigma a_i x_i &= \Sigma a_i(X_i - \bar{X}) = \Sigma a_i X_i - \bar{X} \Sigma a_i \\ &= \Sigma a_i X_i \text{ (since } \Sigma a_i = 0) \\ &= 1 \end{aligned}$$

Substituting $\Sigma a_i x_i = 1$ into (11.39) gives

$$\tilde{\beta}_2 = \beta_2 + \Sigma a_i u_i \tag{11.40}$$

Now, the variance of $\tilde{\beta}_2$ is given by

$$\text{var}(\tilde{\beta}_2) = E[\tilde{\beta}_2 - E(\tilde{\beta}_2)]^2 = E(\tilde{\beta}_2 - \beta_2)^2$$

since $\tilde{\beta}_2$ is assumed to be an unbiased estimator. Therefore, substituting (11.40) into this expression we have

$$\begin{aligned} \text{var}(\tilde{\beta}_2) &= E(\beta_2 + \Sigma a_i u_i - \beta_2)^2 = E(\Sigma a_i u_i)^2 \\ &= E(a_1 u_1 + a_2 u_2 + \ldots + a_n u_n)^2 \\ &= E(a_1^2 u_1^2 + a_2^2 u_2^2 + \ldots + a_n^2 u_n^2 + 2a_1 a_2 u_1 u_2 \\ &\quad + 2a_1 a_3 u_1 u_3 + \ldots + 2a_{n-1} a_n u_{n-1} u_n) \end{aligned}$$

Therefore,

$$\text{var}(\tilde{\beta}_2) = \sigma^2 \Sigma a_i^2 \tag{11.41}$$

using assumptions (A.4) and (A.5), as we did when developing equation (11.34) for the variance of $\hat{\beta}_2$.

We now want to show that $\text{var}(\tilde{\beta}_2)$ is greater than $\text{var}(\hat{\beta}_2)$. Let us define the constants a_i in terms of our original OLS weights w_i, i.e.

$$a_i = w_i + k_i$$

where the k_i are another set of constants. From (11.41):

$$\begin{aligned}\text{var}(\tilde{\beta}_2) &= \sigma^2 \Sigma(w_i + k_i)^2 \\ &= \sigma^2 \Sigma w_i^2 + 2\sigma^2 \Sigma w_i k_i + \sigma^2 \Sigma k_i^2 \qquad (11.42)\end{aligned}$$

Let us consider the first two terms on the right-hand side of (11.42) in turn. The first term is

$$\begin{aligned}\sigma^2 \Sigma w_i^2 &= \sigma^2 \Sigma(x_i/\Sigma x_i^2)^2 = \sigma^2 \Sigma x_i^2/(\Sigma x_i^2)^2 = \sigma^2/\Sigma x_i^2 \\ &= \text{var}(\hat{\beta}_2) \text{ from (11.34)}\end{aligned}$$

The second term is

$$\begin{aligned}2\sigma^2 \Sigma w_i k_i &= 2\sigma^2 \Sigma(x_i/\Sigma x_i^2)k_i \\ &= 2\sigma^2 \Sigma[(X_i - \bar{X})/\Sigma x_i^2]k_i \\ &= 0\end{aligned}$$

since $\Sigma(X_i - \bar{X}) = 0$. Thus (11.42) can be written

$$\text{var}(\tilde{\beta}_2) = \text{var}(\hat{\beta}_2) + \sigma^2 \Sigma k_i^2 \qquad (11.43)$$

but $\sigma^2 \Sigma k_i^2$ is greater than 0; therefore,

$$\text{var}(\tilde{\beta}_2) \geqslant \text{var}(\hat{\beta}_2)$$

and

$$\text{var}(\tilde{\beta}_2) = \text{var}(\hat{\beta}_2)$$

only if the weights k_i were zero, i.e. if $a_i = w_i$ and the weights are the OLS weights. Thus the variance of any estimator other than the OLS estimator will have a larger variance: the OLS estimator is therefore the best estimator. A similar result can be derived for the variance of β_1.

Thus we have shown that if the estimators are derived using the OLS method these estimators turn out to be *best linear unbiased* estimators.

Now we want to consider the *shape* of the sampling distribution of the OLS estimators. We have shown, in equation (11.27), that $\hat{\beta}_2$ is a linear function of the observed values of Y, using assumption (A.6). Also, we know that Y_i is a linear function of u_i, assumption (A.1), and that u_i is itself normally distributed, assumption (A.2). Since a linear function of a normally distributed random variable is itself normally distributed, it follows that $\hat{\beta}_2$

will be normally distributed. The same argument can be deployed to show that $\hat{\beta}_1$ is also normally distributed.

It will be useful to list the assumptions necessary to demonstrate the above properties of the OLS estimators; this is done in Table 11.7.

Table 11.7 Properties of the OLS estimators and the necessary assumptions

Property of OLS estimators	Necessary assumptions
Linear	(A.1), (A.6)
Unbiased	(A.1), (A.3), (A.6)
Best linear unbiased	(A.1), (A.3), (A.4), (A.5), (A.6)
Normally distributed	(A.1), (A.2), (A.6)

We can also conveniently summarise the distributional characteristics of $\hat{\beta}_1$ and $\hat{\beta}_2$ thus:

$$\hat{\beta}_2 \sim N(\beta_2, \sigma^2/\Sigma x_i^2) \tag{11.44}$$

$$\hat{\beta}_1 \sim N(\beta_1, \sigma^2 \Sigma X_i^2/n\Sigma x_i^2) \tag{11.45}$$

The problem with (11.44) and (11.45) is that σ^2 will invariably be unknown and it will be impossible to calculate the value of either $\text{var}(\hat{\beta}_1)$ or $\text{var}(\hat{\beta}_2)$. We shall return to this problem in a moment.

Finally, it is possible to demonstrate (see, for example, Kmenta, 1971, pp. 213–15) that the OLS estimators are the same as the *maximum likelihood* estimators, referred to briefly in section 8.2. Since the maximum likelihood estimators have all the desirable large sample or asymptotic properties, it follows that the OLS estimators are themselves *asymptotically unbiased, asymptotically efficient*, and *consistent*.

The OLS estimators $\hat{\beta}_1$ and $\hat{\beta}_2$ are *point* estimators of the population parameters β_1 and β_2. We now want to discuss the construction of interval estimators of these parameters.

11.6 Statistical inference in the regression model

Interval estimation

We base our discussion here on the development of interval estimators in Chapter 8 and the results contained in equations 8.17 and 8.18. For this reason we limit ourselves to derivation of the important results. If we standardise the expressions in equations (11.44) and (11.45), we have:

$$\frac{\hat{\beta}_2 - \beta_2}{\sqrt{\sigma^2/\Sigma x_i^2}} = \frac{\hat{\beta}_2 - \beta_2}{s_{\hat{\beta}_2}} \sim N(0, 1) \tag{11.46}$$

$$\frac{\hat{\beta}_1 - \beta_1}{\sqrt{\sigma^2 \Sigma X_i^2/n\Sigma x_i^2}} = \frac{\hat{\beta}_1 - \beta_1}{s_{\hat{\beta}_1}} \sim N(0, 1) \tag{11.47}$$

and we can write the following probability statements with regard to $\hat{\beta}_1$ and $\hat{\beta}_2$:

$$\left.\begin{aligned} &P(\hat{\beta}_2 - z_{\alpha/2}s_{\hat{\beta}_2} \leqslant \beta_2 \leqslant \hat{\beta}_2 + z_{\alpha/2}s_{\hat{\beta}_2}) = 1 - \alpha \\ &P(\hat{\beta}_1 - z_{\alpha/2}s_{\hat{\beta}_1} \leqslant \beta_1 \leqslant \hat{\beta}_1 + z_{\alpha/2}s_{\hat{\beta}_1}) = 1 - \alpha \end{aligned}\right\} \quad (11.48)$$

where $s_{\hat{\beta}_2}$, the standard error of $\hat{\beta}_2$, is given by (11.35) and $s_{\hat{\beta}_1}$ by (11.37).

The $100(1 - \alpha)$ per cent confidence intervals are thus:

$$\left.\begin{aligned} &(\hat{\beta}_2 - z_{\alpha/2}s_{\hat{\beta}_2} \leqslant \beta_2 \leqslant \hat{\beta}_2 + z_{\alpha/2}s_{\hat{\beta}_2}) \\ &(\hat{\beta}_1 - z_{\alpha/2}s_{\hat{\beta}_1} \leqslant \beta_1 \leqslant \hat{\beta}_1 + z_{\alpha/2}s_{\hat{\beta}_1}) \end{aligned}\right\} \quad (11.49)$$

As we have pointed out, σ^2 in the standard error equations (11.35) and (11.37) will generally be unknown because the u_i will be unknown. However, we do have estimators of the u_i, i.e. the $\hat{u}_i$, since from (11.18)

$$\hat{u}_i = Y_i - \hat{Y}_i$$

It seems plausible, therefore, to consider the use of $\hat{u}$ as an estimator of u which in turn might enable us to find a suitable estimator for σ^2. The variance of the $\hat{u}_i$ is given by

$$\text{var}(\hat{u}) = \frac{1}{n}\Sigma(\hat{u}_i - \bar{\hat{u}})^2 \quad (11.50)$$

but

$$\bar{\hat{u}} = 0$$

i.e. the mean of the sample of residuals or estimated disturbances is zero. This can be shown as follows:

$$\hat{u}_i = Y_i - \hat{Y}_i = Y_i - \hat{\beta}_1 - \hat{\beta}_2 X_i$$

$$\Sigma\hat{u}_i = \Sigma Y_i - n\hat{\beta}_1 - \hat{\beta}_2 \Sigma X_i$$

(note that this right-hand side is the first of the normal equations)

$$\begin{aligned} &= n\bar{Y} - n(\bar{Y} - \hat{\beta}_2\bar{X}) - \hat{\beta}_2 n\bar{X} \\ &= n\bar{Y} - n\bar{Y} + \hat{\beta}_2 n\bar{X} - \hat{\beta}_2 n\bar{X} \\ &= 0 \end{aligned}$$

and therefore

$$\bar{\hat{u}} = \Sigma\hat{u}_i/n = 0/n = 0$$

This result will always be true for all the estimation procedures mentioned in this book. However, $\bar{u}$, the sample mean of the true disturbances, is typically *not* zero.

It follows from (11.50) that

$$\text{var}(\hat{u}) = \frac{1}{n} \Sigma \hat{u}_i^2 \tag{11.51}$$

and we could use this as an estimator of σ^2. Unfortunately, it can be demonstrated that var($\hat{u}$) is a biased estimator of σ^2 (see, for example, Kmenta, 1971, pp. 222–4), in particular:

$$\text{var}(\hat{u}) = \sigma^2 (n-2)/n \tag{11.52}$$

It follows from (11.51) and (11.52) that an unbiased estimator of σ^2, which we denote by $\hat{\sigma}^2$, is given by

$$\hat{\sigma}^2 = \frac{1}{n-2} \Sigma \hat{u}_i^2 \tag{11.53}$$

where the $\hat{u}_i^2$ may be calculated using (11.18), i.e.

$$\hat{u}_i = Y_i - \hat{Y}_i$$

$\hat{\sigma}^2$ is often referred to as the *standard error of the regression*.

Alternatively, we can avoid the need to calculate the individual $\hat{u}_i$s by rearranging (11.53). From (11.18)

$$\Sigma \hat{u}_i^2 = \Sigma(Y_i - \hat{Y}_i)^2 = \Sigma(Y_i - \hat{\beta}_1 - \hat{\beta}_2 X_i)^2$$
$$= \Sigma[Y_i - (\bar{Y} - \hat{\beta}_2 \bar{X}) - \hat{\beta}_2 X_i]^2$$

(substituting for $\hat{\beta}_1$ from (11.24))

$$= \Sigma(y_i - \hat{\beta}_2 x_i)^2 = \Sigma(y_i^2 - 2\hat{\beta}_2 y_i x_i + \hat{\beta}_2^2 x_i^2)$$
$$= (\Sigma y_i^2 - 2\hat{\beta}_2 \Sigma y_i x_i + \hat{\beta}_2^2 \Sigma x_i^2) \tag{11.54}$$

but from (11.23)

$$\hat{\beta}_2 \Sigma x_i^2 = \Sigma x_i y_i$$

and multiplying both sides by $\hat{\beta}_2$ gives

$$\hat{\beta}_2^2 \Sigma x_i^2 = \hat{\beta}_2 \Sigma x_i y_i$$

Therefore, substituting this last result into (11.54) gives

$$\Sigma \hat{u}_i^2 = (\Sigma y_i^2 - 2\hat{\beta}_2^2 \Sigma x_i^2 + \hat{\beta}_2^2 \Sigma x_i^2) = (\Sigma y_i^2 - \hat{\beta}^2 \Sigma x_i^2)$$

Substituting this last result into (11.53) gives

$$\hat{\sigma}^2 = \frac{1}{n-2} (\Sigma y_i^2 - \hat{\beta}_2^2 \Sigma x_i^2) \tag{11.55}$$

Thus when σ^2 is unknown we can estimate its value using (11.53) or (11.55) and substitute into (11.34) to (11.37) to produce the estimated variances and standard errors of the OLS estimators (estimated because we are replacing σ^2 by its estimator $\hat{\sigma}^2$), i.e.

$$\text{estimated var}(\hat{\beta}_2) = \hat{\sigma}^2/\Sigma x_i^2 \tag{11.56}$$

$$\text{estimated } s_{\hat{\beta}_2} = \sqrt{\hat{\sigma}^2/\Sigma x_i^2} \tag{11.57}$$

$$\text{estimated var}(\hat{\beta}_1) = \hat{\sigma}^2 \Sigma X_i^2/n\Sigma_i^2 \tag{11.58}$$

$$\text{estimated } s_{\hat{\beta}_1} = \sqrt{\hat{\sigma}^2 \Sigma X_i^2/n\Sigma x_i^2} \tag{11.59}$$

To avoid needless repetition we shall miss out the word 'estimated' from now on, but we must remember that since σ^2 will only rarely be known we shall normally be dealing only with estimated measures of dispersion.

Thus when n is large enough we can use $\hat{\sigma}^2$ in place of σ^2 in the confidence interval equations of (11.48) and (11.49). When n is small, however, we face the same problem which we encountered in Chapter 8, which is that particular sample values of $\hat{\sigma}^2$ may be considerably different from the true variance σ^2. When confronted with this difficulty in Chapter 8 we examined the distribution of $\hat{\sigma}^2$. If we do the same thing here, we find that χ^2 is again involved, i.e.

$$(n-2)\hat{\sigma}^2/\sigma^2 \sim \chi^2_{n-2} \tag{11.60}$$

The $n-2$ degrees of freedom arise out of the need to use up two of the n degrees of freedom provided by the data in fixing the values of $\hat{\beta}_1$ and $\hat{\beta}_2$.

If we first divide (11.60) by its number of degrees of freedom $(n-2)$ take the square root and then divide the result into the standard normal distribution in (11.46), say, we have

$$\frac{\hat{\beta}_2 - \beta_2/\sqrt{\sigma^2/\Sigma x_i^2}}{\sqrt{[(n-2)\sigma^2/\hat{\sigma}^2]/(n-2)}} \tag{11.61}$$

but, by the definition in equation (8.37), a standard normal distribution divided by the square root of an χ^2-distribution, itself divided by its number of degrees of freedom, defines a t-distribution with the same number of degrees of freedom.

By rearranging and cancelling terms we can express (11.61) as

$$\frac{\hat{\beta}_2 - \beta_2}{\sqrt{\hat{\sigma}^2/\Sigma x_i^2}} = \frac{\hat{\beta}_2 - \beta_2}{s_{\hat{\beta}_2}} \sim t_{n-2} \tag{11.62}$$

In the same way we can show that

$$\frac{\hat{\beta}_1 - \beta_1}{\sqrt{\hat{\sigma}^2 \Sigma X_i^2/n\Sigma x_i^2}} = \frac{\hat{\beta}_1 - \beta_1}{s_{\hat{\beta}_1}} \sim t_{n-2} \tag{11.63}$$

and we can use (11.62) and (11.63) to make the following probability statements:

$$P\left(t_{n-2,\alpha/2} \leqslant \frac{\hat{\beta}_2 - \beta_2}{s_{\hat{\beta}_2}} \leqslant t_{n-2,\alpha/2}\right) = 1 - \alpha \tag{11.64}$$

$$P\left(t_{n-2,\alpha/2} \leqslant \frac{\hat{\beta}_1 - \beta_1}{s_{\hat{\beta}_1}} \leqslant t_{n-2,\alpha/2}\right) = 1 = 1 - \alpha \tag{11.65}$$

These expressions can be rearranged in the same way as equations (11.46) and (11.47) to give the following probability statements with regard to β_1 and β_2:

$$P(\hat{\beta}_2 - t_{n-2,\alpha/2}s_{\hat{\beta}_2} \leqslant \beta_2 \leqslant \hat{\beta}_2 + t_{n-2,\alpha/2}s_{\hat{\beta}_2}) = 1 - \alpha \tag{11.66}$$

$$P(\hat{\beta}_1 - t_{n-2,\alpha/2}s_{\hat{\beta}_1} \leqslant \beta_1 \leqslant \hat{\beta}_1 + t_{n-2,\alpha/2}s_{\hat{\beta}_1}) = 1 - \alpha \tag{11.67}$$

and the corresponding confidence intervals are therefore

$$(\hat{\beta}_2 - t_{n-2,\alpha/2}s_{\hat{\beta}_2} \leqslant \beta_2 \leqslant \hat{\beta}_2 + t_{n-2,\alpha/2}s_{\hat{\beta}_2}) \tag{11.68}$$

$$(\hat{\beta}_1 - t_{n-2,\alpha/2}s_{\hat{\beta}_1} \leqslant \beta_1 \leqslant \hat{\beta}_1 + t_{n-2,\alpha/2}s_{\hat{\beta}_1}) \tag{11.69}$$

where the standard errors $s_{\hat{\beta}_2}$ and $s_{\hat{\beta}_1}$ are given by (11.57) and (11.59). We can illustrate the use of these equations by constructing the 95 per cent confidence intervals for the OLS estimators calculated earlier in the union–real-wage example. The data and calculations associated with Table 11.2 provide us with the following summary measures:

$$\Sigma x_i y_i = 11.8240; \quad \Sigma y_i^2 = 40.1240; \quad \Sigma x_i^2 = 4.0290$$

and we can further calculate $\Sigma X_i^2 = 1216.2300$. From the above regression example

$$\hat{\beta}_1 = -9.6510; \quad \hat{\beta}_2 = 2.9347$$

The first step is to calculate $\hat{\sigma}^2$ using either (11.53) or (11.55). Using (11.55) we have

$$\hat{\sigma}^2 = \left(\frac{1}{10-2}\right) [40.1240 - 2.9347^2 \times 4.0290]$$

$$= 0.6780$$

The second step is to obtain the value of t, appropriate to a 95 per cent confidence level, from the t-table, i.e.

$$100(1 - \alpha) = 95$$

Therefore,

$$(1 - \alpha) = 0.95; \quad \alpha = 0.05; \quad \alpha/2 = 0.025$$

From the t-table the appropriate value of t is

$$t_{n-2,\alpha/2} = t_{8,0.025} = \pm 2.306$$

The 95 per cent confidence interval for β_2 is found by substituting into (11.68); first, we must calculate $s_{\hat{\beta}_2}$ using (11.57), i.e.

$$s_{\hat{\beta}_2} = \sqrt{\hat{\sigma}^2/\Sigma x_i^2}$$

$$= \sqrt{0.678/4.0290} = 0.4107$$

Substituting into (11.68) gives the 95 per cent confidence interval for β_2 as

$$[2.9347 - (2.306 \times 0.4107) \leqslant \beta_2 \leqslant 2.9347 + (2.306 \times 0.4107)]$$

which resolves to

$$(1.9887 \leqslant \beta_2 \leqslant 3.8807)$$

or, in terms of a probability statement,

$$P(1.9887 \leqslant \beta_2 \leqslant 3.8807) = 0.95$$

i.e. there is a probability of at least 0.95 that the interval from 1.9887 to 3.8807 will contain the true population parameter β_2. Notice that this interval does *not* contain zero. This implies that there is a probability of 0.95 that the true value of β_2 is different from zero; we return shortly to a further interpretation of this important result.

By using (11.69) we can construct a similar confidence interval for β_1; first, the estimated standard error of $\hat{\beta}_1$ is, from (11.59),

$$s_{\hat{\beta}_1} = \sqrt{\hat{\sigma}^2 \Sigma X_i^2 / n \Sigma x_i^2}$$

$$= \sqrt{(0.6780 \times 1216.2300)/(10 \times 4.0290)}$$

$$= 4.5240$$

Substituting into (11.69) gives us the 95 per cent confidence interval, i.e.

$$[-9.6510 - (2.306 \times 4.5240) \leqslant \beta_1 \leqslant -9.6510 + (2.306 \times 4.5240)]$$

or

$$(-20.0834 \leqslant \beta_1 \leqslant 0.7814)$$

The corresponding probability statement is therefore

$$P(-20.0834 \leqslant \beta_1 \leqslant 0.7814) = 0.95$$

Notice that this interval *does* embrace zero, and accordingly we can interpret this probability statement as meaning that there is a probability of at least 0.95 that β_1 will *not* be significantly different from zero. We shall consider this result again before too long.

Goodness of fit

If we glance again at Figure 11.3, we can see that there are many lines we could have drawn through the scatter of observations. The reason that we derived a line using the method of ordinary least squares is because estimators obtained in this way turn out to be the best linear unbiased ones.

What we want to discuss now is how *well* the estimated line fits the data. For example, if all the points lie exactly on the estimated line all the deviations or disturbance terms would be zero, in which case $\Sigma\hat{u}_i^2 = 0$, and we could say that the line *fits* the data perfectly. Of course, this situation will rarely arise in practice (it would imply a completely deterministic relationship between the variables), but the closer all the points are to an estimated regression line the better would we think the fit of the line to be.

In this section we want to develop a qualitative measure of *goodness of fit* which will enable us to judge how well a line fits a given set of data and to compare the goodness of fit of one regression line with another.

We know that if a line fits the data exactly, then $\Sigma\hat{u}_i^2 = 0$, and this would be the best line that could be found. The worst line would be one drawn horizontally parallel to the X-axis, since this would imply *no* relationship whatsoever between Y and X (in these circumstances Y would assume a fixed value regardless of the value of X). If this is the case, then the slope of a horizontal line is zero, i.e. $\hat{\beta}_2 = 0$, and from (11.24)

$$\hat{\beta}_1 = \bar{Y}$$

but from (11.18)

$$\hat{u}_i = Y_i - \hat{Y}_i = Y_i - \hat{\beta}_1 - \hat{\beta}_2 X_i$$

Therefore, when $\hat{\beta}_2 = 0$;

$$\hat{u}_i = Y_i - \bar{Y} = y_i \qquad (11.70)$$

and

$$\Sigma\hat{u}_i^2 = \Sigma y_i^2$$

Thus the worst line will produce a value for the error sum of squares of Σy_i^2. Therefore, $\Sigma\hat{u}_i^2$ will vary from 0 for a perfect 'goodness of fit' to Σy_i^2 for the worst 'goodness of fit', i.e.

$$0 \leqslant \Sigma\hat{u}_i^2 \leqslant \Sigma y_i^2$$

Dividing through by Σy_i^2 gives

$$0 \leqslant \Sigma\hat{u}_i^2 / \Sigma y_i^2 \leqslant 1 \qquad (11.71)$$

Since a measure which gets *larger* as the goodness of fit improves appeals more than (11.71), we can rearrange it thus:

$$\text{Goodness of fit} = 1 - \Sigma\hat{u}_i^2/\Sigma y_i^2 \tag{11.72}$$

The right-hand side of (11.72) is more familiar than it looks, as we shall demonstrate. From (11.18) we have

$$\begin{aligned}\hat{u}_i &= Y_i - \hat{Y}_i \\ &= y_i + \bar{Y} - (\hat{y}_i + \bar{\hat{Y}}) = y_i + \bar{Y} - \hat{y}_i - \bar{Y}\end{aligned}$$

since $\bar{\hat{Y}} = \bar{Y}$. Therefore,

$$\hat{u}_i = y_i - \hat{y}_i \tag{11.73}$$

from which

$$y_i = \hat{y}_i + \hat{u}_i \tag{11.74}$$

$$y_i^2 = \hat{y}_i^2 + \hat{u}_i^2 + 2\hat{y}_i\hat{u}_i$$

and summing throughout gives

$$\Sigma y_i^2 = \Sigma\hat{y}_i^2 + \Sigma\hat{u}_i^2 + 2\Sigma\hat{y}_i\hat{u}_i \tag{11.75}$$

Now consider the last term on the right-hand side of this expression. We have

$$\hat{y}_i = \hat{Y}_i - \bar{\hat{Y}} = \hat{\beta}_1 + \hat{\beta}_2 X_i - \bar{Y}$$

but from (11.24) $\bar{Y} = \hat{\beta}_1 + \hat{\beta}_2\bar{X}$; therefore,

$$\hat{y}_i = \hat{\beta}_1 + \hat{\beta}_2 X_i - (\hat{\beta}_1 - \hat{\beta}_2\bar{X}) = \hat{\beta}_2(X_i - \bar{X})$$

Thus

$$\hat{y}_i = \hat{\beta}_2 x_i \tag{11.76}$$

(provided there is a constant in the equation). Substituting this result into (11.73) gives

$$\hat{u}_i = y_i - \hat{\beta}_2 x_i \tag{11.77}$$

and multiplying (11.76) by (11.77) gives

$$\hat{y}_i\hat{u}_i = \hat{\beta}_2 x_i(y_i - \hat{\beta}_2 x_i) = \hat{\beta}_2 x_i y_i - \hat{\beta}_2^2 x_i^2$$

and summing gives

$$\Sigma\hat{y}_i\hat{u}_i = \hat{\beta}_2\Sigma x_i y_i - \hat{\beta}_2^2\Sigma x_i^2 = \hat{\beta}_2(\Sigma x_i y_i - \hat{\beta}_2\Sigma x_i^2) \tag{11.78}$$

but from (11.23)

$$\hat{\beta}_2 = \Sigma x_i y_i/\Sigma x_i^2$$

and substituting (11.23) into (11.78) gives

$$\begin{aligned}\Sigma\hat{y}_i\hat{u}_i &= \hat{\beta}_2\left(\Sigma x_i y_i - \frac{\Sigma x_i y_i}{\Sigma x_i^2}\Sigma x_i^2\right) \\ &= \hat{\beta}_2(\Sigma x_i y_i - \Sigma x_i y_i)\end{aligned}$$

Therefore,

$$\Sigma \hat{y}_i \hat{u}_i = 0 \tag{11.79}$$

Substituting (11.79) into (11.75) gives

$$\Sigma y_i^2 = \Sigma \hat{y}_i^2 + \Sigma \hat{u}_i^2 \tag{11.80}$$

i.e.

$$\Sigma \hat{u}_i^2 = \Sigma y_i^2 - \Sigma \hat{y}_i^2 \tag{11.81}$$

and substituting (11.81) into (11.72) gives

$$\begin{aligned}\text{Goodness of fit} &= 1 - (\Sigma y_i^2 - \Sigma \hat{y}_i^2)/\Sigma y_i^2 \\ &= 1 - \Sigma y_i^2/\Sigma y_i^2 + \Sigma \hat{y}_i^2/\Sigma y_i^2 \\ &= 1 - 1 + \Sigma \hat{y}_i^2/\Sigma y_i^2\end{aligned}$$

i.e.

$$\text{Goodness of fit} = \Sigma \hat{y}_i^2/\Sigma y_i^2 \tag{11.82}$$

but, using (11.76),

$$\begin{aligned}\Sigma \hat{y}_i^2/\Sigma y_i^2 &= \Sigma(\hat{\beta}_2 x_i)^2/\Sigma y_i^2 \\ &= \hat{\beta}_2^2 \Sigma x_i^2/\Sigma y_i^2 = (\Sigma x_i y_i/\Sigma x_i^2)^2 \Sigma x_i^2/\Sigma y_i^2 \\ &= (\Sigma x_i y_i)^2/\Sigma x_i^2 \Sigma y_i^2\end{aligned} \tag{11.83}$$

and substituting (11.83) into (11.82) gives

$$\text{Goodness of fit} = (\Sigma x_i y_i)^2/\Sigma x_i^2 \Sigma y_i^2 = r^2 \tag{11.84}$$

from (11.7).

In other words, the measure of goodness of fit is identical to the square of the correlation coefficient discussed in section 11.2. We shall follow convention and label the goodness of fit R^2. Now, we saw that

$$-1 \leqslant r \leqslant 1$$

Therefore,

$$0 \leqslant R^2 \leqslant 1 \tag{11.85}$$

Thus we have a measure of goodness of fit R^2 which varies between $R^2 = 0$, the worst fit, when $\Sigma \hat{u}_i^2 = \Sigma y_i^2$, and $R^2 = 1$, the best fit, when $\Sigma \hat{u}_i^2 = 0$. We have shown that goodness of fit can be expressed by

$$R^2 = 1 - \Sigma \hat{u}_i^2/\Sigma y_i^2 \tag{11.86}$$

or by

$$R^2 = \Sigma \hat{y}_i^2/\Sigma y_i^2 \tag{11.87}$$

R^2 is more properly known as the *coefficient of determination*, but this full title is not often used. Returning now to (11.80) we have

$$\Sigma y_i^2 = \Sigma \hat{y}_i^2 + \Sigma \hat{u}_i^2$$

This expression, we see, is made up of three separate 'sums of squares', which can be described as follows:

1. $\Sigma y_i^2 = \Sigma(Y_i - \bar{Y})^2$ measures the total *variation*[1] of the observed Y values around their sample mean $\bar{Y}$. It is usually referred to as the *total sum of squares*, and denoted by SST.
2. $\Sigma \hat{y}_i^2 = \Sigma(\hat{Y}_i - \bar{\hat{Y}})^2$ measures the total variation of the *estimated* Y values around their mean $\bar{\hat{Y}} = \bar{Y}$, and is referred to as the 'explained' or *regression sum of squares*, usually denoted SSR.
3. $\Sigma \hat{u}_i^2 = \Sigma(Y_i - \hat{Y}_i)^2$ measures the total variation in the estimated disturbance or error terms around the estimated line, and is referred to as the 'unexplained' or *error sum of squares*, usually denoted by SSE.

Thus (11.80) can be expressed in sums of squares form as

$$\text{SST} = \text{SSR} + \text{SSE} \tag{11.88}$$

In other words, the total variation in the dependent variable can be partitioned into two separate components, one due to or explained by the estimated regression line and one due to the disturbance term (i.e. unexplained).

Equation (11.80) can be expressed in a (sometimes) more convenient form, since, using (11.76),

$$\text{SSR} = \Sigma \hat{y}_i^2 = \Sigma(\hat{\beta}_2 x_i)^2 = \hat{\beta}_2^2 \Sigma x_i^2 \tag{11.89}$$

Therefore, (11.80) becomes

$$\Sigma y_i^2 = \hat{\beta}_2^2 \Sigma x_i^2 + \Sigma \hat{u}_i^2 \tag{11.90}$$

Finally, by substituting from (11.89) into (11.87) we can express R^2 thus:

$$R^2 = \frac{\hat{\beta}_2^2 \Sigma x_i^2}{\Sigma y_i^2} = \frac{\text{SSR}}{\text{SST}} \tag{11.91}$$

This last result states that goodness of fit is the ratio of the variation in the observed Y which is explained by the estimated regression line to the total variation in Y. The more of the variation in

[1] 'Variation' and 'variance' must be distinguished. Variation means the change in Y from one sample observation to another; it is measured by the sum of squares of the deviations of Y from its mean. Variance refers to the dispersion of Y corresponding to some fixed value of X and is calculated by dividing variation by the appropriate number of degrees of freedom.

Y which is attributable to the estimated regression, the higher will R^2 be, and the better the measured goodness of fit. In other words, the goodness of fit referred to is the fit of the estimated line to the observed values of Y, *not* the fit of the estimated line to the true population line. Since both $\Sigma\hat{u}_i^2$ and Σy_i^2 are random variables in (11.86) it follows that R^2 is also a random variable whose value will (like the Y and u values) vary from sample to sample. It is possible to devise a test of significance of R^2 but we leave this to the next chapter. It is enough to note at this point that if the confidence interval for β_2 does *not* embrace zero, then R^2 will be significant, i.e. a significant relationship exists between X and Y.

Because R^2 is a random variable varying from sample to sample it does not necessarily follow that a high R^2 value *definitely* means a significant relationship between X and Y; a high R^2 value might simply be the product of a particular sample. A low R^2 value might be caused by three different reasons. First, X may be a poor 'explainer' of Y, i.e. changes in X have little effect on Y. Second, X may be a relevant variable in terms of explaining the variation in Y but its role might be overwhelmed by the influence of the stochastic disturbances. Third, it might be that we have not specified the regression model properly. In the latter sense R^2 is frequently used as an indication of the correctness of the specification of the model.

We can now return to our earlier regression example and calculate the value of R^2 as a measure of the goodness of fit of our regression line. To use (11.91) we need to know that

$$\hat{\beta}_2^2 = 2.9347^2 = 8.6125; \quad \Sigma x_i^2 = 4.0290; \quad \Sigma y_i^2 = 40.1240$$

Therefore,

$$\begin{aligned} R^2 &= \hat{\beta}_2^2 \Sigma x_i^2 / \Sigma y_i^2 \\ &= (8.6215 \times 4.0290)/40.1240 \\ &= 0.8657 \end{aligned}$$

This value is in fact the square of the value of 0.93 obtained for the correlation coefficient in section 11.2.

Let us now examine the value of the three sum of squares terms, Σy_i^2, $\Sigma\hat{y}_i^2$ and $\Sigma\hat{u}_i^2$, in (11.80). We already know that $\Sigma y_i^2 = 40.1240$. Thus

$$\text{SST} = 40.1240$$

We can obtain SSR by substituting directly into (11.89), i.e.

$$\begin{aligned} \text{SSR} &= \hat{\beta}_2^2 \Sigma x_i^2 \\ &= 2.9347^2 \times 4.0290 \\ &= 34.7360 \end{aligned}$$

We can now rearrange (11.88) to determine SSE, i.e.

$$\begin{aligned} \text{SSE} &= \text{SST} - \text{SSR} \\ &= 40.1240 - 34.7360 \\ &= 5.3880 \end{aligned}$$

which is the value we previously calculated for $\Sigma \hat{u}_i^2$.

Hypothesis testing

We base this section on the assumption, as will invariably be the case, that σ^2 is unknown and is estimated using $\hat{\sigma}^2$. The concepts discussed in Chapter 10 can readily be applied in the context of the classical normal linear regression model, though (as with interval estimates) they are crucially dependent on the normality assumption embodied in (A.2).

The most frequent application of hypothesis testing in the economic and social sciences is the investigation of whether the true regression parameters β_1 and β_2, but particularly β_2, can be held to be significantly different from zero. If a β_2 cannot statistically speaking be distinguished from zero, then it follows that X and Y can have no relationship as specified in (11.13). Clearly, whatever value X has, once it is multiplied by $\beta_2 = 0$ it, too, is reduced to zero and will not contribute anything to the value of Y.

We discuss this test first and then tests as to whether or not β_2 is smaller or larger than some *a priori* value. This discussion assumes a familiarity with the material covered in Chapter 10; it is therefore quite brief.

Let us start with tests on β_2. We know from (11.62) that

$$\frac{\hat{\beta}_2 - \beta_2}{s_{\hat{\beta}_2}} \sim t_{n-2}$$

where $s_{\hat{\beta}_2}$ is calculated from (11.57).

The null hypothesis is

$$H_0: \beta_2 = 0$$

If we are uncertain as to what alternative value β_2 might take, a suitable alternative hypothesis would be

$$H_1: \beta_2 \neq 0 \tag{11.92}$$

If economic theory indicates that β_2 is likely to be positive, then the alternative hypothesis would be

$$H_1: \beta_2 > 0 \tag{11.93}$$

and if negative would be

$$H_1: \beta_2 < 0 \tag{11.94}$$

The practical difference in these various alternative hypotheses is that the first, $H_1: \beta_2 \neq 0$, is resolved with a two-tailed test, while the remaining two, (11.93) and (11.94), with a one-tailed test. In regression analysis the alternative hypothesis of (11.92) is the most commonly occurring, so we will consider that first. Thus we have:

$$\left.\begin{aligned} H_0&: \beta_2 = 0 \\ H_1&: \beta_2 \neq 0 \end{aligned}\right\} \quad (11.95)$$

Under the null hypothesis $\beta_2 = 0$ and (11.62) can be written

$$\frac{\hat{\beta}_2}{s_{\hat{\beta}_2}} \sim t_{n-2} \quad (11.96)$$

which is of course $\hat{\beta}_2$ divided by its own standard error, calculated from (11.57).

To perform the test we compute the value of (11.96) and compare it with some critical value of t, say $t^*_{n-2,\alpha/2}$, taken from the t-table using the chosen α significance level. The situation is illustrated in Figure 11.14.

The decision-rule for β_2 is thus:

$$\left.\begin{aligned} \text{reject } H_0 \text{ if: } & \hat{\beta}_2/s_{\hat{\beta}_2} > t^*_{n-2,\alpha/2} \\ \text{or if: } & \hat{\beta}_2/s_{\hat{\beta}_2} < -t^*_{n-2,\alpha/2} \\ \text{do not reject } H_0 \text{ if: } & -t^*_{n-2,\alpha/2} \leqslant \hat{\beta}_2/s_{\hat{\beta}_2} \leqslant t^*_{n-2,\alpha/2} \end{aligned}\right\} \quad (11.97)$$

Applying the same test to β_1, we have

$$H_0: \beta_1 = 0$$

$$H_1: \beta_1 \neq 0$$

Under the null hypothesis $\beta_1 = 0$ and (11.63) can be written

$$\hat{\beta}_1/s_{\hat{\beta}_1} \sim t_{n-2} \quad (11.98)$$

which again is $\hat{\beta}_1$ divided by its own standard error, computed from (11.59). To perform the test we calculate the ratio in (11.98) and compare it with the critical value of t, say $t^*_{n-2,\alpha/2}$, using the appropriate significance level α. The situation is shown in Figure

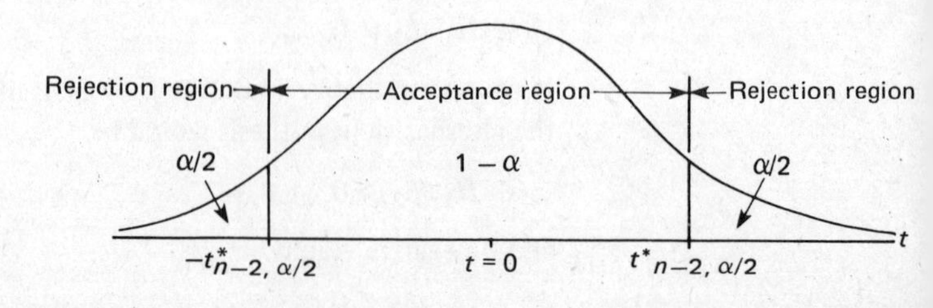

Figure 11.14
Hypothesis testing of OLS estimators

11.14. The decision-rule for β_1 is therefore:

$$\left.\begin{array}{ll} \text{reject } H_0 \text{ if: } & \widehat{\beta}_1/s_{\widehat{\beta}_1} > t^*_{n-2,\alpha/2} \\ \text{or if: } & \widehat{\beta}_1/s_{\widehat{\beta}_1} < -t^*_{n-2,\alpha/2} \\ \text{do not reject } H_0 \text{ if: } & -t^*_{n-2,\alpha/2} \leqslant \widehat{\beta}_1/s_{\widehat{\beta}_1} \leqslant t^*_{n-2,\alpha/2} \end{array}\right\} \quad (11.99)$$

To illustrate the use of these equations let us use our union-membership/real-wages and income example and test whether $\widehat{\beta}_2$ and $\widehat{\beta}_1$ are significantly different from zero. We shall assume that economic theory does not indicate what the sign of β_1 and β_2 is likely to be, and so our hypotheses are as in (11.95) (and similarly for β_1).

We choose a significance level of 5 per cent, i.e.

$$\alpha = 0.05; \quad \alpha/2 = 0.025$$

Since $n = 10$ the critical value of t is found to be

$$t^*_{n-2,\alpha/2} = t^*_{8,0.025} = \pm 2.306$$

We have already calculated $s_{\widehat{\beta}_2}$, using (11.57), to be $s_{\widehat{\beta}_2} = 0.4107$. Substituting into (11.96) gives

$$\widehat{\beta}_2/s_{\widehat{\beta}_2} = 2.9347/0.4107 = 7.1540$$

Applying the decision-rule of (11.97) gives

$$7.1540 > 2.306$$

i.e.

$$\widehat{\beta}_2/s_{\widehat{\beta}_2} > t^*_{n-2,\alpha/2}$$

Thus we reject H_0 in favour of H_1, i.e. there is a probability of at least 0.95 (a probability of a type-I error of 0.05) that β_2 is significantly different from zero. This implies that X is significantly related to Y. In view of the high value of R^2 we obtained above, this result should not surprise us. Notice that this result confirms the confidence interval result, which was that the 95 per cent confidence interval did not embrace zero. As we saw at the end of Chapter 10, confidence interval estimation and hypothesis testing use the same information and lead to the same results.

Now for β_1. We have already calculated $s_{\widehat{\beta}_1}$, using (11.59), to be $s_{\widehat{\beta}_1} = 4.5240$; therefore, substituting into (11.98) gives

$$\widehat{\beta}_1/s_{\widehat{\beta}_1} = -9.6510/4.5240 = -2.1333 \sim t_{n-2}$$

Applying the decision-rule of (11.98) gives

$$-2.306 \leqslant -2.1333 \leqslant 2.306$$

i.e.

$$-t^*_{n-2,\alpha/2} \leqslant \widehat{\beta}_1/s_{\widehat{\beta}_1} \leqslant t^*_{n-2,\alpha/2}$$

Therefore, we accept H_0: the value of $\hat{\beta}_1$ is such as to indicate that there is a probability of at least 0.95 that it was drawn from a population with a mean of zero. However, just because β_1 has been found not to be different from zero does not mean that we can simply set it to zero in the regression equation – remember it has an arithmetic role to play in determining the value of $\hat{Y}$.

Now let us turn to the consideration of one-tailed tests of β_2 where the alternative hypothesis is of the form of (11.93):

$$H_1: \beta_2 > 0$$

As with two-tailed testing, under the null hypothesis

$$H_0: \beta_2 = 0$$

equation (11.62) again reduces to (11.96), i.e.

$$\beta_2/s_{\hat{\beta}_2} \sim t_{n-2}$$

where $s_{\hat{\beta}_2}$ is calculated from (11.57). Now we reject H_0 if the value of the term $\hat{\beta}_2/s_{\hat{\beta}_2}$ exceeds some critical value of t, say $t^*_{n-2,\alpha}$, at some given significance level α. The situation is illustrated in Figure 11.15.

Figure 11.15 Critical value when $H_1: \beta > 0$

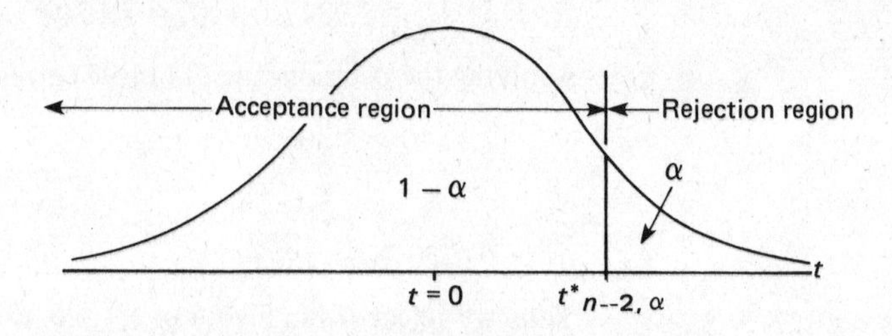

The decision-rule is thus as follows:

$$\left.\begin{aligned} \text{reject } H_0 \text{ if:} \quad & \hat{\beta}_2/s_{\hat{\beta}_2} > t^*_{n-2,\alpha} \\ \text{do not reject } H_0 \text{ if:} \quad & \hat{\beta}_2/s_{\hat{\beta}_2} < t^*_{n-2,\alpha} \end{aligned}\right\} \quad (11.100)$$

For β_1 the decision-rule is similarly:

$$\left.\begin{aligned} \text{reject } H_0 \text{ if:} \quad & \hat{\beta}_1/s_{\hat{\beta}_1} > t^*_{n-2,\alpha} \\ \text{do not reject } H_0 \text{ if:} \quad & \hat{\beta}_1/s_{\hat{\beta}_1} < t^*_{n-2,\alpha} \end{aligned}\right\} \quad (11.101)$$

where $s_{\hat{\beta}_1}$ is calculated using (11.59).

Finally, when the alternative hypothesis is of the form given by (11.94), i.e.

$$H_1: \beta_2 < 0$$

then under the null hypothesis,

$$H_0: \beta_2 = 0$$

(11.62) again reduces to (11.96):

$$\hat{\beta}_2/s_{\hat{\beta}_2} \sim t_{n-2}$$

where $s_{\hat{\beta}_2}$ is calculated from (11.57). Now we will reject H_0 if the value of $\hat{\beta}_2/s_{\hat{\beta}_2}$ is less than some critical value of t, say $-t^*_{n-2,\alpha}$, at some given level of significance α. The situation is illustrated in Figure 11.16.

Figure 11.16 Critical value when $H_1: \beta_2 < 0$

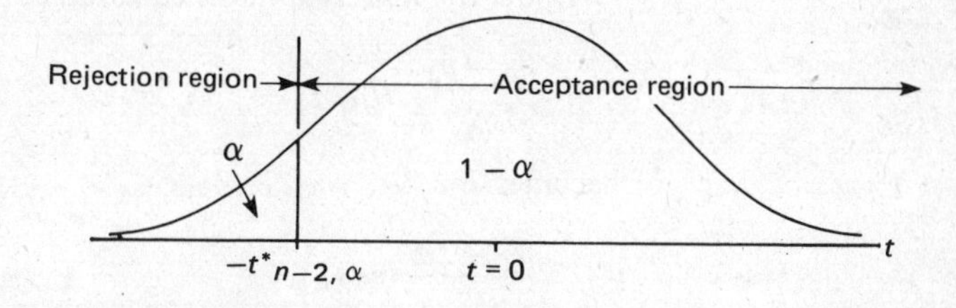

The decision-rule is:

$$\left.\begin{array}{ll} \text{reject } H_0 \text{ if:} & \hat{\beta}_2/s_{\hat{\beta}_2} < -t^*_{n-2,\alpha} \\ \text{do not reject } H_0 \text{ if:} & \hat{\beta}_2/s_{\hat{\beta}_2} > -t^*_{n-2,\alpha} \end{array}\right\} \quad (11.102)$$

For β_1 the decision-rule is similarly:

$$\left.\begin{array}{ll} \text{reject } H_0 \text{ if:} & \hat{\beta}_1/s_{\hat{\beta}_1} < -t^*_{n-2,\alpha} \\ \text{do not reject } H_0 \text{ if:} & \hat{\beta}_1/s_{\hat{\beta}_1} > -t^*_{n-2,\alpha} \end{array}\right\} \quad (11.103)$$

As an illustration of a one-tailed test suppose we have *a priori* information that in the union-membership example β_2 is positive, i.e. $\beta_2 > 0$. Then the hypotheses we wish to test are:

$$H_0: \beta_2 = 0$$

$$H_1: \beta_2 > 0$$

Suppose we again choose a significance level of 0.05; then the critical value of t is

$$t^*_{n-2,\alpha} = t^*_{8,0.05} = \pm 1.860$$

We have already calculated $\hat{\beta}_2/s_{\hat{\beta}_2} = 7.1540$. Therefore, applying rule (11.100), we have

$$7.1540 > 1.860$$

i.e.

$$\hat{\beta}_2/s_{\hat{\beta}_2} > t^*_{n-2,\alpha}$$

Therefore, we reject H_0 in favour of H_1: the sample evidence does support the contention that $\beta_2 > 0$.

Finally, let us suppose that our *a priori* information is such that we believe $\beta_2 = 2.5$ and let us test whether our sample value of $\beta_2 = 2.9347$ is significantly different. Thus our hypotheses are:

$$H_0: \beta_2 = 2.5$$

$$H_1: \beta_2 \neq 2.5$$

although the alternative hypothesis could equally well have been $H_1: \beta_2 > 2.5$.

Under the null hypothesis equation (11.62), i.e.

$$\frac{\hat{\beta}_2 - \hat{\beta}_2}{s_{\hat{\beta}_2}} \sim t_{n-2}$$

becomes

$$\frac{\hat{\beta}_2 - 2.5}{s_{\hat{\beta}_2}} \sim t_{n-2}$$

and substituting $\hat{\beta}_2 = 2.9347$ and $s_{\hat{\beta}_2} = 0.4107$ into this expression gives

$$\frac{\hat{\beta}_2 - 2.5}{s_{\hat{\beta}_2}} = \frac{2.9347 - 2.5}{0.4107} = 1.0584$$

The critical value of t at a significance level of 5 per cent in this two-tailed test case is

$$t^*_{n-2,\alpha/2} = t^*_{8,0.025} = \pm 2.306$$

The decision-rule will be:

$$\left.\begin{aligned} \text{reject } H_0 \text{ if: } & \frac{\hat{\beta}_2 - \beta_2}{s_{\hat{\beta}_2}} < -t^*_{n-2,\alpha/2} \\ \text{or if: } & \frac{\hat{\beta}_2 - \beta_2}{s_{\hat{\beta}_2}} > -t^*_{n-2,\alpha/2} \\ \text{do not reject } H_0 \text{ if: } & -t^*_{n-2,\alpha/2} \leqslant \frac{\hat{\beta}_2 - \beta_2}{s_{\hat{\beta}_2}} \leqslant t^*_{n-2,\alpha/2} \end{aligned}\right\} \quad (11.104)$$

Applying this decision-rule gives

$$-2.306 \leqslant 1.0584 \leqslant 2.306$$

i.e.

$$-t^*_{n-2,\alpha/2} \leqslant \frac{\hat{\beta}_2 - \beta_2}{s_{\hat{\beta}_2}} \leqslant t^*_{n-2,\alpha/2}$$

Accordingly, we accept H_0, i.e. the sample value of $\hat{\beta}_2 = 2.9347$ is *not* sufficiently different from $\beta_2 = 2.5$ to cause us to believe it was taken from a population with a different value.

A rule-of-thumb test

So common is the test that β_1 and β_2 are significantly different from zero that a crude rule-of-thumb test is available which enables us to make a quick but approximate test of the non-zero significance of the OLS estimators.

We have seen above that when the competing hypotheses are of the form:

$$H_0: \beta_2 = 0$$

$$H_1: \beta_2 \neq 0$$

then the value which is of interest is the ratio $\hat{\beta}_2/s_{\hat{\beta}_2}$ or $\hat{\beta}_1/s_{\hat{\beta}_1}$. In most applications, with a significance level of 5 per cent, the critical value of t is *approximately* equal to 2. Thus $\hat{\beta}_1$ or $\hat{\beta}_2$ will be found to be significantly different from zero and we can reject the null hypothesis whenever

$$\hat{\beta}_2/s_{\hat{\beta}_2} > 2 \quad \text{or} \quad \hat{\beta}_1/s_{\hat{\beta}_1} > 2 \tag{11.105}$$

Thus if the standard error of an OLS estimator is *less than one-half* of the value of the estimator, then we can conclude that the corresponding parameter is significantly different from zero. We must emphasise the approximate nature of this test, but it does enable us to examine the results of an OLS regression analysis and to make a rapid appraisal of the significance of the estimator.

Presentation of regression results

It is normal when presenting the results of a regression analysis also to give the standard error or the t-value or both of each estimate below it in brackets and also to give the value of R^2. The results are thus normally presented in the following manner:

$$\underset{(4.5240)}{Y_i = -9.6510} + \underset{(0.4107)}{2.9347X_i} \qquad R^2 = 0.93 \tag{11.106}$$

In this case we have given the standard errors in brackets.

We now want to illustrate all the material covered so far in this chapter with a further example.

Example

Many economists (amongst them Milton Friedman) believe that price inflation in an economy is a function of the money supply, though the relationship is not believed to be instantaneous. In fact various time lags have been postulated – ranging from one year to over two years. In this example we are going to examine this relationship by regressing an index of prices in a given year Y_t against money supply (in £ billion, as measured by M3) in the *previous* year X_t. The subscript t is to indicate that our data are time-series data.

The sample data for the period 1965 to 1976 are given in Table 11.8, together with the calculations necessary to determine the various summary measures. The model whose parameters we wish to estimate is thus

$$Y_t = \beta_1 + \beta_2 X_t + u_t \tag{11.107}$$

The data in columns two and three are plotted on the scatter diagram of Figure 11.17.

Figure 11.17 Scatter diagram of prices Y against money supply X

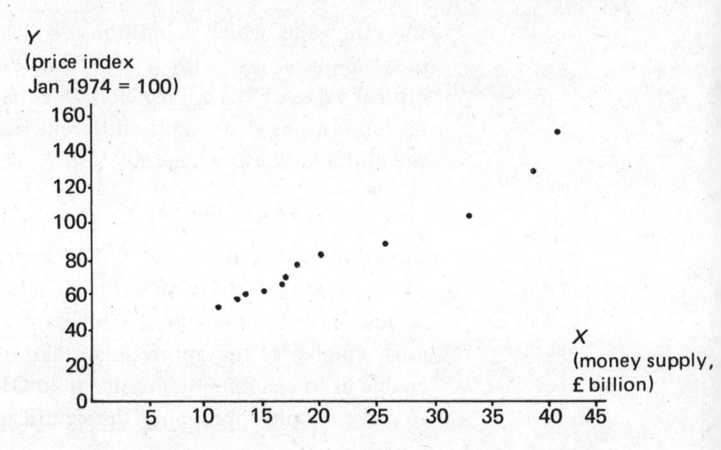

From Table 11.8 we obtain the following measures:

$$\bar{X} = 21.933; \quad \bar{Y} = 87.292; \quad \Sigma X_t^2 = 6891.515; \quad n = 12$$

$$\Sigma x_t y_t = 3430.389; \quad \Sigma x_t^2 = 1118.674; \quad \Sigma y_t^2 = 10889.889$$

First, we calculate the OLS estimators using (11.23) and (11.24), i.e.

$$\hat{\beta}_2 = \Sigma x_t y_t / \Sigma x_t^2 = 3430.389/1118.674 = 3.0665$$

$$\hat{\beta}_1 = \bar{Y} - \hat{\beta}_2 \bar{X} = 87.292 - (3.0665 \times 21.933) = 20.0344$$

Thus the estimated regression line is

$$Y_t = 20.0344 + 3.0665 X_t$$

which can be interpreted as implying that for every unit increase in money supply X_t (where a unit of X_t is £1 billion) there is an increase in the retail price index Y_t of 3.0655 points in the following year. The value of $\hat{\beta}_1$ has little or no economic meaning; literally it means that when money supply is zero, i.e. $X_t = 0$, the retail price index will have a value of 20.0344. It is not possible in reality to have a zero money supply so that $\hat{\beta}_1$ (and the parameter β_1) mean nothing (β_1 might have had more meaning if we had regressed *changes* in X and Y). Of course, as we pointed out

Table 11.8 Calculation of β_1 and β_2 in money supply/price inflation model

Year	X_t (M3, £ billion)	Y_t (Jan 1974 = 100)	X_t^2	x_t	y_t	$x_t y_t$	x_t^2	y_t^2
1965	12.155	58.4	147.744	−9.778	−28.892	282.506	95.609	834.748
1966	13.083	60.7	171.165	−8.850	−26.592	235.339	78.322	707.134
1967	13.555	62.2	183.738	−8.378	−25.092	210.221	70.191	629.608
1968	15.003	65.1	225.090	−6.930	−22.192	153.790	48.025	492.484
1969	16.092	68.7	258.952	−5.841	−18.592	108.596	34.117	345.662
1970	16.596	73.0	275.427	−5.337	−14.292	76.276	28.484	204.261
1971	18.175	79.9	330.331	−3.758	−7.392	27.779	14.122	54.642
1972	20.541	85.6	421.932	−1.392	−1.692	2.355	1.938	2.863
1973	26.245	93.5	688.800	4.312	6.208	26.769	18.593	38.539
1974	33.478	108.5	1120.776	11.545	21.208	244.846	133.287	449.779
1975	37.698	134.8	1421.392	15.765	47.508	748.963	248.534	2257.010
1976	40.573	157.1	1646.168	18.640	69.308	1312.949	347.150	4873.157
Σ	263.194	1047.5	6891.515			3430.389	1118.674	10889.889
	$\bar{X}$ = 263.194/12 = 21.933	$\bar{Y}$ = 1047.5/12 = 87.292						

above, $\hat{\beta}_1$ is still required to play its arithmetic role in determining the value of Y_t for any given value of X_t. The estimated regression line is shown in Figure 11.18. We might note that if we do obtain a value for $\hat{\beta}_1$ which is not feasible or does not make sense this may mean that the model is incorrectly specified. For example, the correct functional form may not be linear, a linear approximation only being a good fit in the sample period.

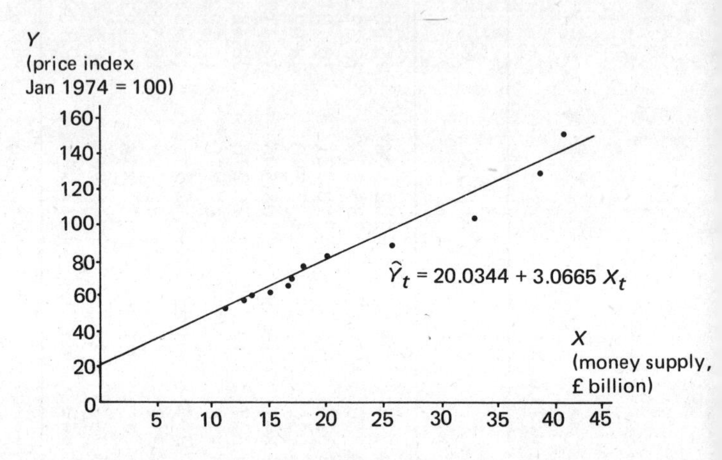

Figure 11.18
Estimated regression line

To calculate the standard error of $\hat{\beta}_1$ and $\hat{\beta}_2$ we must first calculate the estimated value of the estimator $\hat{\sigma}^2$, using (11.55), i.e.

$$\hat{\sigma}^2 = \frac{1}{n-2}(\Sigma y_t^2 - \hat{\beta}_2^2 \Sigma x_t^2)$$

$$= \frac{1}{10}[10889.889 - (3.0665^2 \times 1118.674)]$$

$$= 37.0525$$

By substituting into (11.57) and (11.59) we can calculate the estimated standard errors of $\hat{\beta}_2$ and $\hat{\beta}_1$, i.e.

$$s_{\hat{\beta}_2} = \sqrt{\hat{\sigma}^2/\Sigma x_t^2}$$

$$= \sqrt{37.0525/1118.674}$$

$$= 0.1820$$

$$s_{\hat{\beta}_1} = \sqrt{\hat{\sigma}^2 \Sigma X_t^2 / n\Sigma x_t^2}$$

$$= \sqrt{(37.0525 \times 6891.515)/(12 \times 1118.674)}$$

$$= 4.3614$$

To calculate R^2 we use (11.91), i.e.

$$R^2 = \hat{\beta}_2^2 \Sigma x_t^2 / \Sigma y_t^2$$
$$= 3.0665^2 \times (1118.674/10889.889)$$
$$= 0.9660$$

i.e. 96.6 per cent of the variation in Y is 'explained' by the explanatory variable X, while 3.4 per cent is due to random disturbances.

The estimated regression equation may thus be presented as follows:

$$\hat{Y}_t = \underset{(4.3614)}{20.0344} + \underset{(0.1820)}{3.0665}X_t \qquad R^2 = 0.9660$$

Let us now test, using a 5 per cent significance level, whether β_1 and β_2 can, on the basis of the sample regression results, be said to be significantly different from zero. The rule-of-thumb test described above would suggest that they both are, since

$$s_{\hat{\beta}_1} < \tfrac{1}{2}\hat{\beta}_1 \quad \text{and} \quad s_{\hat{\beta}_2} < \tfrac{1}{2}\hat{\beta}_2$$

However, we can test these results more rigorously. Starting with β_2 we have:

$$H_0: \beta_2 = 0$$
$$H_1: \beta_2 \neq 0$$

Thus substituting into (11.96) gives

$$\hat{\beta}_2 / s_{\hat{\beta}_2} \sim t_{n-2}$$
$$\hat{\beta}_2 / s_{\hat{\beta}_2} = 3.0665/0.1820 = 16.8489$$

The critical value of t with $\alpha = 0.05$ is

$$t^*_{n-2,\alpha/2} = t^*_{10,0.025} = \pm 2.228$$

Applying the decision-rule of (11.97) gives

$$16.8489 > 2.228$$

i.e.

$$\hat{\beta}_2 / s_{\hat{\beta}_2} > t^*_{n-2,\alpha/2}$$

and therefore we reject H_0 in favour of H_1; this result confirms the rule-of-thumb test.

For β_1 we have:

$$H_0: \beta_1 = 0$$
$$H_1: \beta_1 \neq 0$$

Substituting into (11.98) gives

$$\hat{\beta}_1/s_{\hat{\beta}_1} \sim t_{10,0.025}$$

i.e.

$$\hat{\beta}_1/s_{\hat{\beta}_1} = 20.0344/4.3614 = 4.5936$$

but the critical value of t is, from the above,

$$t^*_{10,0.025} = \pm 2.228$$

and applying the decision-rule of (11.99) gives

$$4.5936 > 2.228$$

i.e.

$$\hat{\beta}_1/s_{\hat{\beta}_1} > t^*_{n-2,\alpha/2}$$

and therefore we reject H_0.

Thus the value of both the OLS estimators support the contention that each is significantly different from zero. Now let us confirm these results by constructing the 95 per cent confidence intervals for β_2 and β_1.

Substituting into (11.66) gives

$$P[3.0665 - (2.228 \times 0.1820) \leqslant \beta_2 \leqslant 3.0665 + (2.228 \times 0.1820)] = 0.95$$

which reduces to

$$P(2.6610 \leqslant \beta_2 \leqslant 3.4720) = 0.95$$

i.e. there is a probability of 0.95 that the interval from 2.6610 to 3.4720 will contain the true population parameter β_2. Notice that this interval does *not* contain zero, which implies that β_2 is different from zero; this confirms the result of the hypothesis test.

By substituting into (11.67) we have

$$P[20.0344 - (2.228 \times 4.3614) \leqslant \beta_1 \leqslant 20.0344 + (2.228 \times 4.3614)] = 0.95$$

which is

$$(10.3172 \leqslant \beta_1 \leqslant 29.7516) = 0.95$$

This interval does *not* contain zero, which implies that β_1 in the population is not zero, and this confirms the result of the hypothesis test.

Finally, let us assume that *a priori* information indicated that $\beta_2 = 2.0$. Let us use a hypothesis test to determine whether the sample regression results confirm this value. The hypotheses are now:

$$H_0: \beta_2 = 2.0$$
$$H_1: \beta_2 \neq 2.0$$

Under the null hypothesis (11.62) becomes

$$(\hat{\beta}_2 - \beta_2)/s_{\hat{\beta}_2} = (\hat{\beta}_2 - 2.0)/s_{\hat{\beta}_2} \sim t_{n-2}$$

and substituting the values of $\hat{\beta}_2$ and $s_{\hat{\beta}_2}$ gives

$$(3.0665 - 2.0)/0.1820 = 5.8600$$

The critical value of t is still ±2.228, so that applying the decision-rule of (11.97) gives

$$5.8600 \geqslant 2.228$$

i.e.

$$-t^*_{n-2,\alpha/2} \leqslant (\hat{\beta}_2 - \beta_2)/s_{\hat{\beta}_2} \leqslant t^*_{n-2,\alpha/2}$$

and accordingly we reject H_0. There *is* a statistically significant difference between $\beta_2 = 2.0$ and $\beta_2 = 3.0665$ at a 5 per cent significance level. In other words, it is unlikely that we would draw a sample yielding a $\hat{\beta}_2 = 3.0665$ from a population with a true value for β_2 of 2.0.

11.7 Prediction

Suppose we were interested in forecasting or predicting the level of inflation, i.e. the value of the retail price index Y_0 for some given level of money supply X_0. As we shall see, there are two kinds of prediction which we might make, i.e.

(a) Prediction of the *mean* value of Y_0 corresponding to a given X_0.
(b) Prediction of a *particular* value of Y_0 corresponding to a given X_0.

We can refer to the former as *mean-value prediction* and to the latter as *particular-value prediction*. Let us suppose that, given some value of the explanatory variable, say X_0, we wish to predict the corresponding *mean* value of the dependent variable, say Y_0. Now we know that Y_0 will be a random normally distributed variable because of the disturbance term u in the population regression equation, and for the same reason Y_0 will have a variance of σ^2.

Predicting the mean value of Y_0 means determining the value of $E(Y_0 \mid X_0)$. Now we know that

$$E(Y_0 \mid X_0) = \beta_1 + \beta_2 X_0 \tag{11.108}$$

and if we know β_1 and β_2 we could predict the mean value $E(Y_0 \mid X_0)$ *exactly* simply by substituting the appropriate value of X_0 into (11.108); this amounts to extending the true regression line towards X_0, as in Figure 11.19. Furthermore, if we wish to predict the value of a particular value of Y_0 corresponding to a given X_0 we can do so by forming the appropriate $100(1-\alpha)$ per cent confidence interval for Y_0 provided that, as well as β_1 and β_2, σ is also known, i.e.

$$[E(Y_0 \mid X_0) - z_{\alpha/2}\sigma \leqslant Y_0 \leqslant E(Y_0 \mid X_0) + z_{\alpha/2}\sigma] \quad (11.109)$$

which is, using (11.108),

$$(\beta_1 + \beta_2 X_0 - z_{\alpha/2}\sigma \leqslant Y_0 \leqslant \beta_1 + \beta_2 X_0 - z_{\alpha/2}\sigma) \quad (11.110)$$

Unfortunately in reality we do not know β_1, β_2 or σ^2, and we have to estimate these values.

Let us consider estimation of the mean first.

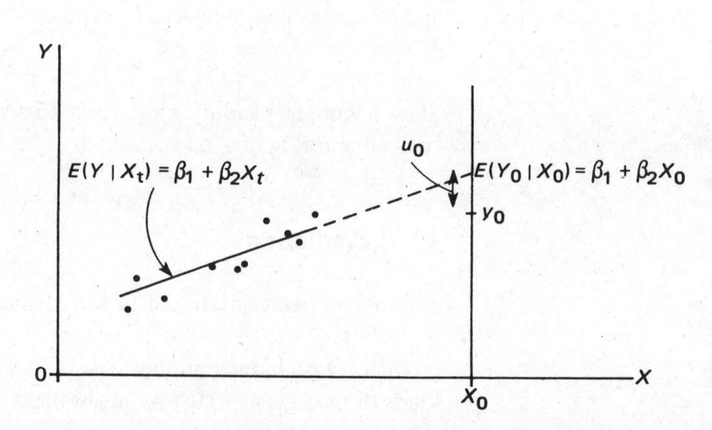

Figure 11.19
Prediction of Y_0 when β_1, β_2 and σ^2 are known

Mean-value prediction

The best point predictor of

$$E(\hat{Y}_0 \mid X_0) = \beta_1 + \beta_2 X_0$$

is

$$\hat{Y}_0 = \hat{\beta}_1 + \hat{\beta}_2 X_0 \quad (11.111)$$

(Note that $\hat{Y}_0$ is also a function of two normally distributed random variables and, it can be shown, is itself therefore normally distributed.)

So in practice instead of the population line in Figure 11.19 we might have that shown in Figure 11.20 (on which we have marked the value of $E(Y_0 \mid X_0)$, though this will of course be unknown).

Now $\hat{Y}_0$ is a random variable because it is a function of the two random variables $\hat{\beta}_1$ and $\hat{\beta}_2$ (whose values will not be known in advance of the sample being taken) and the mean value of $\hat{Y}_0$

Figure 11.20
Prediction of Y_0 when β_1, β_2 and σ^2 are unknown

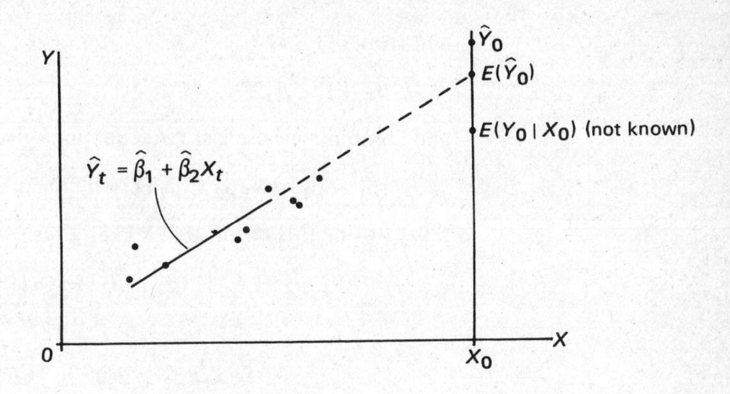

is therefore $E(\hat{Y}_0)$. Thus from (11.111)

$$E(\hat{Y}_0) = E(\hat{\beta}_1 + \hat{\beta}_2 X_0)$$
$$= \beta_1 + \beta_2 X_0$$
$$= E(Y_0 \mid X_0) \tag{11.112}$$

so that $\hat{Y}_0$ is an unbiased estimator of $E(Y_0 \mid X_0)$. In general, $\hat{Y}_0$ will not be the same as $E(Y_0 \mid X_0)$ because the estimated line and the true line will not coincide.

We can form confidence intervals for $\hat{Y}_0$ if we know its variance. From (11.111) we have

$$\text{var}(\hat{Y}_0) = \text{var}(\hat{\beta}_1 + \hat{\beta}_2 X_0) \tag{11.113}$$

where, as we have seen, $\hat{\beta}_1$ and $\hat{\beta}_2$ are random variables and X_0 is a constant. We want to make use of the following theorem:

If P and Q are random variables and a and b are constants and if a third variable Z is defined by

$$Z = aP + bQ$$

then

$$\text{var}(Z) = a^2 \text{ var}(P) + b^2 \text{ var}(Q) + 2ab \text{ cov}(P, Q) \tag{11.114}$$

Applying this result to (11.113) means that

$$\text{var}(\hat{Y}_0) = \text{var}(\hat{\beta}_1) + X_0^2 \text{ var}(\hat{\beta}_2) + 2X_0 \text{ cov}(\hat{\beta}_1, \hat{\beta}_2) \tag{11.115}$$

Since we already know $\text{var}(\hat{\beta}_1)$ and $\text{var}(\hat{\beta}_2)$ we need to find $\text{cov}(\hat{\beta}_1, \hat{\beta}_2)$. From equation (6.14) we have

$$\text{cov}(\hat{\beta}_1, \hat{\beta}_2) = E[\hat{\beta}_1 - E(\hat{\beta}_1)][\hat{\beta}_2 - E(\hat{\beta}_2)]$$
$$= E(\hat{\beta}_1 - \beta_1)(\hat{\beta}_2 - \beta_2) \tag{11.116}$$

but from (11.29)

$$\bar{Y} = \beta_1 + \beta_2 \bar{X} + \bar{u}$$

and from (11.24)

$$\hat{\beta}_1 = \bar{Y} - \hat{\beta}_2 \bar{X}$$

and combining these last two equations gives

$$(\hat{\beta}_1 - \beta_1) = -(\hat{\beta}_2 - \beta_2)\bar{X} + \bar{u}$$

Substituting this result in (11.116) gives

$$\begin{aligned}
\text{cov}(\hat{\beta}_1, \hat{\beta}_2) &= E\ [-(\hat{\beta}_2 - \beta_2)\bar{X} + \bar{u}](\hat{\beta}_2 - \beta_2) \\
&= E[-(\hat{\beta}_2 - \beta_2)^2 \bar{X} + (\hat{\beta}_2 - \beta_2)\bar{u}] \\
&= -\bar{X}E(\hat{\beta}_2 - \beta_2)^2 = -\bar{X}E[\hat{\beta}_2 - E(\beta_2)]^2 \\
&= -\bar{X}\ \text{var}(\hat{\beta}_2) \\
&= -\bar{X}\sigma^2/\Sigma x_t^2 \qquad (11.117)
\end{aligned}$$

which implies that when $\bar{X}$ is positive then $\hat{\beta}_1$ and $\hat{\beta}_2$ are inversely related.

We can now substitute (11.117) into (11.115) to give

$$\begin{aligned}
\text{var}(\hat{Y}_0) &= \sigma^2 \Sigma X_t^2/n\Sigma x_t^2 + X_0^2\sigma^2/\Sigma x_t^2 - 2X_0\bar{X}\sigma^2/\Sigma x_t^2 \\
&= (\sigma^2/n\Sigma x_i^2)(\Sigma X_t^2 + nX_0^2 - 2nX_0\bar{X})
\end{aligned}$$

but

$$\Sigma X_t^2 = \Sigma(x_t + \bar{X})^2 = \Sigma x_t^2 + n\bar{X}^2$$

Therefore,

$$\begin{aligned}
\text{var}(\hat{Y}_0) &= (\sigma^2/n\Sigma x_t^2)(\Sigma x_i^2 + n\bar{X}^2 - 2n\bar{X}X_0 + nX_0^2) \\
&= \sigma^2\,[1/n + (X_0 - \bar{X})^2/\Sigma x_t^2] \qquad (11.118)
\end{aligned}$$

We can thus describe the distribution of $\hat{Y}_0$ as

$$\hat{Y}_0 \sim N\{\beta_1 + \beta_2 X_0, \sigma^2\,[1/n + (X_0 - \bar{X})^2/\Sigma x_t^2]\} \qquad (11.119)$$

Equation (11.118) implies that the variance of $\hat{Y}_0$ gets larger, the further apart $\bar{X}$ and X_0 are. In a model in which variables are measured in successive time periods this implies that the further into the future we want to predict (i.e. the further away X_0 is from $\bar{X}$), the larger will var($\hat{Y}_0$) be.

We can determine the *standard error of* $\hat{Y}_0$ by taking the square root of (11.18), i.e.

$$\text{s.e.}(\hat{Y}_0) = \sqrt{\sigma^2\,[1/n + (X_0 - \bar{X})^2/\Sigma x_t^2]} \qquad (11.120)$$

We have seen that $\hat{Y}_0$ has a mean or expected value given by (11.111), i.e.

$$E(\hat{Y}_0) = E(Y_0 \mid X_0)$$

and has a standard error given by (11.120). Thus we can standardise $\hat{Y}_0$ in the usual way by subtracting its mean and dividing by its standard error, i.e.

$$\hat{Y}_0 - E(Y_0 \mid X_0)/\text{s.e.}(\hat{Y}_0) \sim N(0, 1) \tag{11.121}$$

One problem is that we need to know σ^2 to calculate s.e. $(\hat{Y}_0)$, and of course we do not. However, if n is large enough we can use $\hat{\sigma}^2$ as an estimator of σ^2 in (11.118) and (11.120) and use the standard-normal distribution to make probability statements and form confidence intervals about the mean of $\hat{Y}_0$, i.e. about $E(\hat{Y}_0)$ or $E(Y_0 \mid X_0)$. From (11.121) we can write

$$-z_{\alpha/2} \leqslant [\hat{Y}_0 - E(Y_0 \mid X_0)]/\text{s.e.}(\hat{Y}_0) \leqslant z_{\alpha/2} \tag{11.122}$$

and we can rearrange this in the same way as on previous occasions to give

$$P[\hat{Y}_0 - z_{\alpha/2}\,\text{s.e.}(\hat{Y}_0) \leqslant E(Y_0 \mid X_0) \leqslant \hat{Y}_0 + z_{\alpha/2}\,\text{s.e.}(Y_0)] = 1 - \alpha \tag{11.123}$$

where $\hat{Y}_0$ is calculated using (11.111), with $\hat{\sigma}^2$ replacing σ^2, and s.e. $(\hat{Y}_0)$ calculated in the same way, using (11.120).

In the event that n is small, we must use the t-distribution to make the appropriate probability statement, i.e.

$$P[\hat{Y}_0 - t_{n-2,\alpha/2}\,\text{s.e.}(\hat{Y}_0) \leqslant E(Y_0 \mid X_0) \leqslant \hat{Y}_0 + t_{n-2,\alpha/2}\,\text{s.e.}(\hat{Y}_0)] = 1 - \alpha \tag{11.124}$$

and form the corresponding confidence interval, which is therefore

$$[\hat{Y}_0 - t_{n-2,\alpha/2}\,\text{s.e.}(\hat{Y}_0) \leqslant E(Y_0 \mid X_0) \leqslant \hat{Y}_0 + t_{n-2,\alpha/2}\,\text{s.e.}(\hat{Y}_0)] \tag{11.125}$$

Particular-value prediction

We have just discussed prediction of the *mean* of Y_0, *i.e.* $E(Y_0 \mid X_0)$; we now want to consider prediction of a *particular* value of Y marked Y_0 on the graph of Figure 11.19. Now from the diagram it is clear that

$$Y_0 = \beta_1 + \beta_2 X_0 + u_0$$

The best point predictor of Y_0 is clearly $\hat{Y}_0$, where

$$\hat{Y}_0 = \hat{\beta}_1 + \hat{\beta}_2 X_0$$

In general, of course, $\hat{Y}_0$ will not be the same as Y_0: first, because the estimated line and the true line will not coincide, and second because Y_0 will not equal $E(Y_0 \mid X_0)$. That is, the value will not be on the true line because of the influence of u_0, and to form confidence intervals for Y_0 we need to determine its variance.

Now:

$$\begin{aligned}\operatorname{var}(\hat{Y}_0) &= E[\hat{Y}_0 - E(\hat{Y}_0)]^2 \\ &= E(\hat{Y}_0 - Y_0)^2 \\ &= E[(\hat{\beta}_1 + \hat{\beta}_2 X_0) - (\beta_1 + \beta_2 X_0 + u_0)]^2 \\ &= E[(\hat{\beta}_1 - \beta_1) + (\hat{\beta}_2 - \beta_2)X_0 - u_0]^2\end{aligned}$$

Squaring the right-hand side, taking expected values and using the fact that

$$E(\hat{\beta}_1 - \beta_1)u_0 = E(\hat{\beta}_2 - \beta_2)u_0 = 0$$

(since u_0 and the u_i which are used to form $\hat{\beta}_1$ and $\hat{\beta}_2$ are independent) gives:

$$\begin{aligned}\operatorname{var}(\hat{Y}_0) &= E[(\hat{\beta}_1 - \beta_1)^2 + 2X_0(\hat{\beta}_1 - \beta_1)(\hat{\beta}_2 - \beta_2) \\ &\quad + [X_0(\hat{\beta}_2 - \beta_2)]^2 + u_0^2] \\ &= E(\hat{\beta}_1 - \beta_1)^2 + 2X_0 E(\hat{\beta}_1 - \beta_1)(\hat{\beta}_2 - \beta_2) \\ &\quad + X_0^2 E(\hat{\beta}_2 - \beta_2) + E(u_0^2) \\ &= \operatorname{var}(\hat{\beta}_1) + 2X_0 \operatorname{cov}(\hat{\beta}_1, \hat{\beta}_2) + X_0^2 \operatorname{var}(\hat{\beta}_2) + \sigma^2\end{aligned} \tag{11.126}$$

using (11.116).

Following the same procedure as in the mean-value case we can express (11.26) as

$$\operatorname{var}(\hat{Y}_0) = \sigma^2 [1 + 1/n + (X_0 - \bar{X})^2/\Sigma x_t^2] \tag{11.127}$$

which differs from the corresponding expression in the mean-value case only by the additional term '1' inside the square bracket, i.e. the variance of $\hat{Y}_0$ in (11.127) is larger than in (11.118). The standard error of $\hat{Y}_0$ is found by square rooting (11.128), i.e.

$$\text{s.e.}(\hat{Y}_0) = \sqrt{\sigma^2 [1 + 1/n + (X_0 - \bar{X})^2/\Sigma x_t^2]} \tag{11.128}$$

which means that the appropriate probability statements and confidence intervals, derived in the usual way, are

$$P[\hat{Y}_0 - t_{n-2,\alpha/2}\text{s.e.}(\hat{Y}_0) \leqslant Y_0 \leqslant \hat{Y}_0 + t_{n-2,\alpha/2}\text{s.e.}(\hat{Y}_0)] = 1 - \alpha \tag{11.129}$$

where again $\hat{Y}_0$ is calculated using (11.111), with σ^2 replaced by $\hat{\sigma}^2$ and s.e.$(\hat{Y}_0)$ calculated similarly, using (11.128). The corresponding confidence interval for a *particular* value Y_0 is thus

$$[\hat{Y}_0 - t_{n-2,\alpha/2}\text{s.e.}(\hat{Y}_0) \leqslant Y_0 \leqslant \hat{Y}_0 + t_{n-2,\alpha/2}\text{s.e.}(\hat{Y}_0)] \tag{11.130}$$

The fact that, as we have already noted, var($\hat{Y}_0$) increases as X_0 moves away from $\bar{X}$ (indeed, as the square of this difference) means that the confidence intervals for the particular-value predictor of Y_0 are narrowest at the point where $X_0 = \bar{X}$ and widen at an increasing rate for other values of X_0. The fact that the variance for the mean predictor is smaller than the variance for the particular predictor means that the confidence 'envelope' for the former is narrower than the latter. These points are illustrated in Figure 11.21.

Figure 11.21
Confidence interval 'envelope' for mean- and particular-value predictors

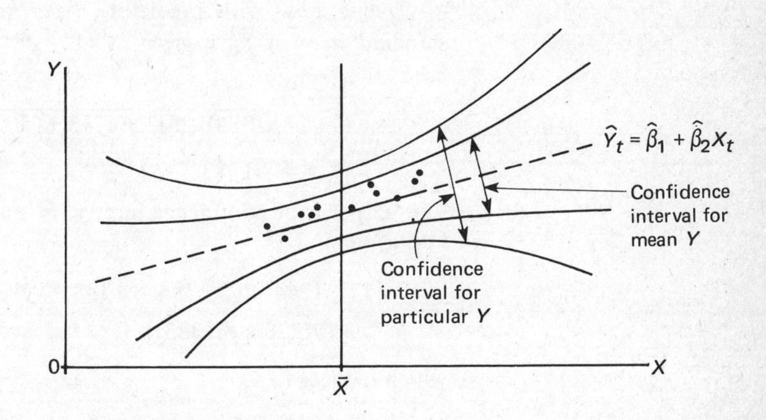

Example

Let us illustrate the above discussion by constructing 95 per cent confidence intervals for the mean value $E(Y_0 \mid X_0)$ and the particular value Y_0 of the price index when money supply reaches £50 billion, i.e. $X_0 = 50$.

Starting with the prediction of $E(Y_0 \mid X_0)$, the point predictor of $E(Y_0 \mid X_0)$ is $\hat{Y}_0$, where

$$\hat{Y}_0 = \hat{\beta}_1 + \hat{\beta}_2 X_0$$

$$= 20.0344 + (3.0665 \times 50)$$

$$= 173.3594$$

To construct confidence intervals we must first calculate s.e.($\hat{Y}_0$) by substituting into (11.120) with σ^2 replaced by $\hat{\sigma}^2$ = 37.0525. Therefore,

$$\text{s.e.}(\hat{Y}_0) = \sqrt{37.0525[1/12 + (50 - 21.933)^2/1118.674]}$$

$$= 5.4017$$

The appropriate value of t is

$$t^*_{n-2,\alpha/2} = t^*_{10,0.025} = \pm 2.228$$

as before. Therefore, substituting into (11.125) gives the 95 per cent confidence interval as

$$(173.3594 - 2.228 \times 5.4017 \leqslant E(Y_0 \mid X_0) \leqslant 173.3594 + 2.228 \times 5.4017)$$

which reduces to

$$(161.3244 \leqslant E(Y_0 \mid X_0) \leqslant 185.3944)$$

In other words, we can be 95 per cent confident that the interval from 161.3244 to 185.3944 will contain the true *mean* value.

Dealing now with predicting a *particular* value of Y_0, the standard error of $\hat{Y}_0$ is given by (11.128) with σ^2 again replaced by $\hat{\sigma}^2$, i.e.

$$\text{s.e.}(\hat{Y}_0) = \sqrt{37.0525[1 + 1/12 + (50 - 21.933)^2/1118.674]} = 8.1383$$

The 95 per cent confidence interval is found by substituting into (11.30), i.e.

$$[173.3244 - (2.228 \times 8.1383) \leqslant Y_0 \leqslant 173.3244 + (2.228 \times 8.1383)]$$

which reduces to

$$155.1923 \leqslant Y_0 \leqslant 191.4565$$

In other words, we can be 95 per cent certain that the interval from 155.1923 to 191.4565 will contain the true *particular* value of the price index.

We indicate both results in Figure 11.22. As we expected, the interval for the predictor of the individual $Y_0 \mid X_0$ is wider than that for the predictor of $E(Y_0 \mid X_0)$ because of the extra error introduced by the disturbance term u_0.

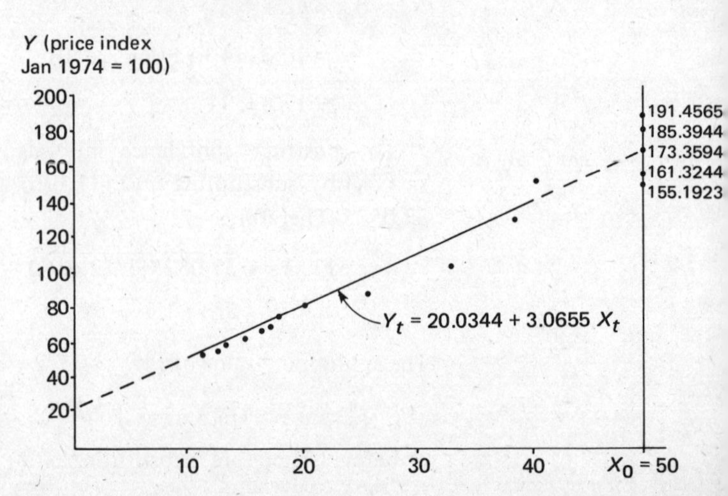

Figure 11.22 95 per cent confidence intervals for $E(Y_0 \mid X_0)$ and $Y_0 \mid X_0$

11.8 Summary

At the beginning of this chapter we introduced the concept of bivariate statistics and in particular the idea of the linear model, distinguishing as we did so between deterministic and stochastic models. We saw that the former have little real use in the economic or social sciences because of their unrealistic nature. We then moved on to introduce the notion of correlation and saw how the correlation coefficient was a useful if limited measure of the strength of the association between two variables.

The shortcomings of correlation when it comes to examining the relationship between two variables, its inability to predict or to say anything about the *nature* of the relationship, caused us to seek a more useful tool. This we found in the method of simple regression analysis: 'simple' because the method we discussed was limited to a model with only two variables, one dependent variable and one independent or explanatory variable. Regression analysis is an extremely powerful and versatile method for examining the nature of the relationship between two variables and for prediction and finds application in a great many different disciplines within as well as beyond the social sciences.

We have seen how to estimate optimally the parameters of a hypothesised linear model, how to test the statistical significance of these parameters individually, and how to measure the goodness of fit of the model as a whole. The estimation procedure we described, known as ordinary least squares, leads to estimators which, in certain circumstances, have all the desirable properties, i.e. they are best linear unbiased or BLU estimators. But these properties only obtain if the set of basic assumptions constituting what is known as the classical normal linear regression model are all satisfied. These assumptions relate principally to the distributional properties of the disturbance term but also encompass the non-stochastic nature of the explanatory variable.

Two questions arise out of the above discussion. First, do theoretical considerations imply that one explanatory variable is adequate in terms of accounting for the behaviour of the dependent variable? Second, what happens to the OLS estimators if one or more of the basic assumptions is violated? We shall answer these questions in the next chapter.

EXERCISES

11.1 Derive equations (11.23) and (11.24) from (11.21) and (11.22).

11.2 With the help of the rules governing the summation operator, equation (4.4), prove that

$$\Sigma x_i y_i / \Sigma x_i^2 = (n\Sigma X_i Y_i - \Sigma X_i \Sigma Y_i)/[n\Sigma X_i^2 - (\Sigma X_i)^2]$$

11.3 Prove that the OLS estimator $\widehat{\beta}_1$ is an unbiased estimator of β_1.

11.4 In Chapter 1 it was reported that a firm of stockbrokers and the Treasury had predicted decreases of 15 per cent and 10 per cent respectively in the number of cigarettes to be sold in 1981 following a 29 per cent increase in prices. Describe how you would go about determining who was more likely to be correct (assuming that the actual figure of sales in 1981 is not available).

11.5 The data in Table 11.9 refer to the number of industrial stoppages (i.e. strikes) per annum, measured in thousands, and the corresponding level of unemployment measured as a percentage of the work-force, for the years 1968 to 1978.

(a) Graph the data in the form of a scatter diagram and comment on what this reveals about the strength and direction of the relationship between the two variables.

(b) Calculate the correlation coefficient for these two variables and comment on its value.

Table 11.9

	1968	1969	1970	1971	1972	1973	1974	1975	1976	1977	1978
Stoppages	2.38	3.12	3.91	2.23	2.50	2.87	2.92	2.28	2.02	2.70	2.47
Unemployment	2.5	2.4	2.6	3.5	3.8	2.7	2.6	4.1	5.7	6.2	6.1

11.6 Using the data in Table 11.9 regress the number of stoppages on the unemployment level using OLS, and so obtain the estimated regression equation. Calculate $\widehat{\sigma}$ and the standard errors of the OLS estimators and hence examine the significance of the parameters of the model using an appropriate significance level. Confirm your findings by constructing the corresponding confidence intervals. Interpret your results. Calculate the error sum of squares and the value of R^2 and comment on the goodness of fit of the estimated regression line to the data.

11.7 Suppose that for the data in Table 11.9 unemployment was expected to be 8 per cent in 1980. Calculate the point predictor for the number of stoppages and the 95 per cent confidence intervals for the mean and particular values. Display your results as in Figure 11.22.

11.8 On a certain day of the week the price of a particular foodstuff, denoted by X and measured in pence per lb, is observed to differ in fifteen major food outlets chosen at random throughout the country. The quantity sold during the day at each of these outlets, denoted by Y and measured in lb, is also observed. Data on X and Y are given in Table 11.10. Plot a scatter diagram of the data and estimate the consumption function using OLS. Draw this estimated regression line on the scatter diagram. Examine the significance and goodness of fit of this line and calculate the price elasticity of demand at the point of means.

Table 11.10

Y lb	28	38	36	42	51	31	49	64	62	52	44	26	38	32	46
X p/lb	22	19	18	17	17	21	16	12	14	15	17	24	20	18	16

11.9 Using the data in Table 11.10 regress Y on X using the method of semi-averages, i.e. calculate the average X and Y values for the first seven observations and for the last seven values and join these two points together to give the semi-average regression line. Compare the efficiency of the two procedures (e.g. by comparing the values of R^2, SSE and $\hat{\sigma}$ in each case.

12 Multiple Regression

OBJECTIVES

After reading this chapter and working through the examples students should understand the meaning of the following terms:

multiple regression model
partial regression parameter
adjusted coefficient of determination $\bar{R}^2$
F-distribution
partial correlation
partial correlation coefficient
zero-order partial correlation coefficient
first-order partial correlation coefficient
SPSS
homoskedasticity
heteroskedasticity
generalised least squares
autocorrelation
autocorrelation coefficient
first-order autoregression
Durbin–Watson statistic
multicollinearity

Students should be able to:

discuss the estimation of the parameters in a multiple regression model
prepare a multiple regression problem for solution by the SPSS computer program
interpret the output from an SPSS program
discuss the consequences for the OLS estimation procedure of a breakdown in any one of the basic assumptions

---

12.1 Introduction

In the last chapter we considered the simple regression model:

$$Y_i = \beta_1 + \beta_2 X_i + u_i \tag{12.1}$$

in which the single explanatory variable X is held to account for *all* of the deterministic variation in the dependent variable Y. The remaining stochastic variation is taken care of by the random disturbance term u. The parameters β_1 and β_2 are unknown, and in Chapter 11 we discussed their estimation using ordinary least squares (OLS) and the application of hypothesis testing to examine their statistical significance.

In reality it would be unusual if the behaviour of any economic or socio-economic variable could be adequately explained by a single explanatory variable. For example, the demand for some good will be a function not only of the price of that good but also of the price of competing goods and of income, and so on. A household's expenditure will be a function not only of its income but also of its size (i.e. number of persons), its age characteristics, its social class, and so on. To try to explain the variation in a dependent variable with only one explanatory variable is thus unnecessarily limiting and fundamentally (in most practical cases) theoretically unsound. For these reasons we want to consider models which contain several explanatory variables. Such models are a logical extension of the simple regression models considered in the preceding chapter.

The principles underlying the estimation and testing of *multiple regression* models are exactly the same as in the simple regression case except that, because of the existence of more than one explanatory variable, the conventional arithmetic approach quickly becomes very unwieldy as the number of variables increases. For this reason many textbooks consider multiple regression via a more advanced form of mathematics known as matrix algebra. This, however, is beyond the scope of an introductory text such as this but a good starting-point for those interested is to be found in Johnston (1972).

There is a second reason why we can safely eschew the mathematical complexities associated with the estimation and testing of multiple regression models. This is because most readers of this book, i.e. those who wish to become actively involved in the analysis of multiple regression models, will have access to computers, almost all of which will have a pre-programmed facility for performing multiple regression programs (for example, SPSS, TSP, HASH, etc.). It is therefore simply not necessary (and for models with more than two explanatory variables it is almost practically impossible) for readers to be able to estimate and test multiple regression models manually, as this can all be done quite satisfactorily by machines. However, it is important to understand the underlying principles governing the multiple regression models, e.g. the basic assumptions, the meaning of significance tests, interpreting the results correctly, and so on.

In this chapter, therefore, we shall concentrate our discussion on the broader features of multiple regression rather than on the arithmetic minutiae.

The simplest type of multiple regression model, i.e. one containing *two* explanatory variables, is conventionally written as

$$Y_i = \beta_1 + \beta_2 X_{2i} + \beta_3 X_{3i} + u_i \tag{12.2}$$

Notice the slight change in notation between (12.2) and (12.1). The two explanatory variables are distinguished by subscripting them as X_2 and X_3 respectively. (In this regard we could think of there being another variable X_1 associated with the constant term β_1 which always assumes the value 1.0 and thus need not be written into the equation explicitly. In this sense there are three explanatory variables, X_1 to X_3, except that X_1 is always equal to 1.0.)

In this chapter we first of all examine the general specification of the multiple regression model, together with the associated set of basic assumptions which (apart from slight modification) are identical to the basic assumptions of the simple regression model set out in assumptions (A.1) to (A.6) of section 11.4. We then consider the desirable properties of the ordinary least squares estimators in the context of multiple regression and look at estimation in a very broad and general sense. We then spend a little time considering the meaning of the parameters in a multiple regression model. Following this we consider the consequences of a breakdown in each of the basic assumptions of the classical normal linear regression model, how such a breakdown might be detected and what remedial measures (if any) are available. Of necessity we shall be able to consider only the barest bones of such a discussion in view of the complexity and volume of existing material in this area.

12.2 The basic model and assumptions

The multiple regression model in its general form is specified by the following population equation:

$$Y_i = \beta_1 + \beta_2 X_{2i} + \beta_3 X_{3i} + \ldots + \beta_k X_{ki} + u_i \tag{12.3}$$

In other words, variation in the dependent variable Y_i is due to a deterministic element, explained by the several variables X_2 to X_k, and a stochastic element due to the disturbance term u_i. The expected value of (12.3), i.e.

$$E(Y_i) = \beta_1 + \beta_2 X_{2i} + \beta_3 X_{3i} + \ldots + \beta_k X_{ki} \tag{12.4}$$

represents not a straight line in two-dimensional space, as was the case with the simple regression model, but a figure in k-dimensional space, known as a *hyperplane*. Our task is to estimate the multiple regression equation of (12.3) by estimating the values of the individual parameters β_1, β_2 to β_k. We shall examine this problem in a moment. First of all let us consider more closely the meaning of the parameters in a multiple regression model.

Each β_j still represents the change in the value of the dependent variable Y for a unit change in the variable X_j, i.e. if X_j increases by one unit, then Y will increase by β_j units (if β_j is positive) or will decrease by β_j units (if it is negative). In this sense the β_js can still be interpreted as 'slope' parameters. However, β_j represents the change in Y_i for unit change in X_j with all the other explanatory variables *held constant*. For this reason the parameters β_2, $\beta_3, \ldots, \beta_k$ are sometimes known as the *partial regression parameters*.

It will perhaps help the reader to see why this interpretation of the role of the regression parameters can be made if we assume for a moment that all the explanatory variables *except* X_j, say, are held constant. Since they are constant they could all be lumped into the intercept term. Thus the model of (12.4) reduces to

$$E(Y_i) = \text{a constant term} + \beta_j X_j \tag{12.5}$$

This has the same form as the simple regression model, and β_j can be interpreted as a slope parameter only if all the other explanatory variables can be considered to be constants. An example might help to clarify the above discussion.

Suppose we had a model which purported to relate the annual sales of coal, SCOAL, to the price of coal, PCOAL, and the price of a substitute fuel, say electricity, PELEC, i.e.

$$\text{SCOAL}_t = \beta_1 + \beta_2 \text{PCOAL}_t + \beta_3 \text{PELEC}_t + u_t$$

In terms of the multiple regression analysis if PCOAL increased by one unit, SCOAL would decrease by β_2 units (assuming a negative value for β_2) and we would be making the assumption that PELEC remains constant. In practice we might well expect that if the price of coal increases so also will the price of electricity since coal is a raw material used in electricity generation. Thus when PCOAL increases so also will PELEC, and this increase in PELEC will also cause a change in coal sales. The effect on coal sales of the increase in electricity prices is an *indirect* effect brought about initially by the increase in the price of coal. Thus there are two influences on coal sales: a direct effect due to the change in PCOAL and an *indirect* effect caused by the influence of PCOAL on PELEC and the subsequent influence of PELEC on SCOAL.

In interpreting the regression parameter β_2 we are measuring only the *direct* effect of changes in PCOAL on SCOAL and assuming implicitly that PELEC remains constant and any indirect effect of it on SCOAL is absent.

This leads us to an important point. If we had started with the simple regression model:

$$\text{SCOAL}_t = \beta_1 + \beta_2 \text{PCOAL}_t$$

the value obtained for β_2 in this model would in general be different from the value obtained for β_2 in the multiple regression model when PELEC is included. This is because in simple regression we make no allowance for the influence of other variables not present, i.e. we do not have to assume they remain constant, and β_2 in this case measures the *total* change in SCOAL, i.e. *both* the direct change and any possible indirect change due to variables not included in the model. In the multiple regression model, as we have seen, β_2 measures only the *direct* effect. This of course enables us to measure the specific contribution of each individual variable. Thus, to sum up, when we estimate (12.3) and obtain numeric values for each of the regression estimators these will each represent the expected change in Y_i for a unit change in the appropriate X_j on the assumption that all the other explanatory variables are being held constant.

The basic assumptions of the classical normal linear regression model in the context of *multiple regression* are as follows:

the model is linear in form (A.1)

u_i is normally distributed (A.2)

the mean of u_i is zero (A.3)

the variance of u_i is a constant, i.e. $E(u_i^2) = \sigma^2$ (A.4)

the u_is are independent, i.e. $E(u_i u_j) = 0 \quad i \neq j]$ (A.5)

the X_js are non-stochastic variables with fixed values (A.6)

no exact linear relationship exists between any of the explanatory variables (A.7)

Assumptions (A.1) to (A.5) are identical to the basic assumptions of the simple regression model set out in section 11.4. Assumption (A.6) simply extends the non-stochastic assumption to include *all* the explanatory variables in the model. Assumption (A.7) is new and states that there should be no exact linear relationship (i.e. no correlation) between the explanatory variables or any sub-set of them. For example, explanatory variable relationships such as

$$X_{3i} = X_{5i}$$

or

$$X_{4i} = -X_{2i} + 1.5X_{3i}$$

etc., would violate assumption (A.7). Breakdown of this assumption brings about what is known as *multicollinearity* in the model, which can have serious consequences when it comes to estimation of the model. We shall deal with this problem more fully in section 12.14.

The complete set of assumptions (A.1) to (A.7) constitutes what is known as the classical normal linear regression model in the context of multiple regression.

12.3 Estimation in the multiple regression model

Following the notation of section 11.3 the estimated or sample regression equation is written

$$Y_i = \hat{\beta}_1 + \hat{\beta}_2 X_{2i} + \hat{\beta}_3 X_{3i} + \ldots + \hat{\beta}_k X_{ki} + \hat{u}_i \tag{12.6}$$

while the corresponding estimated regression 'line' is written

$$\hat{Y}_i = \hat{\beta}_1 + \hat{\beta}_2 X_{2i} + \hat{\beta}_3 X_{3i} + \ldots + \hat{\beta}_k X_{ki} \tag{12.7}$$

The values of the estimators $\hat{\beta}_1$ to $\hat{\beta}_k$ are determined such that the value of the sum of the squares of the estimated disturbance terms (or the error sum of squares), i.e.

$$\Sigma \hat{u}_i^2$$

is minimised. In the multiple regression model, however,

$$\begin{aligned}\Sigma \hat{u}_i^2 &= \Sigma (Y_i - \hat{Y}_i)^2 \\ &= \Sigma [Y_i - (\hat{\beta}_1 + \hat{\beta}_2 X_{2i} + \ldots + \hat{\beta}_k X_{ki})]^2 \end{aligned} \tag{12.8}$$

and the increased complexity of this last expression compared with (11.20) causes there to be a corresponding increase in the complexity of the arithmetic. For this reason, as we stated at the beginning of this chapter, we do not intend to give here a detailed algebraic exposition of estimation in the multiple regression case. It is *just* feasible to do this for the least complicated multiple regression model, i.e. one with only two explanatory variables, but even so the resulting equations are very unwieldy and would occupy several rather uninviting pages of formulae. (Any reader wishing to perform a two-variable regression by hand is referred to, for example, Kmenta (1971, pp. 353–62) or Koutsoyiannis (1977, pp. 119–23).) What we propose to do here is look at estimation from a broader, less-detailed viewpoint. Consider estimation of the parameter β_j. It can be shown that the value of $\hat{\beta}_j$ is given by regressing $\hat{w}_i$ on $\hat{v}_i$, i.e.

$$\hat{\beta}_j = \Sigma \hat{v}_i \hat{w}_i / \Sigma \hat{v}_i^2 \tag{12.9}$$

where the $\hat{v}_i$ are the error terms obtained by regressing X_j on all the other explanatory variables, and the $\hat{w}_i$ are the error terms obtained by regressing Y on all the explanatory variables *except* X_j. That is, we first of all estimate the regression equation:

$$\begin{aligned} X_{ji} = \beta_1 + \beta_2 X_{2i} + \ldots + \beta_{j-1,i} + \beta_{j+1} X_{j+1,i} + \ldots \\ + \beta_k X_{ki} + v_i \end{aligned}$$

and obtain

$$\begin{aligned} \hat{X}_{ji} = \hat{\beta}_1 + \hat{\beta}_2 X_i + \ldots + \hat{\beta}_{j-1} X_{j-1,i} + \hat{\beta}_{j+1} X_{j+1,i} + \ldots \\ + \hat{\beta}_k X_{ki} \end{aligned}$$

from which

$$\hat{v}_i = X_{ji} - \hat{X}_{ji} \tag{12.10}$$

We then estimate the regression equation

$$Y_i = \beta_1 + \beta_2 X_{2i} + \ldots + \beta_{j-1} X_{j-1,i} + \beta_{j+1} X_{j+1,i} + \ldots$$
$$+ \beta_k X_{ki} + w_i$$

and obtain

$$\widehat{Y}_i = \widehat{\beta}_1 + \widehat{\beta}_2 X_{2i} + \ldots + \widehat{\beta}_{j-1} X_{j-1,i} + \ldots \beta_{j+1}, X_{j+1} \ldots + \widehat{\beta}_k X_{ki}$$

from which

$$\widehat{w}_i = Y_i - \widehat{Y}_i \tag{12.11}$$

Finally, we take the appropriate sums of squares from equations (12.10) and (12.11) and combine them, according to (12.9), to obtain the value of $\widehat{\beta}_j$.

This process may appear more complicated than it is. The important point is that we calculate each $\widehat{\beta}_j$ using the error terms $\widehat{v}_i$ which are derived from that part of X_j which is not 'explained' by the other explanatory variables; it is that part of X_j which is unique to X_j and which is not *common* to any of the other explanatory variables.

Thus, as far as the $\widehat{\beta}_j$ are concerned, any change in X_j is a change *only* in X_j. There are no *indirect* changes in any of the other variables. In other words, we can think of the other explanatory variables as being held constant.

It might help to clarify the discussion if we consider the following example. Suppose we specify the following model to explain the sales of butter in the various shops of a supermarket chain during a certain week of the year:

$$\text{BUT}_i = \beta_1 + \beta_2 \text{PBUT}_i + \beta_3 \text{PMAR}_i + \beta_4 \text{PBRE}_i + u_i$$

where

BUT_i = sales of butter in shop i

PBUT_i = price of butter in shop i

PMAR_i = price of margarine in shop i

PBRE_i = price of bread in shop i

Let us suppose we want to estimate β_3, the parameter associated with the variable PMAR.

First, we regress PMAR on all the other explanatory variables, i.e.

$$\text{PMAR}_i = \beta_1 + \beta_2 \text{PBUT} + \beta_4 \text{PBRE}_i + v_i$$

and obtain the estimated relationship

$$\widehat{\text{PMAR}}_i = \widehat{\beta}_1 + \widehat{\beta}_2 \text{PBUT}_i + \widehat{\beta}_4 \text{PBRE}_i$$

from which

$$\hat{v}_i = \text{PMAR}_i - \widehat{\text{PMAR}}_i$$

The residual term $\hat{v}_i$ is that part of the variation in PMAR which is not explained by the remaining explanatory variables PBUT and PBRE; it is variation unique to PMAR.

Second, we regress the dependent variable BUT on to all the other explanatory variables *except* PMAR, i.e.

$$\text{BUT}_i = \beta_1 + \beta_2\text{PBUT}_i + \beta_4\text{PBRE}_i + w_i$$

and obtain the estimated relationship

$$\widehat{\text{BUT}}_i = \hat{\beta}_1 + \hat{\beta}_2\text{PBUT}_i + \hat{\beta}_4\text{PBRE}$$

from which

$$\hat{w}_i = \text{BUT}_i - \widehat{\text{BUT}}_i$$

The residual term $\hat{w}_i$ is that part of the variation in BUT which is not explained by the explanatory variables PBUT and PBRE; it is variation unique to BUT.

Thus in the residual $\hat{v}_i$ we have the variation in PMAR which is not shared by the other two explanatory variables and in $\hat{w}_i$ we have the variation in BUT which is not shared by the other two explanatory variables. The influence of the variables PBUT and PBRE has thus been excluded. This is illustrated schematically in (a) and (b) respectively of Figure 12.1.

Having discounted or controlled for the influences of the other two explanatory variables we can regress $\hat{w}_i$ on to $\hat{v}_i$, i.e.

$$\hat{w}_i = \epsilon_1 + \epsilon_2\hat{v}_i$$

to find out how much of the variation in $\hat{w}_i$ is explained by $\hat{v}_i$ (i.e. how much of the variation in BUT which is *not* explained by PBUT and PBRE *is* explained by PMAR). Thus we have

$$\hat{\epsilon}_2 = \Sigma\hat{v}_i\hat{w}_i/\Sigma v_i^2 \qquad (12.12)$$

The rationale underlying estimation of the parameter β_3, say,

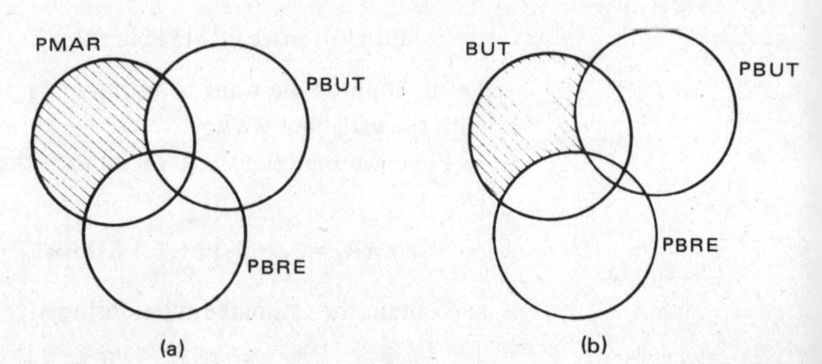

Figure 12.1 Unique variation in PMAR and BUT shown shaded

in the general multiple regression model is to calculate $\widehat{\beta}_3$ using (12.12), in other words by setting $\widehat{\beta}_3 = \epsilon_2$. In this way $\widehat{\beta}_3$ estimates the *direct* influence of PMAR on Y since the process has eliminated any indirect influences due to PBUT and PBRE. Because we do not measure indirect influences the parameters in a multiple regression model are often referred to as *partial regression parameters*.

This is the general principle underlying estimation in the multiple regression model. As we noted above, however, the algebra needed to carry out this procedure is in practice extremely cumbersome. Fortunately, however, such calculations are invariably carried out with the aid of suitable computer programs (one example of which we shall discuss shortly).

Properties and distribution of the OLS estimators

We can use the basic assumptions, as we did in section (11.5), to show that the OLS estimators are best linear unbiased estimators (and the same as the maximum likelihood estimators); they all have the same desirable small sample and asymptotic properties as in the simple regression model.

The distribution of the estimators is normal (or approximately so by the *central limit theorem*) as in the simple regression model and because of their unbiased property they have a mean equal to the true value of the population parameter, i.e.

$$E(\widehat{\beta}_j) = \beta_j \qquad (j = 1 \text{ to } k) \tag{12.13}$$

Without the use of matrix algebra it is not feasible to present the formulae for the variance of the OLS estimators in the general k-variable model because of the complexity of the algebra. Thus we do not give here formulae equivalent to (11.34) to (11.37) of the simple regression model. At this stage let us simply state that the variance of the $\widehat{\beta}_j$ is a function of the variance of the disturbance term and of $\widehat{v}_i$, the error terms from the regression of X_j on the other explanatory variables, given by (12.10), i.e.

$$\text{var}(\beta_j) = \sigma^2/\Sigma\widehat{v}_i^2 \tag{12.14}$$

and the corresponding standard error is thus given by the square root of (12.14), i.e.

$$s_{\beta_j} = \sqrt{\sigma^2/\Sigma\widehat{v}_i^2} \tag{12.15}$$

We can thus summarise the sampling distribution of any $\widehat{\beta}_j$ as follows:

$$\widehat{\beta}_j \sim N(\beta_j, \sigma^2/\Sigma\widehat{v}_i^2) \tag{12.16}$$

The true value of σ^2 will of course not generally be known, and to use (12.14) or (12.15) we have to determine a suitable estimator for it. We can show that an unbiased estimator of σ^2 in a model with k explanatory variables is given by

$$\hat{\sigma}^2 = \frac{1}{n-k} \Sigma \hat{u}_i^2 \tag{12.17}$$

where $\Sigma \hat{u}_i^2$ is given by (12.8). Notice that the number of degrees of freedom has increased from $n - 2$ to $n - k$ between the simple and multiple regression models. In the simple regression case we had two restrictions to satisfy $\Sigma \hat{u}_i = 0$ and $\Sigma \hat{u}_i X_{2i} = 0$ (from the two equations preceding the normal equations (11.21) and (11.22)). In the three-variable model we have three such restrictions to satisfy $\Sigma \hat{u}_i = 0$, $\Sigma \hat{u}_i X_{2i} = 0$ and $\Sigma \hat{u}_i X_{3i} = 0$. In the k-variable model we have k such restrictions and hence the number of degrees of freedom is reduced to $n - k$.

Interval estimation

In the preceding discussion we have shown that the $\hat{\beta}_j$s can be used as *point estimators* of the unknown β_j. As far as *interval estimation* is concerned, no new ideas are involved beyond those expounded in the simple regression model. The general expression for the probability statements about the β_j and corresponding interval estimators parallel those for the simple model given in equations (11.66) to (11.69). Thus in the multiple regression model, for the parameter β_j we have the following probability statement:

$$P(\hat{\beta}_j - t_{n-k,\alpha/2} s_{\hat{\beta}_j} \leqslant \beta_j \leqslant \hat{\beta}_j + t_{n-k,\alpha/2} s_{\hat{\beta}_j}) = 1 - \alpha \tag{12.18}$$

and the corresponding interval estimate is

$$(\hat{\beta}_j - t_{n-k,\alpha/2} s_{\hat{\beta}_j} \leqslant \beta_j \leqslant \hat{\beta}_j + t_{n-k,\alpha/2} s_{\hat{\beta}_j}) \tag{12.19}$$

where $s_{\hat{\beta}_j}$ is calculated using (12.15).

Goodness of fit

The treatment of goodness of fit again parallels that in the simple regression case. The coefficient of determination R^2 is computed as a measure of how well the fitted hyperplane fits the observations in k-dimensional space. Recall that one expression for goodness of fit is given by equation (11.72), i.e.

$$R^2 = 1 - \Sigma \hat{u}_i / \Sigma y_i^2 \tag{12.20}$$

and the same expression applies in the multiple regression case, except of course that $\Sigma \hat{u}_i^2$ is calculated using (12.8) rather than (11.54). The worst fit is when

$$\Sigma \hat{u}_i^2 = \Sigma y_i^2$$

i.e.

$$R^2 = 0$$

and the best fit is when

$$\Sigma \hat{u}_i^2 = 0$$

in which case

$$R^2 = 1$$

R^2 can be alternatively expressed (see equation (11.87)) as

$$R^2 = \Sigma \hat{y}_i^2 / \Sigma y_i^2 \tag{12.21}$$

and using the ideas of section 11.6 we can again decompose SST, the 'total sum of squares' or Σy_i^2, as follows:

$$\text{SST} = \text{SSR} + \text{SSE} \tag{12.22}$$

i.e.

$$\Sigma y_i^2 = \Sigma \hat{y}_i^2 + \Sigma \hat{u}_i^2 \tag{12.23}$$

In the multiple regression case, of course, the term $\Sigma \hat{y}_i^2$, the regression sum of squares, is no longer calculated using (11.89) but by a similar, more complex expression involving *all* the explanatory variables. R^2 can still be expressed as the ratio of the regression sum of squares to the total sum of squares, i.e.

$$R^2 = \text{SSR}/\text{SST} \tag{12.24}$$

In practice all computer regression programs print out the value of R^2, and hand computation is not necessary. As promised in the previous chapter, there is a test of the significance of R^2 but we will postpone this for a little while.

Before we move on to consider hypothesis testing of the multiple regression model there is one final aspect of goodness of fit to consider.

$\bar{R}^2$, the adjusted coefficient of determination

We have defined a measure of goodness of fit R^2 in three ways, equations (11.86), (11.87) and (11.91). Consider (11.91), i.e.

$$R^2 = 1 - \Sigma \hat{u}_i^2 / \Sigma y_i^2$$

Suppose we start with a model with two explanatory variables, X_2 and X_3, and then add a third explanatory variable X_4. The value of SST, Σy_i^2, will remain unchanged (in no way does it depend upon the number of explanatory variables); however, we can see intuitively that the addition of a further explanatory variable will in general cause a decrease in the value of $\Sigma \hat{u}_i^2$ (at least it will not increase). With fixed Σy_i^2 the value of R^2 is almost bound to show an increase (even if only by a small amount) between the two- and three-variable models, and this would seem to indicate an improvement in the goodness of fit or explanatory power of the model. Suppose that the true value of the parameter β_4 is actually zero (i.e. X_4 does not influence Y), In general, because of sampling, the estimated value of β_4 will not be exactly zero and its non-zero arithmetic value will contribute to a decrease in $\Sigma \hat{u}_i^2$ and a consequent increase in R^2. However, this is not a 'real' increase in the explanatory power of the model because β_4 is statistically not significant; its origin lies in the fact that $\hat{\beta}_4$ is a sample result and we would not expect it to be exactly equal to β_4 (i.e. to equal zero).

Thus we are faced with the question whether any increase in R^2 between one model and another (with one or more *extra* explanatory variables in it) is a significant increase in the goodness of fit or explanatory power of the model or whether it is due more to the sort of sampling phenomena described above. With a single extra explanatory variable this amounts to asking whether the extra variable is significant or not. With several extra explanatory variables it amounts to asking whether the several extra variables taken together are significant.

The conclusion to this discussion is that we cannot simply compare R^2 values for models with different numbers of explanatory variables and conclude that because the R^2 of a larger model is bigger than that for a smaller model the former is a better model. We need to take into account the effect on $\Sigma \hat{u}_i^2$ of the extra variables and adjust R^2 accordingly. A measure has been devised to deal with this problem, known as the *adjusted coefficient of determination*. It is calculated using

$$\bar{R}^2 = 1 - (1 - R^2)[(n - 1)/(n - k)] \qquad (12.25)$$

Any increase in the value of R^2 is adjusted by the presence of k in this expression. $\bar{R}^2$ in fact measures the proportion of estimated *variance* which is explained.

Examination of (12.25) will reveal that

1. if $k = 1$, then $R^2 = \bar{R}^2$
2. if $k > 1$, then $R^2 > \bar{R}^2$ (unless $R^2 = 1$, in which case $\bar{R}^2 = \bar{R}^2$)

In general if n is large compared with k then R^2 and $\bar{R}^2$ will not differ very much, but if n is small and k is large the two measures may differ considerably (indeed $\bar{R}^2$ might take a negative value).

The value of $\bar{R}^2$ is usually included in the print-out from regression programs and we can use it as an aid in choosing between models. For example, suppose we had the following regression results from two models:

Model 1		*Model 2*	
with X_2 X_3 X_4		with X_2 X_3 X_4 X_5	
$R^2 = 0.91$	$\bar{R}^2 = 0.87$	$R^2 = 0.92$	$\bar{R}^2 = 0.85$

The addition of X_5 to the model improves the goodness of fit, as measured by R^2 (increasing it from 0.91 to 0.92) and would suggest that model 2 is better because of its greater explanatory power. However, the value of $\bar{R}^2$ actually drops, suggesting that the addition of X_5 does not significantly increase the explanatory power of the model.

We may ask: 'How can $\bar{R}^2$ fall when an extra explanatory is added to the model?' To see why, let us rewrite (12.25) as

$$\bar{R}^2 = 1 - \frac{\text{SSE}/(n-k)}{\text{SST}/(n-1)} \tag{12.26}$$

using (11.88) and (11.91), which is equal to

$$\bar{R}^2 = 1 - \frac{\Sigma\hat{u}_i^2/(n-k)}{\Sigma y_i^2/(n-1)} \tag{12.27}$$

When an extra variable is added $\bar{R}^2$ can only decrease if the second term on the right-hand side increases.

The term Σy_i^2 is fixed, so for the second term on the right-hand side to increase $\Sigma\hat{u}_i^2/(n-k)$ must increase. However, as we saw above, $\Sigma\hat{u}_i^2$ will usually decrease when an extra variable is added. This leaves $(n-k)$, which will of course decrease when k increases by 1, and if this decrease relative to the decrease in $\Sigma\hat{u}_i^2$ is big enough then the term $\Sigma\hat{u}_i^2/(n-k)$ will as a whole increase and so $\bar{R}^2$ will decrease.

$\bar{R}^2$ will only *increase* when $\hat{\beta}_j$, the estimated value of the parameter, is greater than its estimated standard error, $s_{\hat{\beta}_j}$, i.e. when t is greater than 1. None the less it is still possible for the value of $\bar{R}^2$ to increase when an extra variable is added to the model even though the estimated value of its parameter is not significantly different from zero. Although these circumstances are not usual, it indicates that the values of $\bar{R}^2$ and R^2 should be treated with some caution. The safest way of testing whether an extra variable adds significantly to the explanatory power of a model is to apply the conventional t-test to its estimated parameter and ascertain whether it is significantly different from zero.

12.4 Hypothesis testing

Tests of the individual parameters

To test the significance of any one of the estimators the appropriate test-statistic takes the same form as (11.62), i.e.

$$\frac{\hat{\beta}_j - \beta_j}{s_{\hat{\beta}_j}} \sim t_{n-k,\alpha/2} \tag{12.28}$$

This is the same as in the simple regression case except that the number of degrees of freedom is now $n - k$ (as explained above).

In the usual event that we wish to test whether the β_j is significantly different from zero, then the competing hypotheses are:

$$H_0: \beta_j = 0$$

$$H_1: \beta_j \neq 0$$

In this case the test-statistic of (12.28) reduces to

$$\frac{\hat{\beta}_j}{s_{\hat{\beta}_j}} \sim t_{n-k,\alpha/2} \tag{12.29}$$

Following the usual hypothesis test procedure, the numeric value of the test-statistic (12.29) is compared with the theoretical value from the t-tables at the chosen level of significance and the decision to accept or not to accept the null hypothesis is taken.

There may be occasions when we wish to test not that some β_j is significantly different from zero but that it is equal to some given value, say $\beta_j = b$. In these circumstances the competing hypotheses are:

$$H_0: \beta_j = b$$

$$H_1: \beta_j \neq b$$

and the appropriate test-statistic, from (12.28), is

$$\frac{\hat{\beta}_j - b}{s_{\hat{\beta}_j}} \sim t_{n-k,\alpha/2} \tag{12.30}$$

Testing the over-all significance of the estimated regression – the F-test

Let us ask the following question. Suppose we have tested the individual significance of the k parameters in a regression equation and found none of them to be significantly different from zero. Now, is there anything more that we can do? The answer, perhaps surprisingly, is yes.

It is quite possible that while none of the explanatory variables significantly influences Y *individually*, it is conceivable that some of the explanatory variables *acting together* have a joint influence on Y which is statistically significant. If this were the case, we would obviously be interested in knowing about it. How, then, can we detect such a situation?

Essentially, this question relates to whether or not, despite the zero significance of the *individual* explanatory variables, the proportion of the total variation in Y_i 'explained' by the *whole* model is a significant proportion. Thus the above question relates to the size of R^2, and effectively asks: 'Is R^2 big enough to indicate that the proportion of SST explained by all the explanatory variables in the model is a statistically significant propertion?'

In other words, the test of the over-all significance of an estimated regression can be viewed as a test of the significance of R^2. Without further ado we can state that there is a distribution called the F-distribution which enables us to perform a test of the over-all significance of the regression model, i.e. a test of the significance of the value of R^2. It can be shown that

$$\frac{\text{SSE}/(k-1)}{\text{SSR}/(n-k)} \sim F_{k-1,n-k} \tag{12.31}$$

or equivalently and perhaps more conveniently

$$\frac{R^2/(k-1)}{(1-R^2)/(n-k)} \sim F_{k-1,n-k} \tag{12.32}$$

where F has $k - 1$ degrees of freedom in the numerator and $n - k$ in the denominator.

We shall describe the test procedure in a moment, but first a few words about the F-distribution. The F-distribution is similar to the chi-distribution in being skewed to the right, but as $k - 1$ and $n - k$ increase in size the distribution approaches the normal distribution. This is shown in Figure 12.2.

The F-distribution has a mean equal to $(n - k)/(n - k - 2)$. Finally, the square of a t-distributed random variable with k

Figure 12.2
The F-distribution

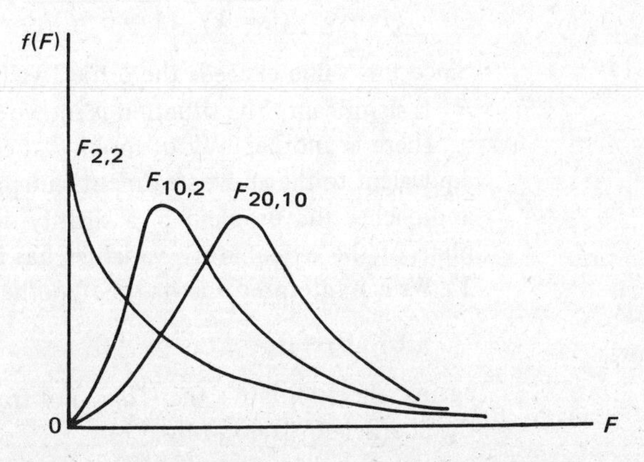

degrees of freedom has an F-distribution with 1 and k degrees of freedom, i.e.

$$t_k^2 = F_{1,k} \tag{12.33}$$

Let us now see how the critical value of F is obtained. Tables of values for the F-distribution are usually given at 5 per cent and 1 per cent significance levels, and such tables are given in the appendix. We reproduce a small portion of such a table in Table 12.1.

Table 12.1 Portion of a 5 per cent F-table

Values of $F_{0.05,k-1,n-k}$					
		$k-1$ = degrees of freedom in the numerator			
		1	2	3	4
$n-k$ = degrees of freedom in the denominator	1	161	200	216	225
	2	18.5	19.0	19.2	19.2
	3	10.1	9.55	9.28	9.12
	4	7.71	6.94	6.59	
	5	6.61	5.79	5.41	
	6	5.99	5.14	4.76	
	7	5.99			

Suppose a regression analysis is based on ten observations in the following model:

$$Y_i = \beta_1 + \beta_2 X_{2i} + \beta_3 X_{3i} + \beta_4 X_{4i} + u_i$$

and yields a value of $R^2 = 0.76$. We wish to test whether the model explains a significant proportion of the variation in Y.

The number of degrees of freedom in the numerator is $k - 1 = 3$, and in the denominator is $n - k = 6$. We thus look in column 3 and row 6 and find the critical value of F is 4.76.

The value of the left-hand side of (12.32) is

$$\frac{R^2/(k-1)}{(1-R^2)/(n-k)} = \frac{0.76/3}{(1-0.76)/6} = 6.33$$

Since this value exceeds the critical value of 4.76, we accept that R^2 is significant. The situation is shown in Figure 12.3.

There is another way of looking at the F-test which, although equivalent to the above treatment (and producing the same result), approaches the problem in a slightly different way. This is that none of the explanatory variables has a significant influence on Y_i. We can interpret this via the hypothesis test:

$$H_0: \beta_2 = \beta_3 = \ldots = \beta_k = 0 \tag{12.34}$$

against the alternative that H_0 is not true, i.e. that at least one of the coefficients is non-zero.

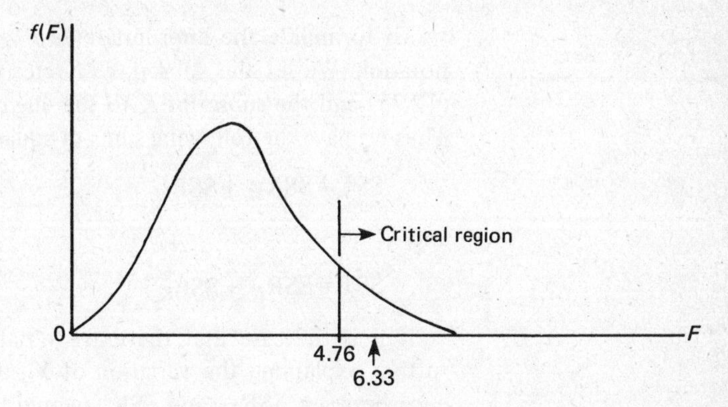

Figure 12.3 Showing critical region of F-distribution

If H_0 is found to be true, then the variation of Y from one observation to the next is not influenced by changes in any of the explanatory variables but is due to purely random factors. In this case a value of R^2 different from zero is due entirely to these random factors.

We should note that in general if a t-test applied separately to each parameter yields even one significant parameter, then the F-test will also give a significant result, and there is no point in performing the test. However, as we pointed out above, if the t-test reveals no significant parameters it is still worth while to proceed to the F-test to examine the *joint* influence of the variables. (For a further discussion on this see Maddala (1977, pp. 122–4).)

Testing the significance of additional variables

A somewhat different, perhaps more formal approach to the addition of extra variables to a regression model is possible. As we have seen, one possible drawback with the $\bar{R}^2$ approach is that even if $\bar{R}^2$ increases when an additional variable is included in the model, is the increase in $\bar{R}^2$ (and conversely the decrease in SSE) *statistically* meaningful? To answer this question in the general situation of adding a number of explanatory variables, suppose we start with the model in m variables, i.e.

$$Y_i = \beta_1 + \beta_2 X_{2i} + \ldots + \beta_m X_{mi} + u_i \tag{12.35}$$

We then add further variables to obtain the model

$$Y_i = \beta_1 + \beta_2 X_{2i} + \ldots + \beta_m X_{mi} + \beta_{m+1} X_{m+1,i} + \ldots + \beta_k X_{k1} + w_i \tag{12.36}$$

We can test the statistical significance of the extra variables by testing the null hypothesis

$$H_0\colon \beta_{m+1} = \beta_{m+2} = \ldots = \beta_k = 0 \tag{12.37}$$

against H_1: that H_0 is not true.

To formulate the appropriate test we must introduce a new notation. We let the subscript M refer to the original model of (12.35) and the subscript K to the augmented model of (12.36). Thus we have the following sums-of-squares relationships:

$$\text{SST} = \text{SSR}_M + \text{SSE}_M$$

and

$$\text{SST} = \text{SSR}_K + \text{SSE}_K$$

If it is the case that the extra variables are not pertinent in further explaining the variation of Y_i, then any observed difference between SSR_M and SSR_K would be due merely to chance (sampling) factors.

If the null hypothesis of (12.37) is true, then it can be shown that the appropriate test-statistic is

$$\frac{(\text{SSR}_K - \text{SSR}_M)/(K - M)}{\text{SSE}_K/(n - K)} \sim F_{K-M,\, n-K} \tag{12.38}$$

which can be written as

$$\frac{(R_K^2 - R_M^2)/(K - M)}{(1 - R_K^2)/(n - K)} \sim F_{K-M,\, n-K} \tag{12.39}$$

The procedure is thus one of performing two regressions, the first on the original model with M variables, the second on the augmented model with K variables. From the first regression we would obtain SSR_M, SSE_M and R_M^2 and from the second SSR_K, SSE_K and R_K^2. By substituting into either (12.38) or (12.39) we can perform the desired test.

12.5 Partial correlation

In the multiple regression model

$$Y_i = \beta_1 + \beta_2 X_{2i} + \beta_3 X_{3i} + \ldots + \beta_k X_{ki} + u_i$$

we have seen that each parameter β_j measures the effect on Y of a unit change in the associated variable X_j, on the assumption that any indirect effect via the other explanatory variables is discounted. That is, we take it that if X_j changes by one unit, all the other explanatory variables are held constant. For this reason, as we have previously noted, the β_j are sometimes referred to as the *partial* regression parameters.

It is perhaps natural to wonder if we can extend this notion of partial regression to include a measure of the 'direct' correlation between the variables.

We can always calculate the simple correlation between Y and any *one* of the explanatory variables. For example, the simple correlation between Y and X_3 could be determined thus:

$$r_{YX_3} = \frac{\Sigma x_3 y}{\sqrt{\Sigma x_3^2 \Sigma y^2}}$$

And in general

$$r_{YX_j} = \frac{\Sigma x_j y}{\sqrt{\Sigma x_j^2 \Sigma y^2}} \tag{12.40}$$

would denote the simple correlation between Y and the explanatory variable X_j.

One question which may now occur to us is this: 'Does r_{YX_3} measure the true strength of the relationship between Y and X_3?' What if the apparent correlation between Y and X_3 is due to the fact that both of them are being influenced by some other variable, say X_4? In other words, if X_4 is a strong influence on Y as well as on X_3, then changes in X_4 will cause both Y and X_3 to change simultaneously. If we are unaware of the presence or influence of X_4, we shall observe only the fact that Y and X_3 appear to be strongly related in their movements. We might conclude, not without reason in the circumstances, that it is Y and X_3 that are influencing each other, i.e. that it is these two variables that are correlated.

What we would like to know is whether there is any 'pure' relationship between Y and X_3 after any common influence which X_4 may have on both of them has been removed or netted out. We illustrate the position schematically in Figure 12.4.

In Figure 12.4 the region of influence or activity of each variable is represented by a circle. In Figure 12.4(a) the three variables are completely uncorrelated and the value of any correlation coefficient between any pair of them would be zero. In Figure 12.4(b) X_4 is correlated with both Y and X_3 but the latter two are not correlated. However, if we were to calculate a correlation coefficient between Y and X_3 it would not be zero, because

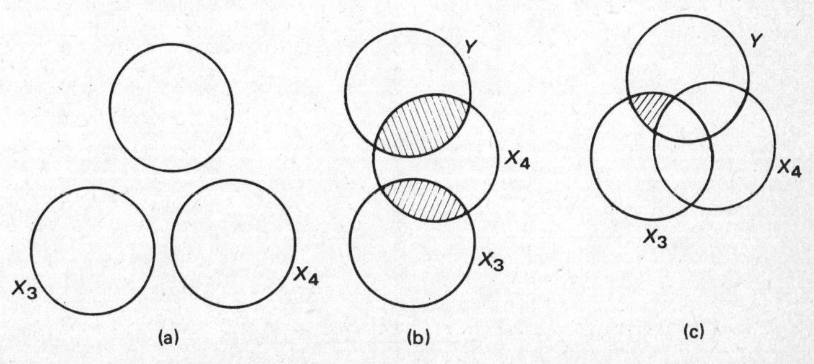

Figure 12.4 Schematic representation of correlation between variables

of the common influence of X_4, as explained above. Such a result would obviously be misleading. In Figure 12.4(c) the most likely situation is illustrated: X_3 is correlated with Y, and X_4 is correlated with Y, while X_3 and X_4 are themselves correlated. We are interested in the correlation between Y and X_3 after any common influence due to X_4 has been removed. In other words, we are concerned with the shaded portion in Figure 12.4(c).

Such a correlation coefficient exists and is known as the *partial correlation coefficient*. It measures the 'pure' relationship between two variables after the common influence of all other variables has been netted out, i.e. with all other variables being held constant or controlled for.

How, then, are we to derive the value of the partial correlation coefficients (because, of course, there will be as many of them as there are different pairwise combinations from among all the variables)? One way would be as follows. Consider the data shown in Table 12.2 which relate to a model with two explanatory variables. Suppose we wished to determine the partial correlation between Y and X_2 with the effect of X_3 being discounted, i.e. with X_3 being held constant. One way would be to consider only those pairs of values of Y and X_2 for all of which X_3 had the same value. Thus, using the data in Table 12.2 as an example we would choose observations 2, 3, 6, 8 and 10 (for all of which $X_3 = 3$). If we calculated the simple correlation coefficient between Y and X_2 for these five observations, we would effectively be holding X_3 constant, i.e. we would obtain the partial correlation coefficient between Y and X_2.

The obvious drawback of this approach is that in practice we have few enough observations in total to start off with. The chances of finding enough of them for which the third variable has the same (i.e. constant) value, *if any*, are slender, and of

Table 12.2

Observation number	Y_i	X_{2i}	X_{3i}
1	1	4	5
2	2	6	3
3	2	5	3
4	3	7	4
5	3	8	5
6	4	9	3
7	4	9	2
8	5	10	3
9	6	11	6
10	8	12	3

course the problem becomes much worse when we have more than two explanatory variables.

However, things are not as black as they would at first appear, for there is an alternative approach which is much more promising.

We need first of all to establish a new notation. Let us define

$$r_{12.34\ldots k} \tag{12.41}$$

as the partial correlation between variable 1 and variable 2, with all the variables to the right of the first dot, i.e. variable 3, variable 4 up to variable k, being held constant.

In regression, variable 1 corresponds to the independent variable Y, variable 2 to X_2, etc., and the number of variables to the right of the dot represents the 'order' of the partial correlation coefficient. Thus $r_{12.3}$ would be a first-order partial correlation coefficient between Y and X_2 with X_3 held constant; $r_{12.34}$ would be a second-order partial correlation coefficient between Y and X_2 with X_3 and X_4 held constant; and so on.

For a model with two explanatory variables there would be three first-order coefficients, defined as follows: $r_{12.3}$ = the partial correlation between Y and X_2 with X_3 held constant; $r_{13.2}$ = the partial correlation between Y and X_3 with X_2 held constant; $r_{23.1}$ = the partial correlation between X_2 and X_3 and Y held constant. (There can of course be no higher-order coefficients than first-order in a model with only two explanatory variables.) For this model we could also quite easily calculate the three simple correlation coefficients, i.e.

r_{12} = the simple correlation between Y and X_2

r_{13} = the simple correlation between Y and X_3

r_{23} = the simple correlation between X_2 and X_3

The 'simple' correlation coefficient is usually referred to as the *zero*-order coefficient, a nomenclature we shall use from now on.

Now although, as we have seen, it is difficult to calculate the partial correlation coefficients of first-order and above, it is a straightforward matter to calculate the zero-order correlations, i.e. we simply apply equation (12.40). Fortunately, there is a direct relationship between the first-order correlation coefficients and the zero-order coefficients. These are

$$r_{12.3} = \frac{r_{12} - r_{13}r_{23}}{\sqrt{(1 - r_{13}^2)(1 - r_{23}^2)}} \tag{12.42}$$

$$r_{13.2} = \frac{r_{13} - r_{12}r_{23}}{\sqrt{(1 - r_{12}^2)(1 - r_{23}^2)}} \tag{12.43}$$

$$r_{23.1} = \frac{r_{23} - r_{12}r_{13}}{\sqrt{(1 - r_{12}^2)(1 - r_{13}^2)}} \tag{12.44}$$

Thus having calculated the zero-order correlation coefficients it is a simple matter to determine the value of any of the first-order correlations.

Furthermore, similar (albeit somewhat more cumbersome) expressions can be derived which relate first-order coefficients to second-order coefficients, second to third, and so on. In other words, there is a hierarchy of partial correlation coefficients, all of which can be calculated by starting with the zero-order correlations and proceeding step by step to the higher orders.

In the next section we are going to put some flesh on to the above bare bones of the multiple regression model by considering a numeric example whose regression analysis has been performed by a computer. This will also give us the opportunity to describe and discuss a typical regression program output.

12.6 SPSS: a computer program for regression analysis

We have already stated that in practice estimation of multiple regression models is undertaken using one of the recognised computer programs which have been written especially for this task. There are a large number of such programs, e.g. HASH, TSP, and they all differ in their complexity and range of analysis.

In this section we want to discuss one of the most popular programs, the Statistical Package for the Social Sciences (or SPSS) developed in Chicago. This is a package of statistical programs, offering an analysis of the many different statistical features embodied in a set of data. One part of this package is a powerful regression analysis with a variety of options in terms of the desired output. In the first part of this section we list the program together with a commentary on the various steps and a discussion of the options available. We then give an example of the program applied to a two-variable multiple regression model which seeks to examine the causes of inflation in terms of money supply, as measured by M3, and an earnings variable, and use this to describe and discuss the output from the SPSS program.

A complete and detailed description of this program is to be found in the second edition of the SPSS manual, Nie *et al.* (1975) together with a volume of more recent procedures and facilities (Hull and Nie, 1979).

The program

The SPSS program takes the following general form:

Run cards
Data definition cards
Task definition cards for first task
Data cards

Task definition cards for any subsequent tasks
Finish card

The run card identifies the particular program to the computer and initiates the program.

The data definition cards provide the computer with information on the structure and contents of the model to be analysed, e.g. the number of variables (with their chosen names), the number of observations, and how the data are to be arranged on the cards or on the tape. The task definition cards tell the computer which type of analysis is to be performed on the data. In SPSS there are several available types of analysis, any one or more of which may be used, e.g. descriptive statistics, bivariate correlation (Pearson correlation, rank correlation and scatter diagrams), partial correlation, multiple regression and a number of more advanced programs.

This section is followed by the data. If further analysis is required, the appropriate task definition card(s) is included here. The program is terminated with a finish card.

The following program is designed to perform, first, a regression analysis, and then a partial correlation analysis, on the price–inflation model:

$$\text{PRICES}_t = \beta_1 + \beta_2\text{M3} + \beta_3\text{EARNINGS}_t$$

The data consist of seventeen annual observations, from 1963 to 1979, on UK data, as shown in Table 12.3.

Table 12.3 Annual data for price–inflation model

Year	Prices	M3	Earnings
1963	63.2	11.500	55.19
1964	66.2	12.155	57.31
1965	69.3	13.083	59.96
1966	71.8	13.555	61.96
1967	73.6	14.895	65.61
1968	77.9	16.092	70.30
1969	81.6	16.596	74.35
1970	88.0	18.175	84.41
1971	95.9	20.541	94.83
1972	103.3	26.245	108.10
1973	116.4	33.478	121.40
1974	136.1	37.698	157.10
1975	170.0	40.571	197.00
1976	195.6	45.129	220.20
1977	219.4	49.566	232.90
1978	237.8	56.931	274.80
1979	278.8	63.957	323.40

Prices = the index of retail prices at December (31 July 1972 = 100).
M3 = money stock M3 at year-end, £ billion.
Earnings = index of basic weekly wage rates of manual workers at December (31 July 1972 = 100).

The program is as follows:

```
RUN NAME           PRICE AGAINST MONEY SUPPLY AND
                   EARNINGS
VARIABLE LIST      PRICES, M3, EARNINGS
INPUT MEDIUM       CARD
INPUT FORMAT       FREEFIELD
NO. OF CASES       17
PARTIAL CORR       PRICES WITH M3 BY EARNINGS (1)/
                   PRICES WITH EARNINGS BY M3 (1)
OPTIONS            3
STATISTICS         1, 2, 4, 5, 8, 9
READ INPUT DATA
11.500    55.19    63.2
12.155    57.31    66.2
  .         .        .
  .         .        .
  .         .        .
63.957   323.4    278.8
REGRESSION         VARIABLES=PRICES, M3, EARNINGS/
                   REGRESSION=PRICES WITH M3,
                   EARNINGS/RESIDUALS
OPTIONS            7, 8, 11, 15, 21
STATISTICS         1, 2, 4, 5
FINISH
```

As can be seen, the program is short and simple. It starts with the run card RUN NAME, which is written in columns 1 to 8 (the first 15 spaces are usually known as the *control field*; and from column 16 on, they are known as the *specification field*), and we can identify our program by giving it any name of up to sixty-four spaces long. The run name used here is PRICES AGAINST MONEY SUPPLY AND EARNINGS. There then follow four data definition cards, the first of which is the VARIABLE LIST card. This instructs the computer in the number of variables in the model and what they have been named. In this case we have three variables named PRICES, M3 and EARNINGS. The second data definition card is the INPUT MEDIUM card, which in column 16 on tells the computer that the program will be input on card. The INPUT FORMAT card describes the way in which the data have been entered on to each card (and how, therefore, they are to be read off). The order in which the data are entered on to the card must be the same as on the VARIABLE LIST card, i.e. on each card the first value should be the value of PRICES, followed by M3 and then EARNINGS. There will be one card for each year in forward chronological order. The word FREEFIELD in columns 16 to 24 of the INPUT FORMAT card tells the computer that the data on the cards are not in any pre-set or specific columns but that each value follows the preceding values separated from it by a space. The alternative format, known as FIXED, is more versatile and efficient, though not quite so easy to understand (full details can be found in the SPSS manual).

The last data definition card, NO. OF CASES, simply indicates how many observations there are. The first analysis to be performed on the data is a calculation of the partial correlation coefficients. The first task definition card is thus PARTIAL CORR punched in columns 1 to 12 of the control field. The computer is told exactly which partials to calculate and print out in columns 16 on. PRICES WITH M3 BY EARNINGS(1) means calculate the partial correlation coefficients between PRICES, which is of course the dependent variable, and M3, holding EARNINGS constant; this is $r_{\text{PRICES, M3. EARNINGS}}$. The (1) after EARNINGS indicates the order of the partial coefficient (first in this case) and must be specified. On the same card, separated from the first instruction by a 'slash' (or /), follows the instruction to calculate the coefficient between PRICES and EARNINGS holding M3 constant (also first order).

There then follows the OPTIONS card which indicates which of several options available in the partial correlation program are to be called upon. From these we have specified '3', which instructs the computer to apply a two-tailed significance test to the value calculated for each coefficient. In the absence of a call for this option the computer would automatically calculate a one-tailed test. The next card indicates which of the available statistical analyses we want performed: STATISTICS 1 requests the machine to calculate and print the zero-order correlation coefficients as well as the first-order values; STATISTICS 2 causes the mean and standard deviation of each variable on the PARTIAL CORR card to be printed; and so on (full details of all the available OPTIONS and STATISTICS are given in the SPSS manual).

The next card is the READ INPUT DATA card. This tells the machine that the first set of task definition cards is ended and that the data cards follow immediately. The first data card contains the data for 1963 on PRICES, M3 and EARNINGS. After the data cards the task definition cards for the second task appear. The REGRESSION card is followed, in columns 16 on, by a list of the variables upon which the regression analysis is to be performed. In this case VARIABLES=PRICES, M3, EARNINGS indicates that the regression is to be carried out on these three variables. Following the '/' the computer is told, on the next card, that the regression will have PRICES as the dependent variable and M3 and EARNINGS as the explanatory variables, signified by REGRESSION=PRICES WITH M3, EARNINGS in column 16 on.

Since the standard regression program in the SPSS package does not automatically print out information on the error or residual terms it is necessary to call in the sub-program **RESIDUALS**. This program will calculate and print the values of all the error terms, i.e. the $\hat{u}$s, as well as the estimated values of the dependent variable. By choosing the appropriate **OPTIONS** and **STATISTICS** we can, among other things, plot the error terms, which (as we shall see later) may be of some use in detecting the breakdown of the basic assumptions.

Output from the program

The above program fed into a computer produced the output shown between pages 322 and 326. In the actual output the graph of the $\hat{u}$ values is plotted to the right and alongside the table of residual values. The output from the SPSS regression program is literal and easy to follow; however, we shall go through it to discuss some features in the output and make one or two additional points.

```
VOGELBACK COMPUTING CENTER
NORTHWESTERN UNIVERSITY

S P S S – – STATISTICAL PACKAGE FOR THE SOCIAL SCIENCES

VERSION 7.0 – – JUNE 27 1977

1000 VARIABLE VERSION FOR THE UMRCC CDC 7600 - - - - - - SEPTEMBER 1979
LCM VERSION – MAXIMUM AVAILABLE WORKSPACE = 65536 (200000 B) WORDS

RUN NAME          PRICE AGAINST MONEY SUPPLY AND EARNINGS
VARIABLE LIST     PRICES, M3, EARNINGS
INPUT FORMAT      FREEFIELD
NO. OF CASES      17
PARTIAL CORR      PRICES WITH M3 BY EARNINGS(1)/PRICES WITH EARNINGS BY M3(1)
OPTIONS           3
STATISTICS        1,2,4,5,8,9
READ INPUT DATA

034500 SCM, 001000 LCM NEEDED FOR PARTIAL CORR

OPTION – 1
IGNORE MISSING VALUE INDICATORS

OPTION – 3
TWO-TAILED TEST OF SIGNIFICANCE

- - - - - - - - - - - - - - - - - - - - - - - - - - - - - - - - - - - - - - - - - - - - - -
- - -
```

PRICE AGAINST MONEY SUPPLY AND EARNINGS

FILE NONAME (CREATION DATE = 21/05/81)

VARIABLE	MEAN	STANDARD DEV	CASES
PRICES	126.1706	68.5168	17
M3	28.8672	17.1393	17
EARNINGS	132.8718	85.6117	17
PRICES	126.1706	68.5168	17
EARNINGS	132.8718	85.6117	17
M3	28.8672	17.1393	17

PRICE AGAINST MONEY SUPPLY AND EARNINGS

---------------PARTIAL CORRELATION

ZERO ORDER PARTIALS

	PRICES	M3	EARNINGS
PRICES	1.0000	.9859	.9977
	(0)	(15)	(15)
	S = .001	S = .001	S = .001
M3	.9850	1.0000	.9903
	(15)	(0)	(15)
	S = .001	S = .001	S = .001
EARNINGS	.9977	.9903	1.0000
	(15)	(15)	(0)
	S = .001	S = .001	S = .001

(COEFFICIENT / (D.F.) / SIGNIFICANCE)

PRICE AGAINST MONEY SUPPLY AND EARNINGS

---------------PARTIAL CORRELATION

CONTROLLING FOR.. EARNINGS

	M3
PRICES	−.2101
	(14)
	S= .435

(COEFFICIENT / (D.F.) / SIGNIFICANCE)

```
---------------PARTIAL CORRELATION

CONTROLLING FOR..   M3

                    EARNINGS

PRICES                 .9163
                    (    14)
                    S=  .001

(COEFFICIENT / (D.F.) / SIGNIFICANCE)   (A VALUE OF 99.000 IS PRINTED)

---------------------------------------------------------------
---

PRICE AGAINST MONEY SUPPLY AND EARNINGS

REGRESSION      VARIABLES=PRICES,M3,EARNINGS/
                REGRESSION=PRICES WITH M3,EARNINGS/RESIDUALS
OPTIONS         7,8,11,15,21
STATISTICS      1,2,4,5
FINISH

046100 SCM, 001000 LCM NEEDED FOR REGRESSION

OPTION – 1
IGNORE MISSING VALUE INDICATORS

OPTION – 7
SUPPRESS SUMMARY TABLE OUTPUT

OPTION – 8
WRITE CORRELATION MATRIX ON BCDOUT

OPTION – 11
WRITE RESIDUALS ON BCDOUT

OPTION – 15
WRITE MEANS AND STDS ON BCDOUT

OPTION – 21
PRINT T, NOT F, IN THE STEP-BY-STEP OUTPUT

---------------------------------------------------------------
---

PRICE AGAINST MONEY SUPPLY AND EARNINGS

* * * * * * * * * * * * * * * * * * * * * * * * * * MULTIPLE REGR
* * * * * *

DEPENDENT VARIABLE..    PRICES

MEAN RESPONSE       126.17059       STD. DEV.       68.51684

VARIABLE(S) ENTERED ON STEP NUMBER    1..    M3
                                             EARNINGS
```

```
MULTIPLE R          .99776              F
SIGNIFICANCE
R SQUARE            .99552          1557.13115
        .000
ADJUSTED R SQUARE          .99489
STD DEVIATION             4.90011

---------------------- VARIABLES IN THE EQUATION -----------
------------

VARIABLE          B          STD ERROR B        T                BETA
     F
------
                                              --------    --------

                                              SIGNIFICANCE  ELASTICITY

IFICANCE

M3             -.41294174     .51362189      -.80398002       -.1032960
                                                  .435          -.09448
EARNINGS        .88031124     .10282596      8.5611769        1.0999473
                                                  .000           .92707
(CONSTANT)      21.122542     2.5903081      8.1544516
                                                  .000

ALL VARIABLES ARE IN THE EQUATION

--------------------------------------------------------------
---

* * * * * *

OBSERVATION          Y VALUE          Y ESTIMATE          RESIDUAL
        +2SD
      1.            63.20000          64.95809          -1.758090
      2.            66.20000          66.55387          -.3538727
      3.            69.30000          68.50349           .7965125
      4.            71.80000          70.06920          1.730798
      5.            73.60000          72.72900           .8710044
      6.            77.90000          76.36336          1.536636
      7.            81.60000          79.72050          1.879498
      8.            88.00000          87.68696           .3130435
      9.            95.90000          96.12022          -.2202210
     10.            103.3000          105.4465          -2.146531
     11.            115.4000          114.1679          2.232137
     12.            136.1000          143.8524          -7.752360
     13.            170.0000          177.7904          -7.790397
     14.            195.6000          196.3314          -.7314296
     15.            219.4000          205.6792          13.72084
            R
     16.            237.8000          239.5229          -1.722885
     17.            278.8000          279.4047          -.6046825

NOTE - (*) INDICATES ESTIMATE CALCULATED WITH MEANS SUBSTITUTED
       R  INDICATES POINT OUT OF RANGE OF PLOT
```

```
NUMBER OF CASES PLOTTED          17.
NUMBER OF 2 S.D. OUTLIERS         1.  OR  5.88 PERCENT OF THE TOTAL

VON NEUMANN RATIO   1.98886          DURBIN-WATSON TEST   1.87187

NUMBER OF POSITIVE RESIDUALS    8.
NUMBER OF NEGATIVE RESIDUALS    9.
NUMBER OF RUNS OF SIGNS         7.

NORMAL APPROXIMATION TO SIGN DISTRIBUTION IMPOSSIBLE.
USE A TABLE FOR EXPECTED VALUES.

            * * * * * * * * * * * * * * * * * * * *
```

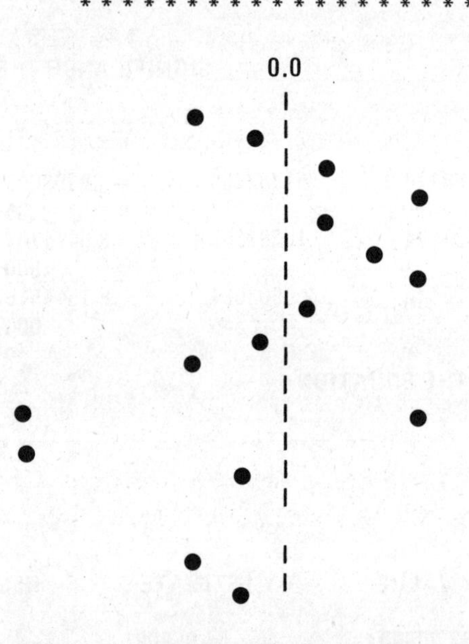

The output begins with a listing of the first part of the program up to the **READ INPUT DATA** card (i.e. after completion of the first task). This is followed by a table of the means and standard deviations of all the variables in the model as listed on the **PARTIAL CORR** card (this is requested by **STATISTICS 2**). The output of the partial correlation analysis follows, starting with a table (actually a matrix) of zero-order correlation coefficients (as requested by **STATISTICS 1**). Note that this table gives not only the value of the coefficients but also the number of degrees of freedom (d.f.) involved, in parentheses ( ) beneath each coefficient, and beneath this the level at which each coefficient is significant. For example, the zero-order correlation coefficient between **PRICES** and **M3** is printed thus:

0.9859
(15)
$s = 0.001$

In other words, $r_{\text{PRICES, M3}} = 0.9859$ calculated on the basis of 15 d.f., and this value is significant at $\alpha = .001$ (i.e. 99.9 per cent).

The two partial correlation coefficients requested are then printed, **PRICES** against **M3** controlling for **EARNINGS**, and **PRICES** against **EARNINGS** controlling for **M3**. The number of d.f. and the significance level is again given. For example,

$$r_{\text{PRICES, M3.EARNINGS}} = -0.2101$$
$$(14)$$
$$s = 0.438$$

means that the partial correlation coefficient between **PRICES** and **M3** with the effect of **EARNINGS** held constant or controlled for is –0.2101 calculated using 14 d.f. and this value is significant at only $\alpha = 0.438$ (i.e. 43.8 per cent). It is interesting to compare this partial value with the simple or zero-order value between the same variables, i.e. $r_{\text{PRICES, M3}} = 0.9858$. Taken at face-value the result could be very misleading as to the nature of the relationship between prices and money supply, when the indirect contribution of the earnings variable is not known or not included.

The value of –0.2101 gives a coefficient of partial determination of

$$r^2_{\text{PRICES, M3.EARNINGS}} = -0.2101^2 = 0.0441$$

In other words, only 4.4 per cent of the variation in **PRICES** which is *not* explained by **EARNINGS** *is* explained by **M3**. Contrast this with the other partial value, i.e.

$$r_{\text{PRICES,EARNINGS.M3}} = 0.9163$$

which implies that 83.9 per cent of the variation in **PRICES**, which is *not* explained by **M3** *is* explained by **EARNINGS**.

Following the partial correlation output there is then a listing of the rest of the program, specifically those task definition cards dealing with the regression analysis. The first part of the regression analysis output begins with the name of the dependent variable **PRICES** together with the values of its mean and standard deviation. There then follows a listing of various summary measures associated with the regression. We see that

R SQUARE = 0.99552

ADJUSTED R SQUARE = 0.99489

STD DEVIATION = 4.90011

$F = 1557.13115$

The value of R^2 tells us that **M3** and **EARNINGS** jointly account for 99.55 per cent of the variation in **PRICES**. The value of R^2

would be of interest if we were comparing this model with a similar one containing fewer or more variables. **STD DEVIATION** gives the value of the estimated standard deviation of the disturbance terms from $\Sigma\hat{u}_t^2/(n-k)$, equation (12.17). This is also known as the *standard error* of the regression. The F-statistic value is significant since $F_{2,14} = 3.74$ (at $\alpha = 0.05$). This indicates that the R^2 value is significant, as is the regression as a whole.

Information on the regression estimates is then printed in tabular form. The first column gives the variable name, the second column gives the value of the corresponding $\hat{\beta}_j$, the third column its standard error $s_{\hat{\beta}_j}$, the fourth column the value of the t-statistic, calculated from $\hat{\beta}_j/s_{\hat{\beta}_j}$. Finally, the last column gives the value of the beta coefficient,[1] and under this the elasticity of the variable with respect to **PRICES** at the point of respective means. From this output we are able to write the estimated regression equation

$$\underset{}{\text{PRICES}_t} = \underset{(2.5903)}{21.1225} - \underset{(0.5136)}{0.4129\text{M3}_t} + \underset{(0.1028)}{0.8803\text{EARNINGS}_t}$$

$$R^2 = 0.9955 \qquad \bar{R}^2 = 0.9948$$

The rule-of-thumb test tells us immediately that the money supply variable is not a significant explainer of prices in the period covered by the data, but that the **EARNINGS** variable is significant. (We must, however be cautious about using the rule-of-thumb test if n is small.)

The final section of the computer print-out is concerned with an analysis of the error terms; this information is given in tabular form. The first column gives the number of the observation, the second column (headed '+2SD') is used to indicate when any particular value of the error term is more than plus two standard deviations away from the mean of u (the letter R is printed in this column to indicate such a value and that it cannot be plotted since its value is outside the range of the graph). The remaining three columns contain the values of $\hat{Y}$, Y and $\hat{u}$. The final column, headed '−2SD', indicates those $\hat{u}$ values which are more than minus two standard deviations from the mean of $\hat{u}$. Beneath this table (but to the right of it in practice) is a graph of the $\hat{u}$ values plotted against the number of observations. The output below the table of residual values relates to the $\hat{u}$s, the number of positive values, and so on. We shall return to discuss the meaning and

[1] We have not explained what beta coefficients are. They are the coefficients which would be obtained if all the observations on the variables were first standardised before the regression was performed. They are therefore 'unit-free' coefficients. In effect they measure the number of standard deviations by which Y changes for one standard deviation change in any one of the X_j.

significance of the Durbin–Watson statistic in a later section (we shall not discuss the meaning of the von Neumann ratio or the reference to the sign distribution since we have not discussed these).

We have not the space here to discuss other programs in the SPSS package which also might be of interest in the context of regression (for example, the non-linear regression or the regression with dummy variables programs), nor have we discussed the full range of **OPTIONS** and **STATISTICS** available with either regression or partial correlation. For this the reader is referred to the SPSS manual.

12.7 Breakdown of the basic assumptions

We have seen that the application of OLS to a model of the general form

$$Y_i = \beta_1 + \beta_2 X_{2i} + \beta_3 X_{3i} + \ldots + \beta_k X_{ki} + u_i \tag{12.45}$$

will yield estimators which are best linear unbiased and normally distributed only if all of the basic assumptions of the classical normal linear regression model, (A.1) to (A.6) of section 12.2, are satisfied. In this section we first examine how we might test whether or not each of the assumptions is in fact satisfied in any particular model, and second, what, if any, are the consequences for the properties of the OLS estimators should any of the assumptions be violated. Since this is not an econometrics book (wherein a full and detailed discussion of what is to follow would be appropriate) we intend to consider these points only briefly.

A.1: linearity

This assumption relates to the linearity of the relationship between the dependent and the explanatory variables, i.e.

$$Y_i = \beta_1 + \beta_2 X_{2i} + \beta_3 X_{3i} + \ldots + \beta_k X_{ki} + u_i$$

We would specify the linearity of such a relationship on the basis of economic theory, i.e. we would hope that economic theory would provide some evidence upholding this assumption. If such evidence is not available, then we can assume a linear relationship by default and then perhaps apply some tests to determine whether our assumption is valid.

Let us assume that we have established beyond doubt that there is a linear relationship between Y and all the explanatory variables *other* than X_2, but we are uncertain as to whether or not the relationship between Y and X_2 is also linear. We want therefore to consider the form of the relationship between Y and

X_2 leaving aside the remaining explanatory variables. In particular we want first to examine the consequences of a non-linear relationship, and then see how we may test for linearity.

Suppose that the true relationship between Y and X_2 is non-linear and is of the form

$$Y_i = \beta_1 + \beta_2 X_{2i} + \beta_3 X_{2i}^2 + u_i \tag{12.46}$$

but we mistakenly assume it is linear and is of the form

$$Y_i = \beta_1 + \beta_2 X_{2i} + u_i \tag{12.47}$$

These two relationships are illustrated in Figure 12.5.

Suppose that the application of OLS to the linear model of (12.47) yields the estimated regression equation

$$\hat{Y}_i = \hat{\beta}_1 + \hat{\beta}_2 X_{2i} \tag{12.48}$$

It is fairly obvious that, if the true relationship is as given by (12.46), the OLS estimators $\hat{\beta}_1$ and $\hat{\beta}_2$ obtained using (12.47) will tend not to be significant. Furthermore, they will also be pretty meaningless in terms of any relationship between Y and X_2.

The lack of significance of the estimators may thus cause us to reject the possibility of a relationship between Y and X_2 even though such a relationship (albeit non-linear) may well exist. Of course, if the departure from linearity is not too severe, or if X_2 is a stronger influence on Y than is X_2^2, we may find $\hat{\beta}_2$ to be significant in (12.48).

This is perhaps better than before, but it is still unsatisfactory. First of all, we are missing out an important variable X_2^2 which may affect the value we obtain for $\hat{\beta}_1$ and $\hat{\beta}_2$. Second, the influence of the missing variable has to be incorporated into the disturbance term in (12.47); this may cause its variance to increase, which will in turn adversely affect the size of the estimated standard errors, with obvious consequences for hypothesis testing. Finally, the disparity between (12.46) and (12.47) may increase as X_2 increases, with consequences for the forecasting efficiency of the model. With these points in mind tests for non-linearity

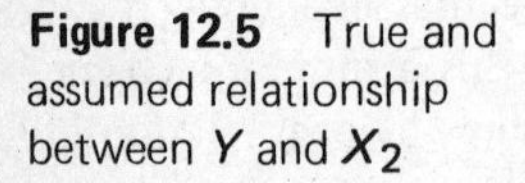

Figure 12.5 True and assumed relationship between Y and X_2

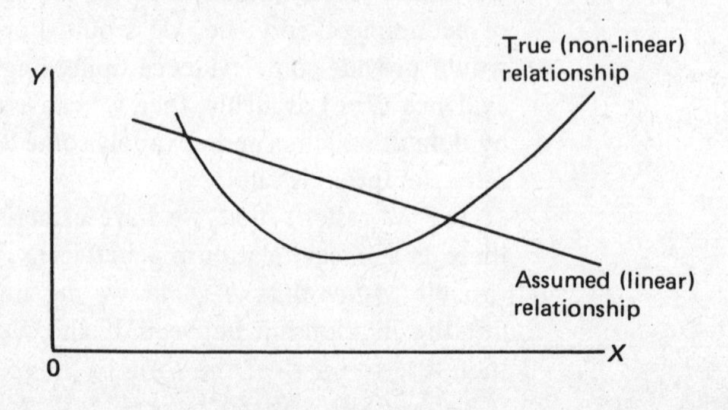

seem advisable. There are several possibilities which we shall discuss briefly – for a more complete discussion see, for example, Kmenta (1971, pp. 466–72).

First, however complex the relationship between Y and X_2 is, we can always represent it by using a polynomial of high enough degree. The degree of a polynomial is equal to the highest power of X_2 in the equation. For example, (12.46) is a polynomial of degree 2 (since X_2^2 is the highest power of X_2 in the equation). In general a polynomial of degree k is written

$$Y_i = \beta_0 + \beta_1 X_{2i} + \beta_2 X_{3i}^2 + \ldots + \beta_k X_{ki}^k + u_i \tag{12.49}$$

One way of testing for linearity, therefore, would be to specify the relationship between Y and X_2 as a polynomial in as high a degree (say 3 or 4) as is thought necessary. The application of OLS and inspection of the significance of the parameters of the higher powers of X_2 will indicate whether or not the assumption of linearity is reasonable.

A second possibility for testing linearity arises out of the fact that if the relationship is linear then the value of the slope parameter β_2 remains constant throughout the range of values of the explanatory variable. For example, suppose a scatter of observations on Y and X_2 is as shown in Figure 12.6. If we divide the observations into two or three sub-groups (ordered by increasing size of X_2) and compute separate regressions on each group, then if the relationship is truly linear and stable each of the slope estimators should have the same value. Of course the values of the $\hat{\beta}$s are unlikely to be exactly the same even if the relationship is linear, and so we must test any difference between the values to determine whether these are significant (indicating non-linearity) or not using the test for the difference between means described in section 10.3 or section 10.5.

The final test relates to the way in which the disturbance terms are scattered randomly around the true regression line. If the true relationship is non-linear and we attempt to force a linear fit, the disturbance will not be randomly scattered around the linear relationship. This is illustrated in Figure 12.7.

Figure 12.6 Testing for non-linearity by dividing into sub-groups

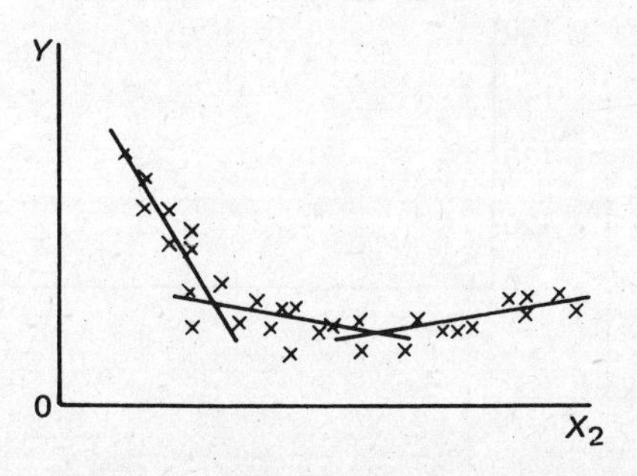

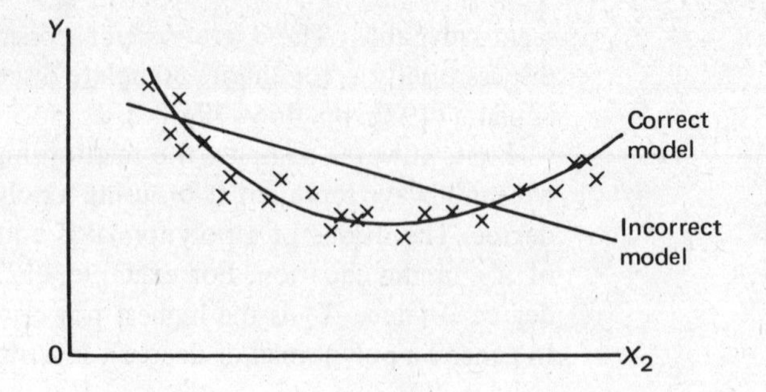

Figure 12.7 Non-random dispersion of disturbance terms around an incorrect linear model

If we were able to observe the disturbance terms (ordered against increasing values of X_2), we would see a first group of positive values, then a negative group, which is followed by another positive group. Of course we cannot observe the disturbance terms, but we do have the error terms available. Visual inspection of the $\hat{u}_i$s (by plotting them against increasing values of X_2) for such non-random patterns can be used as a test for linearity, although unfortunately this test may be confused with the testing of assumption (A.4) (which we shall come to shortly).

As an illustration of these methods consider the scatter diagram shown in Figure 12.8. This relates to the total consumption of coal per annum (in millions of tonnes) in the United Kingdom and to a general index of retail prices of fuel and light (1974 = 100) between 1962 and 1977.

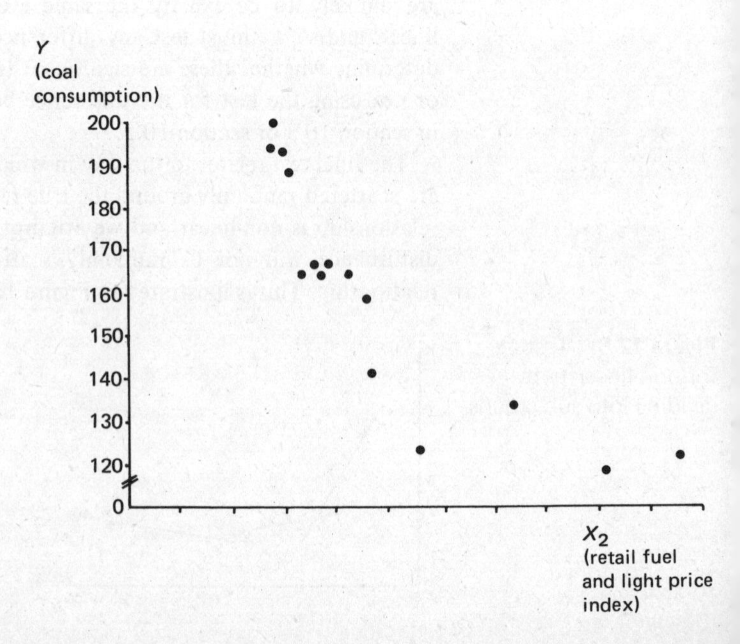

Figure 12.8 Scatter diagram of UK annual coal consumption against price of fuel and light

The relationship displayed is clearly non-linear, but suppose it was specified in linear form, i.e. as

$$Y_t = \beta_1 + \beta_2 X_{2t}$$

If we suspect a non-linear relationship, we could divide the sixteen observations into two groups of eight observations each, the first from 1962 to 1969 and the second from 1970 to 1977. Separate regressions carried out on these two groups produce the following results:

$$1962\text{–}9: \quad \hat{Y}_t = 307.2253 - 1.9986\, X_{2t}$$

$$1970\text{–}7: \quad \hat{Y}_t = 148.6995 - 0.1462\, X_{2t}$$

The value of the slope coefficient $\hat{\beta}_2$ is much lower (i.e. less steep) in the latter period from 1970 to 1977 than in the former, reflecting the flattening out of the relationship in the later years, and clearly visible in Figure 12.8. Whether the difference between the two values of $\hat{\beta}_2$, i.e. -1.9986 and -0.1462, is significant or not will depend upon an application of the test described in section 10.3. In this example the two values are clearly significantly different and we would adduce this as further evidence of a non-linear relationship.

We can substantiate this by examining the pattern of values of the error terms for a regression on all sixteen observations. The estimated regression equation is found to be

$$Y_t = 201.3056 - 0.4817\, X_{2t}$$

and the error terms have the following values (rounded to one decimal place for convenience):

20.0	27.3	19.8	16.2	7.1	−3.9	0.0	−1.5
−6.5	−18.9	−34.3	−21.8	−30.1	−8.1	10.2	24.4

Inspection of these values reveals a pattern of positive and negative values typical of the sort of non-linear relationship described above, a series of positive values followed by a series of negative values, followed by more positive values. We can see this more clearly in Figure 12.9, where the error terms (the $\hat{u}_t$s) are plotted against increasing values of the explanatory variable.

As a second example of plotting the $\hat{u}_t$s to detect non-linearity we can refer back to the computer output example given in section 12.6 and to the plot of the residual terms on page 326, which we reproduce here (Figure 12.10), though it is turned through 90°. The table of residual values is also shown.

We have joined the points in the graph together to make identification of any pattern in it the more easily discernible. However,

Figure 12.9 Plot of error terms as a test for non-linearity

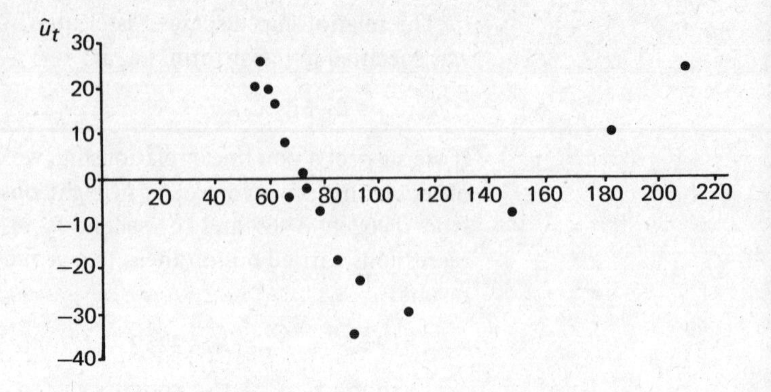

Figure 12.10 does not immediately suggest that non-linearity is present in this model, certainly not one which could be represented by a polynomial of low degree. On the evidence of this graph at least we would be inclined to think that assumption (A.1) is fulfilled.

Given that we discover there to be a non-linear relationship between two variables, is there any way in which the situation can be rectified. One common approach is to *transform* the model to linear form by taking logarithms of both variables and applying OLS to log Y and log X. That is, the model is specified as

$$\log Y_t = \beta_1 + \beta_2 \log X_{2t} + u_t$$

If this does not produce a linearised relationship, we might turn to a more complex polynomial specification of the sort discussed earlier in this section. Once a satisfactory transformation to linear form has been achieved, the application of OLS to the transformed

Figure 12.10

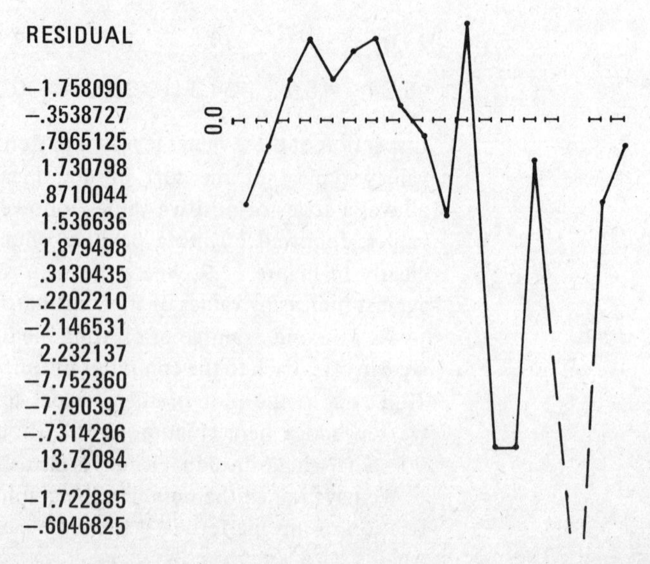

data will yield estimators with all the desirable properties (these will of course be estimators of the parameters in the transformed model).

A.2: normal u

This second assumption relates to the normality of the distribution of the disturbance terms. If this assumption is violated, then the OLS estimators are still best linear unbiased (we can see this from Table 11.7), but are no longer maximum likelihood estimators. They are also consistent and asymptotically efficient.

Thus the breakdown of this assumption has little effect on the desirable properties (both small sample and large sample) of the OLS estimators.

However, in terms of confidence interval estimation and hypothesis testing of the estimators, lack of normality in the distribution of the u_is will cause the OLS estimators not to be normally distributed when n is small. This means that we cannot make use of the z- and t-distributions to construct confidence intervals or to perform tests of significance on the estimates. This means that none of the estimating and testing procedures described in Chapter 11 can be applied. This is a serious consequence, especially in view of the fact that as economic or social scientists we are often obliged to work with small samples. It may be of some comfort to know that for large n the OLS estimators will be approximately normal (because of the central limit theorem), and so the tests may be applied even though the results may not be completely accurate.

However, following the development of the basic idea of the disturbance term in Chapter 11, an assumption of its normality does not seem unreasonable. In any case small departures from normality will not have serious consequences for estimation and hypothesis testing and the results will hold reasonably well.

One possible way of testing the normality of the u_is is to examine the shape of the distribution of the $\hat{u}_i$s after arranging them in frequency table and histogram form. However, the number of observations required for a reliable outcome of this test might be more than is available.

A.3: zero mean of u

This third assumption relates to the zero mean of the disturbance term, i.e.

$$E(u_i) = 0$$

The consequences for the OLS estimators arising out of the breakdown of this assumption are not particularly serious. To see this, suppose that the mean of the u_i is some constant k, i.e.

$$E(u_i) = k$$

Taking expected values of (12.45) gives

$$E(Y_i) = \beta_1 + \beta_2 X_{2i} + \ldots + \beta_k X_{ki} + k$$
$$= (\beta_1 + k) + \beta_2 X_{2i} + \ldots + \beta_k X_{ki} \quad (12.50)$$

and the application of OLS to (12.50) yields an estimate of the composite intercept term $(\beta_1 + k)$, from which there is no way of determining the separate value of β_1. In other words, the estimator of the intercept term is biased. Estimation of the remaining parameters β_2 to β_k is, however, unaffected. In other words, breakdown of this assumption affects only the estimated value of the intercept term, and since, as we have already noted, interpretation of the meaning of the intercept term in most economic and social science applications is often difficult, the consequences may not be too serious.

Is it possible to test the assumption of zero mean? The answer is no. The individual u_i of course are unknown. The error terms $\hat{u}_i$ are our only means of estimating the u_i but, as we have seen, $\Sigma\hat{u}_i = 0$ and thus $\bar{\hat{u}} = \Sigma\hat{u}_i/n = 0$, i.e. the mean of the error terms is defined always to be zero, so we cannot measure this mean and use it as a test of $E(u_i)$.

A.4: constant variance of u

Assumption (A.4) requires that the variance of the disturbance term should be constant, i.e.

$$E(u_i^2) = \sigma^2 \quad (12.51)$$

When this assumption is fulfilled the disturbance terms are said to be *homoskedastic*. When the assumption is violated the disturbance terms are said to be *heteroskedastic*.

We may write the general form of a non-constant disturbance variance by replacing (12.51) with

$$E(u_i^2) = \sigma_i^2 \quad (12.52)$$

The subscript i on the right-hand side of (12.52) indicates that the value of $E(u_i^2)$, i.e. var(u_i), is not constant but varies from one observation to another.

Consequences

The presence of heteroskedastic disturbance terms in a model may have serious consequences for the OLS estimators. From Table 11.7 we can see that the OLS estimators will still be unbiased among all linear estimators since assumption (A.4) is not required for the proof of unbiasedness. Furthermore, it is possible to demonstrate that the OLS estimators are also consistent. However, assumption (A.4) is required if we wish to prove that OLS estimators have the smallest variance among all linear unbiased estimators. It is possible that there exist alternative estimators having a smaller variance than the OLS estimators. Thus the latter are not efficient, nor (as can be shown) are they asymptotically efficient.

These difficulties are compounded if, unknowingly, we apply OLS to a model with heteroskedastic disturbances. The OLS estimators will still have some desirable properties but the estimated variances of these estimators will be biased. To see this suppose we are concerned with the heteroskedastic model

$$Y_i = \beta_1 + \beta_2 X_{2i} + u_i \tag{12.53}$$

where the variance of u is given by (12.52). Now the variance of $\hat{\beta}_2$, say, is given by

$$\text{var}(\hat{\beta}_2) = E(\hat{\beta}_2 - \beta_2)^2 \tag{12.54}$$

from equation (11.33). But from (11.23)

$$\hat{\beta}_2 = \Sigma x_{2i} y_i / \Sigma x_{2i}^2$$

and from (11.30)

$$y_i = \beta_2 x_{2i} + u_i - \bar{u}$$

Therefore,

$$\begin{aligned} \hat{\beta}_2 &= \Sigma x_{2i}(\beta_2 x_{2i} + u_i - \bar{u}) / \Sigma x_{2i}^2 \\ &= \beta_2 + \Sigma x_{2i} u_i / \Sigma x_{2i}^2 \end{aligned}$$

Since $\bar{u}\Sigma x_{2i} = 0$. Substituting this last result (12.54) gives

$$\begin{aligned} \text{var}(\hat{\beta}_2) &= E(\beta_2 + \Sigma x_{2i} u_i / \Sigma x_{2i}^2 - \beta_2)^2 \\ &= E(\Sigma x_{2i} u_i / \Sigma x_{2i}^2)^2 \end{aligned} \tag{12.55}$$

We now need the following theorem:

> The square of a sum is equal to the sum of squares plus twice the sum of cross-products, where the cross-products include all possible pairs for which the first subscript is smaller than the second, i.e.

$$(\Sigma a_i)^2 = \Sigma a_i^2 + 2 \sum_{i<j} a_i a_j \tag{12.56}$$

Applying this result to (12.55) gives

$$\begin{aligned} \text{var}(\hat{\beta}_2) &= E[\Sigma x_{2i}^2 u_i^2 / (\Sigma x_{2i}^2)^2] + 2E\Sigma(x_{2i} u_i x_{2j} u_j / \Sigma x_{2i}^2) \\ &= E[\Sigma x_{2i}^2 u_i^2 / (\Sigma x_{2i}^2)^2] + 0 \end{aligned}$$

Since $E(u_iu_j) = 0$, by assumption (A.5), thus

$$\text{var}(\hat{\beta}_2) = \Sigma x_{2i}^2 E(u_i)^2 / (\Sigma x_{2i}^2)^2$$

i.e.

$$\text{var}(\hat{\beta}_2) = \Sigma x_{2i}^2 \sigma_i^2 / (\Sigma x_{2i}^2)^2 \tag{12.57}$$

using (12.52).

Let us now suppose that in (12.57)

$$\sigma_i^2 = \sigma^2 k_i \tag{12.58}$$

where the k_i are some non-stochastic weights. Substituting from (12.58) into (12.57) gives

$$\text{var}(\hat{\beta}_2) = \Sigma x_{2i}^2 \sigma^2 k_i / (\Sigma x_{2i}^2)^2$$

$$= \frac{\sigma^2}{\Sigma x_{2i}^2} \times \frac{\Sigma x_{2i}^2 k_i}{\Sigma x_{2i}^2} \tag{12.59}$$

Now, the first term on the right-hand side of (12.59) is the variance of $\hat{\beta}_2$ using OLS under the assumption of homoskedasticity. Thus we can write (12.59) as

$$\text{var}(\hat{\beta}_2) = \text{var}(\hat{\beta}_2^{\text{OLS}}) [\Sigma x_{2i}^2 k_i / \Sigma x_{2i}^2] \tag{12.60}$$

or, rearranging, as

$$\text{var}(\hat{\beta}_2^{\text{OLS}}) = \text{var}(\hat{\beta}_2) / [\Sigma x_{2i}^2 k_i / \Sigma x_{2i}^2] \qquad (k_i > 0) \tag{12.61}$$

This last expression is read as 'the variance of $\hat{\beta}_2$ assuming homoskedasticity and applying OLS is equal to the actual variance of $\hat{\beta}_2$ under the assumption of heteroskedasticity divided by the term $\Sigma x_{2i}^2 k_i / \Sigma x_{2i}^2$' (a similar result can be derived for $\hat{\beta}_1$).

Now, in many applications in economics we might expect X_{2i} and k_i to be positively correlated, and if the denominator in (12.61) is greater than 1.0 then (12.61) will *underestimate* the true variance of $\hat{\beta}_2$ as given by (12.57). This last result is of considerable significance for hypothesis testing in the heteroskedastic model. If $\text{var}(\hat{\beta}_2)$ is calculated to be smaller than it actually is, so of course is the standard error of $\hat{\beta}_2$, $s_{\hat{\beta}_2}$, which in turn means that the ratio

$$\hat{\beta}_2 / s_{\hat{\beta}_2} \sim t$$

will be larger than it should otherwise be, and this will cause us to reject H_0 when perhaps we should accept it. Similar consequences follow for the confidence intervals, which will be narrower than they should be. The net result of this is that we tend to accept parameters as being significant when they are not.

The application of OLS to a model with heteroskedasticity may thus cause us to draw erroneous conclusions as to the relevance of certain variables in the model. This is a serious consequence and emphasises the need for us to be able to detect the presence of heteroskedasticity in a model so that, if possible, we can use some alternative approach.

Detection of heteroskedasticity

Let us suppose that in an income–expenditure model the variance of the disturbance terms is a function of the explanatory variable X_2. In this case we can write

$$E(u_i^2) = \sigma X_{2i} \tag{12.62}$$

Thus as the values of X_2 increase (i.e. as income increases) var(u) also increases according to (12.62). If we plot the estimated values of the u_i^2s, i.e. the observable $\hat{u}_i^2$s, against the explanatory variable with which var(u) is suspected of being associated (X_2 in this case), then a visual inspection of the graph might indicate the presence of heteroskedasticity. In Figure 12.11 we show several possibilities resulting from such a plot.

A scatter such as that in Figure 12.11(a) indicates that the variance of the disturbance term is a constant and is independent of X_2. A pattern such as that in Figure 12.11(b) suggests that var($\hat{u}$) and hence var(u) is directly proportional to the value of X_2. Increasing values of X_2 correspond to increasing values of var($\hat{u}$) though the relationship appears to be linear, such as that in (12.5.2). Finally, Figure 12.11(c) suggests that the size of var($\hat{u}$) increases at an ever-quickening rate according to the size of X_2. This implies a non-linear relationship with var($\hat{u}$) proportional to some power of X_2, e.g.

$$E(\hat{u}_i^2) = \sigma^2 X_{2i}^2 \tag{12.63}$$

The usefulness of plotting the error terms as described above should not be underestimated and the method may reveal much about the relationship between var(u) and the explanatory variable with which it is suspected of being related.

However, more rigorous procedures are available. For example, Park (1966) has suggested that if we specify the relationship between var(u) and X_2 in the most general form of equation

Figure 12.11 Testing for heteroskedasticity

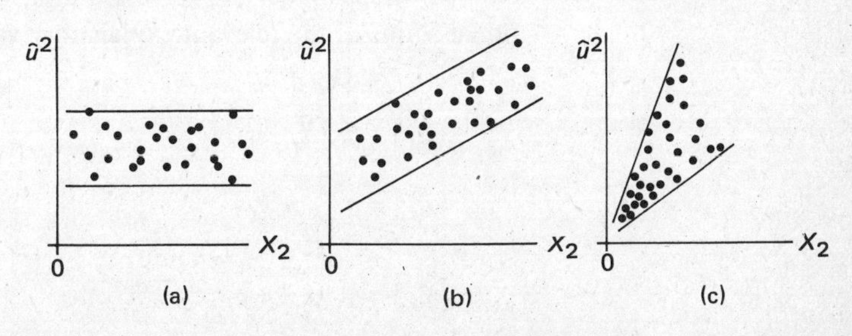

(12.64), i.e.

$$\sigma_i^2 = \sigma^2 X_{2i}^{\gamma} e^{v_i} \tag{12.64}$$

where e is the base of the natural logarithms, then taking natural logarithms of both sides of (12.64) gives

$$\ln \sigma_i^2 = \ln \sigma^2 + \gamma \ln X_{2i} + v_i \tag{12.65}$$

and by regressing $\ln \sigma_i^2$ against $\ln X_{2i}$ and examining whether γ is significantly different from zero, we shall thus be able to determine whether any relationship exists between σ_i^2 and X_{2i}. Of course, we cannot actually use (12.65) since we do not know the σ_i^2. Again we resort to the estimated disturbances and instead use

$$\ln \hat{\sigma}_i^2 = \ln \sigma^2 + \hat{\gamma} \ln X_{2i} \tag{12.66}$$

If $\hat{\gamma}$ is found to be indicative of a non-zero γ, then heteroskedasticity may be said to exist in the model. Furthermore, the actual value of $\hat{\gamma}$ may provide some indication of the nature of the relationship between var(u) and X_2. This test has been criticised as being unreliable on the grounds that the disturbance term in (12.66), i.e. v_i, may itself be subject to heteroskedasticity. However, the test, considered along with the evidence from a plot of $\hat{u}_i^2$ against the relevant explanatory variable, may still provide a useful guide to the presence and form of any heteroskedasticity present in the model.

Several other more rigorous tests are available, e.g. Goldfeld and Quandt (1965) and Glejser (1969). We cannot discuss them in detail here, but we can summarise briefly.

In Goldfeld and Quandt the observations, ordered according to the size of the relevant explanatory variable, are split into two groups (after omitting a small number of central observations): a group corresponding to low values of the appropriate explanatory variable, and a group corresponding to the high values. Separate regressions are performed on these two groups and the value of their sum of squared errors is compared using an F-test. If heteroskedasticity is not present, the ratio of the two SSE values should be unity.

In Glejser's method, the $|\hat{u}_i|$ obtained from an OLS regression on the whole model are regressed against several different functional forms of the relevant explanatory variable, e.g.

$$|\hat{u}_i| = \beta X_{ji} + v_i$$

$$|\hat{u}_i| = \beta \sqrt{X_{ji}} + v_i$$

$$|\hat{u}_i| = \beta \frac{1}{X_{ji}} + v_i$$

$$|\hat{u}_i| = \beta \frac{1}{\sqrt{X_{ji}}} + v_i$$

If any of the coefficients are found to be significant, then heteroskedasticity may be said to be present and its form indicated by the model type found to be most suitable.

A final test involves the use of Spearman's rank correlation coefficient. In this, OLS is applied to the whole model, and the absolute values of the error terms so obtained are correlated with the suspect explanatory variable to produce the rank correlation coefficient r_s. We then calculate the value of the expression

$$r_s\sqrt{n-2}/\sqrt{1-r_s^2} \sim t_{n-k,\,\alpha/2} \tag{12.67}$$

and if this value exceeds the critical value of t we can assume that heteroskedasticity is present.

Before turning to methods for dealing with heteroskedasticity we should note that the condition is more likely to arise in cross-section rather than time-series models. For example, consider once again the relationship between household expenditure and household income. Time-series data on national aggregates of these variables over, say, twenty annual observations might yield in an economy growing at a moderate rate an approximate doubling of income (and thus of expenditure) in real terms. However, a cross-section study of a random sample of all households in a given year will yield perhaps a ten- or twenty-fold difference between the lowest and highest groups of household income. The latitude for increasingly larger variations in expenditure over such an income range is much greater than in the time-series case, i.e. heteroskedasticity is more likely to be present.

Remedial measures

If the application of one of the above tests indicates heteroskedastic disturbances, what remedies are available? If we know the form of the relationship between var(u) and the explanatory variable in question, it is possible to transform the model to a more suitable specification and hence apply OLS to obtain best linear unbiased estimators of the transformed equation. We do not have the space to consider this procedure more than briefly, but suppose heteroskedasticity in a model arises because var(u) is related to X_2, as in (12.63), i.e.

$$E(u_i^2) = \sigma^2 X_{2i}^2 \tag{12.63}$$

If the original model is of the form

$$Y_i = \beta_1 + \beta_2 X_{2i} + u_i \tag{12.68}$$

and if we divide this through by $\sqrt{X_{2i}^2}$ we have

$$Y_i/\sqrt{X_{2i}^2} = \beta_1/\sqrt{X_{2i}^2} + \beta_2 X_{2i}/\sqrt{X_{2i}^2} + u_i/\sqrt{X_{2i}^2} \tag{12.69}$$

which we can write as

$$Y_i/X_{2i} = \beta_1/X_{2i} + \beta_2 + v_i \tag{12.70}$$

where

$$v_i = u_i/X_{2i}.$$

Now,

$$E(v_i^2) = E(u_i/X_{2i})^2 = \frac{1}{X_{2i}^2} E(u_i^2)$$
$$= \sigma^2$$

using (12.63). In other words, v_i in (12.70) is a homoskedastic disturbance.

Furthermore,

$$E(v_i) = E(u_i/X_{2i}) = \frac{1}{X_{2i}} E(u_i)$$

using assumption (A.6) (non-stochastic X_{ji}s), and therefore

$$E(v_i) = 0$$

using assumption (A.3). Moreover,

$$E(v_i v_j) = E\left[\frac{u_i}{X_{2i}} \frac{u_j}{X_{2i}}\right] = \frac{1}{X_{2i}^2} E(u_i u_j)$$

using (A.6), and therefore

$$E(v_i v_j) = 0$$

using (A.4) (independent u_is).

Thus the disturbance term v_i in (12.70) obeys the basic assumptions, and the application of OLS to this equation will produce BLU estimators of its parameters. Note, however, that when we estimate (12.70) the result we obtain for the constant term is in fact an estimate of the slope parameter β_2.

For this approach to the problem of heteroskedasticity to be contemplated we have to know not only which explanatory variable in the model is related to the disturbance terms, but more particularly the exact nature of this relationship. The first of these problems may not be so difficult to overcome (using some of the methods described earlier, e.g. plotting the $\hat{u}_i^2$ against each X_j). Some information as to the form of the relationship between u and the X_j may become available from one or other of the detection methods (e.g. Glejser's approach) but we shall often have no more to go on than an informed guess. The form given by (12.63) is commonly used in the event that no other information is available and frequently gives good results.

Whatever the assumed form of var(u), the general approach is as described above: find an appropriate transformation, apply it to the original model and then apply OLS to the transformed model. This approach is known as the method of *generalised least squares* (or GLS) and has wide application in econometrics as an estimation technique which may be particularly effective when a model, for one reason or another, is not well behaved.

A.5: independence of u

Assumption (A.5) relates to the assumption of no correlation between disturbance terms (taken together with the assumption of normality: (A.4) also implies *independence* between the disturbances), i.e.

$$E(u_i u_j) = 0 \quad (i \neq j) \tag{A.4}$$

When this assumption breaks down the model is said to have autocorrelated disturbances or to suffer from *autocorrelation*. If this is the case, then we can write

$$E(u_i u_j) \neq 0 \quad (i \neq j) \tag{12.71}$$

Equation (12.71) implies (for example) that the value of the disturbance in any particular quarter of a set of quarterly observations is influenced by the value of the disturbance in some previous quarter. Or in the case of a cross-section of household expenditures, if the expenditure of one household is considerably higher than the mean expenditure of all households, then the expenditure of some other household may also be higher.

There are several forms which the relationship between the disturbances might take in the model of

$$Y_t = \beta_1 + \beta_2 X_{2t} + u_t \tag{12.72}$$

For example, it might be a moving-average process, i.e.

$$u_t = v_t + \alpha v_{t-1}$$

which is a moving-order structure of order 1 (the 'order' of the process is determined by the longest lag term on the righthand side of the above expression).

Alternatively, the relationship between the us might be more suitably described by an autoregressive process, i.e.

$$u_t = \rho u_{t-1} + v_t \tag{12.73}$$

where v_t is a random normally distributed and well-behaved disturbance term, i.e.

$$E(v_t) = 0 \tag{12.74}$$

$$E(v_t^2) = \sigma^2 \tag{12.75}$$

$$E(v_t v_s) = 0 \quad (t \neq s) \tag{12.76}$$

$$E(v_t u_{t-1}) = 0 \tag{12.77}$$

and the constant ρ is known as the *autocorrelation coefficient* such that,

$$|\rho| < 1$$

This last expression ensures that the condition is not 'explosive'.

Equation (12.73) implies that the value of the disturbance term in any particular time period is partially determined by the value of u in the previous time period (multiplied by ρ) and partially by the stochastic term ν.

Equation (12.73) is known as a *first-order autoregressive* scheme: 'first-order' because u_t is related to u_{t-1} (i.e. via *one* time lag) and not to u_{t-2} or to u_{t-3}, etc. (although there are indirect connections), and 'autoregressive' because (12.73) is a regression-type relationship and indicates regression by a variable on itself (albeit lagged one period).

In the usual event that we are considering autocorrelated disturbances in the context of a time-series model, such as (12.72) and (12.73), then (12.71) is more properly written as

$$E(u_t u_{t-s}) \neq 0 \quad (t > s) \tag{12.78}$$

Consequences

As with heteroskedastic models, the presence of autocorrelation affects the optimality of the OLS estimators. Inspection of Table 11.7 shows that assumption (A.5) is not required for the estimators to be unbiased, so provided assumptions (A.1), (A.3) and (A.6) are satisfied then the application of OLS to a model having autocorrelated disturbances, as in (12.73), will still yield linear unbiased estimators. They can also be shown to be consistent.

However, assumption (A.5) is required in the proof that the OLS estimators are best, i.e. have a smaller variance than any other linear unbiased estimators. Thus OLS estimators are not efficient in small samples, nor are they asymptotically efficient.

Turning to the loss of 'smallest variance' property it can be shown (see, for example, Kmenta (1971, p. 276)) that when we apply OLS to a model such as (12.72), assumed to be suffering from first-order autocorrelation as in (12.73), then the variance of the estimator $\hat{\beta}_2$, for example, is given by the expression

$$\text{var}(\hat{\beta}_2^{\text{AUTO}}) = \frac{\sigma^2}{\Sigma x_t^2} + \frac{2\sigma^2}{(\Sigma x_t^2)^2}\left[\rho \sum_{t=2}^{n} x_t x_{t-1} + \rho^2 \sum_{t=3}^{n} x_t x_{t-2} + \ldots\right] \tag{12.79}$$

Without the assumption of autocorrelation the normal OLS formula for the variance of $\hat{\beta}_2$ in a model such as (12.72) is given by (11.34), i.e.

$$\text{var}(\hat{\beta}_2) = \sigma^2 / \Sigma x_t^2$$

In most economic situations ρ will usually be positive, and thus

$$\text{var}(\hat{\beta}_2) < \text{var}(\hat{\beta}_2^{\text{AUTO}})$$

In other words, the application of OLS to a model such as (12.72), ignoring the presence of autocorrelation, will generally produce estimated variances of the coefficients which are smaller than the true variances (under the assumption that (12.73) describes the autoregressive structure correctly). It should be emphasised that even $\text{var}(\hat{\beta}_2^{\text{AUTO}})$ is not a minimum variance. The application of OLS to a model containing autocorrelation, even if the form of the autocorrelation is known, does not lead to best estimators (except of course by chance).

The consequences of this for hypothesis testing are the same as in the heteroskedastic case. Underestimates of the $\text{var}(\hat{\beta}_j)$s and hence of the $s_{\hat{\beta}_j}$s will cause us to accept β_js as significant when they are not, i.e. we tend to reject $H_0: \beta_j = 0$ more often than we should. In the same way confidence intervals will be narrower than they should really be, leading to similarly incorrect decisions as to the significance of the parameters.

The adverse effects of autocorrelation do not end here, however. A further problem is that $\hat{\sigma}^2$, which is an unbiased estimator of σ^2 in a well-behaved model, is in fact biased in a model containing autocorrelated disturbance terms. Normally, $\hat{\sigma}^2$ for the model of (12.72) is given by (11.53), i.e.

$$\hat{\sigma}^2 = \Sigma u_t^2 / n - 2$$

In a model whose disturbances follow (12.73) it can be shown (see Kmenta, 1971, p. 281) that

$$E(\hat{\sigma}^2) = \sigma^2 \{ n - [2/(1-\rho)] - 2\rho r \} / n - 2 \qquad (12.80)$$

where

$$r = \sum_{t=2}^{n} x_t x_{t-1} / \Sigma x_t^2$$

In most economic applications both ρ and r will be positive, and so

$$E(\hat{\sigma}^2) < \sigma^2$$

In other words, $\hat{\sigma}^2$ will on average be smaller than the true σ^2, and this effect will be transmitted through to the estimated value of $\text{var}(\hat{\beta}_2)$, say, via equation (11.34) such that it too will underestimate $\text{var}(\hat{\beta}_2)$.

Thus in an autocorrelated model the estimated variance of the OLS estimators will be biased downwards for the two reasons described above. Because of the potentially serious consequences which arise when we apply OLS to an autocorrelated model it becomes imperative to try to detect its presence in a model before estimation is attempted.

Detection

One possible approach to the detection of autocorrelation is by means of a graphical examination of the error terms. If successive disturbance terms are related to each other, then this relationship should manifest itself by means of some noticeably non-random pattern when the disturbance terms are plotted against time (we are confining our analysis to time-series models since autocorrelation is so much more likely in these than in cross-section models).

Since we do not know the values of the u_ts we have to make do, as in the heteroskedastic model, with the error terms, i.e. the $\hat{u}_t$s. Now if positive autocorrelation exists in the model, i.e. if ρ in (12.73) is positive, then successive disturbance terms and consequently successive error terms will be positively related.

Thus we might observe longish strings of error terms of the same sign, i.e. a positive string followed by a negative string followed by a positive string, and so on. The change in string sign is caused by the random term v_t in (12.73) which may occasionally be of the opposite sign to the term ρu_{t-1} and large enough to cause a change in sign of the whole right-hand side of this equation.

When negative autocorrelation is present there will be a tendency for any particular disturbance term to be followed by a disturbance term of opposite sign. Thus when the error terms are plotted we are likely to observe oscillation of the sign of the $\hat{u}_t$s from positive to negative and back again. As we pointed out above, however, most economic applications will tend to exhibit positive autocorrelation rather than negative if autocorrelation is present. Figure 12.12 illustrates these conditions. If such a diagram shows a fairly random displacement of the $\hat{u}_t$s around the horizontal axis, this would be suggestive of no autocorrelation.

One of the advantages of plotting the error terms in this way is that it may help to provide information about missing variables. If the error terms display some systematic pattern over the whole of the time period considered, and if we obtain a significant value, this may suggest the inclusion of some additional explanatory

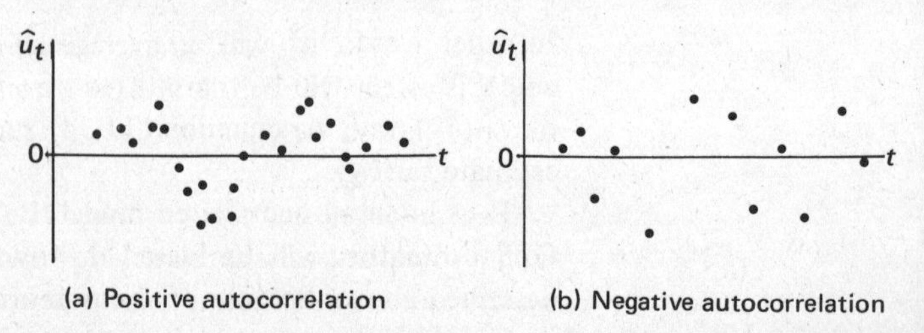

Figure 12.12 Graphical detection of autocorrelation

variable. Examination of the pattern of the error-term graph may be suggestive of the variation in some relevant explanatory variable which has not, for some reason, been initially included in the model, in which case the variation in the error term will tend to reflect this variation. Inclusion of the appropriate variable may remove the autocorrelation problem.

As an example of the use of the error-term graph in the detection of autocorrelation let us consider the graphical output of section 12.6, and in particular the residuals data on page 325 (reproduced here, as Figure 12.13, for convenience with the graph turned through 90°).

Figure 12.13

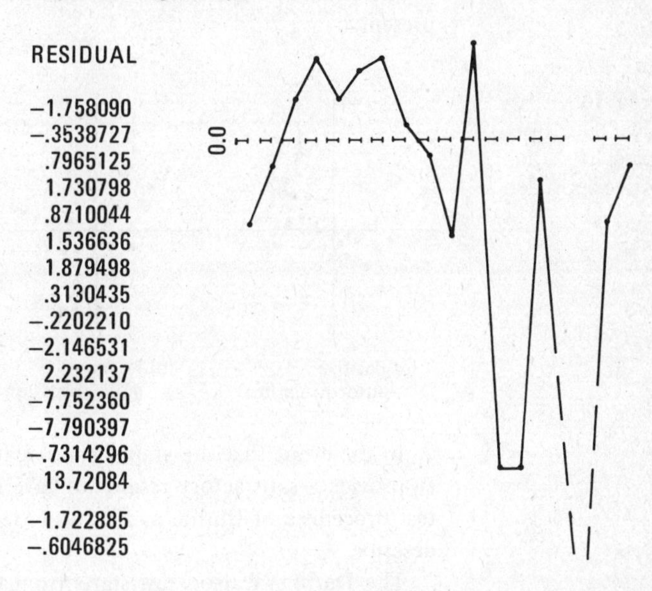

If we look at the signs on the value of the error terms, we see that the series starts with two negative values which are then followed by a longish string of six positive values. This in turn is followed by a string of five negatives broken only by a single positive. Two of the last three values are negative. Although we might normally hope for a few more observations on which to base any decision we make, we might be inclined to view this series of error-term values with some suspicion: negative followed by positive followed by negative is suggestive of positive autocorrelation. On the other hand, we have suggested that such a disposition of the error-term values could be an indication not of autocorrelation but of a specification error, i.e. a missing variable, whose influence is making itself felt via the disturbance term and is thus reflected in the error-term values. If this is the case, we would look for a relevant variable whose variation over the period in question mirrored that of the error term in Figure 12.13.

One difficulty with the above test is that if there is a tendency for values of the explanatory variable(s) to increase with time, as is unfortunately the case with many economic series, then this test for autocorrelation is identical to the test for non-linearity discussed earlier. One alternative test involves plotting the $\hat{u}_t$s against the $\hat{u}_{t-1}$s on a scatter diagram. If there is positive autocorrelation, most of the observations will tend to fall in the lower left-hand and upper right-hand quadrants, as shown in Figure 12.14(a). Negative autocorrelation produces a scatter, as in Figure 12.14(b), while the random scatter of Figure 12.14(c) would suggest that autocorrelation (at least of the first degree) is not present.

Figure 12.14
Graphical detection of autocorrelation: an alternative approach

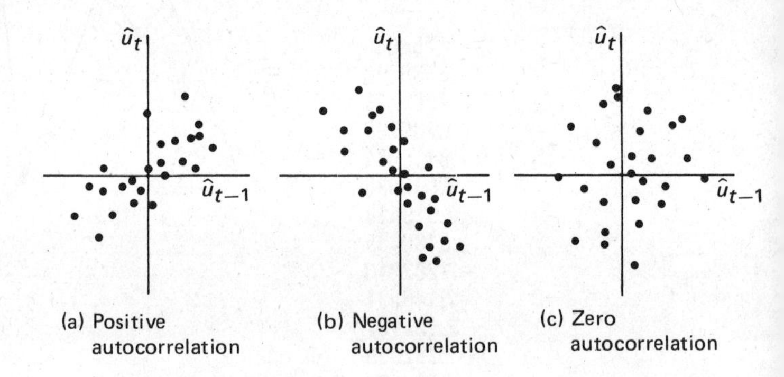

In the event that the graphical procedures fail, for any reason, to provide a satisfactory result, we have recourse to the analytic test procedure of Durbin and Watson (1951) which we shall now describe.

The Durbin–Watson test starts from the null hypothesis that autocorrelation is not present, i.e.

$$H_0: \rho = 0$$

$$H_1: \rho \neq 0$$

and we differentiate between these two using as the test-statistic the term

$$d = \frac{\sum_{t=2} (\hat{u}_t - \hat{u}_{t-1})^2}{\sum_{t=1} \hat{u}_t^2} \tag{12.81}$$

where d is known as the Durbin–Watson statistic and is compared with two critical theoretical values of d derived from a table of the Durbin–Watson statistic (see Table A.5 in the appendix) denoted d_L and d_U (the subscripts indicating 'lower' and 'upper' respectively). Notice that the numerator is a summation of $n - 1$

observations, i.e. from $t = 2$ to $t = n$, made inevitable by the presence of the $\hat{u}_{t-1}$ term.

The appropriate decision-rules are then as follows:

$$\left.\begin{array}{l} \text{if } d < d_L\text{: reject } H_0 \text{ in favour of positive autocorrelation} \\ \text{if } d > 4 - d_L\text{: reject } H_0 \text{ in favour of negative autocorrelation} \\ \text{if } d_U \leqslant d \leqslant 4 - d_U\text{: accept } H_0 \text{ (implies no autocorrelation)} \\ \left.\begin{array}{l}\text{if } d_L \leqslant d \leqslant d_U\text{:} \\ \text{or if } 4 - d_U \leqslant d \leqslant 4 - d_L\text{:}\end{array}\right\} \text{the test is inconclusive} \end{array}\right\} \quad (12.82)$$

These decision-rules are illustrated in Figure 12.15.

Figure 12.15 The Durbin–Watson test for autocorrelation

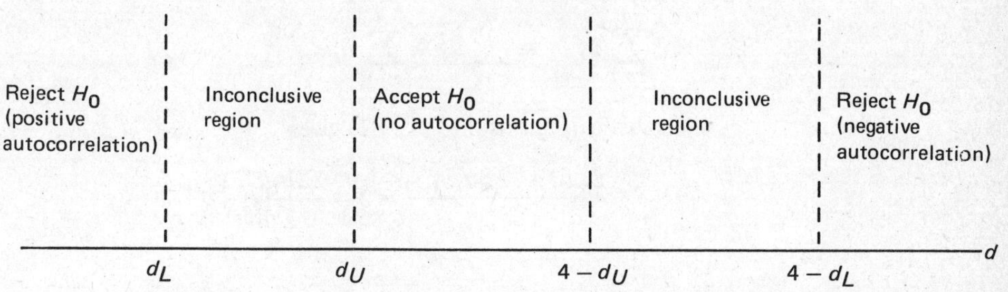

The value of d will normally be printed out automatically in most regression programs and thus does not in practice have to be calculated. We can see immediately that the inconclusive regions present some difficulty since if our value of d falls in one of these regions we can neither reject nor accept H_0, i.e. we cannot say whether or not autocorrelation is present.

Fortunately, Theil and Nagar (1961) have suggested that, provided the explanatory variables move reasonably smoothly, i.e. with no sudden large increases or decreases from one observation to another, then the appropriate boundary value is close to d_U. In this case the decision-rules become:

$$\left.\begin{array}{l} \text{if } d < d_U\text{: reject } H_0 \text{ in favour of positive autocorrelation} \\ \text{if } d > 4 - d_U\text{: reject } H_0 \text{ in favour of negative autocorrelation} \\ \text{if } d_U \leqslant d \leqslant 4 - d_U\text{: accept } H_0\text{, i.e. no autocorrelation} \end{array}\right\} \quad (12.83)$$

These revised decision-rules are illustrated in Figure 12.16.

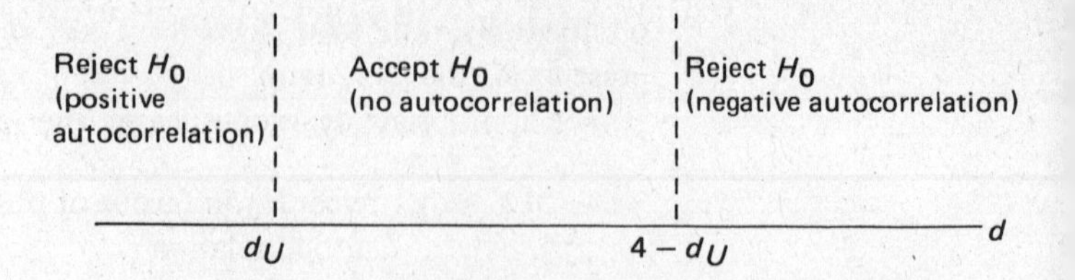

Figure 12.16 Theil and Nagar's approximation of the Durbin–Watson test

Finally, before we consider an example showing the use of these rules, it will be useful to develop a rule-of-thumb test for the detection of autocorrelation. To do this, consider (12.81) again. Expanding the numerator of this expression gives

$$d = \frac{\Sigma \hat{u}_t^2 + \Sigma \hat{u}_{t-1}^2 - 2\Sigma \hat{u}_t \hat{u}_{t-1}}{\Sigma \hat{u}_t^2} \tag{12.84}$$

Now $\Sigma_{t=2}^{n} \hat{u}_t^2$, $\Sigma_{t=1}^{n} \hat{u}_t^2$ and $\Sigma_{t=1}^{n} \hat{u}_{t-1}^2$ differ only by one observation so that (and particularly in large samples) we can consider them to be approximately equal, i.e.

$$\sum_{t=2}^{n} \hat{u}_{t-1}^2 \cong \sum_{t=2}^{n} \hat{u}_t^2 \cong \sum_{t=1}^{n} \hat{u}_t^2$$

and substituting this into (12.81) gives

$$d = \frac{2\Sigma \hat{u}_t^2 - 2\Sigma \hat{u}_t \hat{u}_{t-1}}{\Sigma \hat{u}_t^2} = 2 - \frac{2\Sigma \hat{u}_t \hat{u}_{t-1}}{\Sigma \hat{u}_t^2}$$

i.e.

$$d \cong 2(1 - \Sigma u_t u_{t-1} / \Sigma \hat{u}_t^2 \tag{12.85}$$

which can be written

$$d \cong 2(1 - \hat{\rho}) \tag{12.86}$$

since the second term inside the bracket in (12.85) is the expression for the OLS estimator of ρ in (12.73).

Now if there is no autocorrelation in the model, then $\rho = 0$, and (12.86) becomes

$$d = 2(1 - 0) = 2$$

If there is exact positive autocorrelation, then $\rho = 1$, and (12.86) becomes

$$d = 2(1 - 1) = 0$$

while if there is exact negative autocorrelation, $\rho = -1$, and (12.86) becomes

$$d = 2(1 + 1) = 4$$

Thus, as a rule of thumb, we can establish the following decision-rules. If d is close to 2, then we can assume no autocorrelation; if d is close to zero, we can assume positive autocorrelation; if d is close to 4, we can assume negative autocorrelation.

Let us now illustrate the application of the above rules by returning to the example relating total sales of coal between 1962 and 1977 to a general index of the retail price of fuel and light, first considered when we examined the assumption of linearity. The regression program prints out the following results:

$$Y_t = 201.3056 - 0.4817\, X_{2t} \qquad R^2 = 0.5792$$
$$(11.4577)\quad (0.1097) \qquad d = 0.3587$$

A plot of the error terms given on page 333 against time will yield a graph similar to that of Figure 12.9 (since X_2 increases more or less constantly with time) and is shown in Figure 12.17. (We could of course use the error terms to calculate d, employing (12.81) but since the value of d is printed out we can avoid this chore.)

Figure 12.17 Plot of error terms against time for coal-sales example

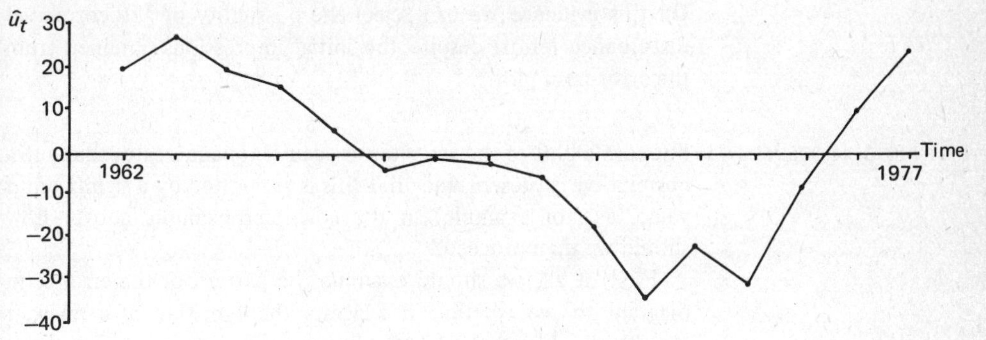

The graph is typical of positive autocorrelation, a string of positive error terms followed by a string of negative terms followed by more positive terms.

This evidence is supported by application of the rule-of-thumb test since $d = 0.3587$ is closer to zero than it is to 2, also suggesting positive autocorrelation. Let us now see whether this is further substantiated by the Durbin–Watson test.

To determine the critical values d_U and d_L from Table A.5 we need to know the number of observations, $n = 16$ in this example, and k', the number of explanatory variables *excluding* the constant term. In this example $k' = 1$, which gives the following critical values, for a significance level of 0.05,

$$d_L = 1.10; \quad d_U = 1.37$$

With $d = 0.3587$ and applying the decision-rules of (12.82) we get

$$0.3587 < 1.10$$

i.e.

$$d < d_L$$

and therefore we reject H_0: $\rho = 0$ in favour of the likelihood of positive autocorrelation. These results underpin the graphical and rule-of-thumb tests.

Finally, let us return once again to the computer output for the price–inflation model of section 12.6. Under the table of residual values the Durbin–Watson statistic is given as equal to 1.87187. We have already seen that graphical analysis of the $\hat{u}$s is suggestive of positive autocorrelation. This possibility is not supported, however, by this value of d (= 1.87187), which being close to 2 is indicative of no autocorrelation (using the rule-of-thumb test). We can confirm this result more formally using the Durbin–Watson tables. With $n = 17$ and $k' = 2$ we get $d_L = 1.02$ and $d_U = 1.54$ at 5 per cent significance. Thus the value of $d = 1.87187$ lies in the accept H_0 region (between d_U and $4 - d_U$). On this evidence we can reject the possibility of autocorrelated disturbance terms, despite the initial impressions obtained from the error-term plot.

Remedial measures

Suppose a plot of the error terms against time indicates that autocorrelation is present and that this is supported by a significant d value, as (for example) in the coal-sales example above: how should we then proceed?

First of all, we should examine the pattern of the error-term diagram to see whether it suggests the omission of a relevant variable. In Figure 12.17, for example, we may realise that the pattern of error terms corresponds closely to the movement in mean annual temperature for the years in question or to the variation in some other relevant variable. If this is the case, it might be profitable to include this variable in the model, repeat the regression analysis, and re-examine the results to ascertain whether the plot of the error terms and the statistic value still suggest autocorrelation. If they do not, we may have cured the problem, otherwise we have to turn to other measures.

The method we describe is another example of the generalised least squares approach we discussed briefly in the context of heteroskedasticity. The rationale of GLS is the same as before, i.e.

(a) Find a suitable transformation and apply it to the original data to remove the autocorrelation.

(b) Apply OLS to the transformed model, which will then yield BLU estimators.

In what follows we emphasise that we are considering first-order autoregression only since more complex relationships present too many difficulties to be considered here. For simplicity of exposition we also restrict ourselves to a model containing only one explanatory variable, though the method we shall discuss is quite general.

Thus we start with the model of (12.72) and (12.73), i.e.

$$Y_t = \beta_1 + \beta_2 X_{2t} + u_t \tag{12.72}$$

$$u_t = \rho u_{t-1} + v_t \tag{12.73}$$

If (12.72) holds in time period t, it holds in any other time period. In particular in time period $t - 1$ we have

$$Y_{t-1} = \beta_1 + \beta_2 X_{2,t-1} + u_{t-1} \tag{12.87}$$

and multiplying through by ρ gives

$$\rho Y_{t-1} = \rho\beta_1 + \rho\beta_2 X_{2,t-1} + \rho u_{t-1} \tag{12.88}$$

Subtracting (12.88) from (12.72) gives

$$Y_t - \rho Y_{t-1} = (\beta_1 - \rho\beta_1) + \beta_2 X_{2t} - \rho\beta_2 X_{2,t-1} + u_t - \rho u_{t-1}$$

or

$$(Y_t - \rho Y_{t-1}) = \beta_1(1 - \rho) + \beta_2(X_{2t} - \rho X_{2,t-1}) + v_t \tag{12.89}$$

using (12.73).

Now (12.89) contains the disturbance term v_t, which by assumption is well behaved (see (12.74) to 12.77)), and the application of OLS to (12.89) will produce BLU estimators of the parameters β_2 and $\beta_1(1 - \rho)$ and, provided ρ is known, of β_1 itself.

Unfortunately there are two difficulties with the transformation procedure described above. First, we do not know ρ (we shall consider this problem shortly), and second we are unable to compute the *first* observation on the transformed variables ($Y_t - \rho Y_{t-1}$) and ($X_{2t} - \rho X_{2,t-1}$) even if ρ is known, because, of course, there is no observation preceding Y_1 and $X_{2,1}$.

In fact, in place of these first observations we can use the following:

$$Y_1 = Y_1\sqrt{1 - \rho^2} \tag{12.90}$$

$$X_{2,1} = X_{2,1}\sqrt{1 - \rho^2} \tag{12.91}$$

It is important to include these first observations to ensure that the application of OLS to the transformed model yields BLU estimators, though in practice, especially if n is large, their omission may make little difference to the numeric values of the estimators obtained.

Now we turn to the second, more difficult problem, i.e. ignorance of the value of ρ. There are two possible approaches. The first is to estimate from (12.73) using

$$\hat{\rho} = \frac{\Sigma \hat{u}_t \hat{u}_{t-1}}{\Sigma \hat{u}_{t-1}^2} \tag{12.92}$$

However, $\hat{\rho}$ estimated thus is biased even though it is consistent.

The second possibility is to derive $\hat{\rho}$ from the expression for the Durbin–Watson statistic given by (12.86), i.e.

$$\hat{\rho} \cong 1 - d/2 \tag{12.93}$$

although, remember, this relationship is only approximate. For small samples Theil and Nagar (1961) have suggested the use of the following more accurate expression:

$$\hat{\rho} = [n^2(1 - d/2) + k^2]/(n^2 - k^2) \tag{12.94}$$

where k is the number of explanatory variables, including the constant term.

In the coal-sales example above $d = 0.3587$, so that using (12.93) gives

$$\hat{\rho} \cong 1 - 0.3587/2 = 0.8206$$

or using (12.94),

$$\hat{\rho} = 0.8495$$

As it happens, the application of this GLS estimation procedure does not in this case produce improved estimators or 'improve' (except marginally) the d-statistic value. This may suggest several things. For example, we might try a higher-order autoregressive model, or it might be a different type of autocorrelation. Alternatively, it might be that the problem is caused by specification error, i.e. the exclusion of a relevant variable, rather than by correlation between the disturbances, and the search for and inclusion of such a variable would be the next step.

A.6: non-stochastic Xs

Assumption (A.6) asserts that the explanatory variables should be non-stochastic and have fixed values in repeated samples. The first part of this assumption means that we should be able to predetermine or fix the values taken by the explanatory variables and then observe the corresponding values taken by the dependent variable. The second part of this assumption requires that we should be able to repeat the analysis as often as we wish using the same values of the X_js each time. In practice in most economic and social science applications we are not able to choose the values of the explanatory variables. If we were examining national income–expenditure relationships, for example, the figures we have to use for our analysis are those which we find in official sources. We take these values as given; we cannot pre-select figures for national income. In other words, the values of the explanatory variable will usually be stochastic in nature. Furthermore, for the same reasons as outlined above, we are usually confined to a single sample. Thus on both counts the reality is that assumption (A.6) will be doubly violated.

However, the fact that we cannot keep the values of the X_js fixed in repeated samples (because we cannot normally take repeated samples) is not of great importance. This part of assumption (A.6) is really for mathematical convenience as much as anything else. It enables us to derive the desirable properties of the OLS estimators rather more simply than would be the case if we had to allow for variation in the values from one sample to another.

The non-stochastic nature of the explanatory variables is, however, rather more important since subsumed by this assumption is the very important notion that *there is no statistical relationship between the explanatory variables and the disturbance terms.* The rationale here is that if the disturbance terms are random, and if the explanatory variables are non-stochastic, then there can be no systematic relationship between them. However, if the X_js are not, then there is a possibility of such a systematic relationship arising between the disturbances and the X_js; such a relationship can have serious consequences, as we shall see.

There are three possibilities:

1. X_{ji} and u_h are statistically independent, for $i = 1, 2, \ldots, n$; $h = 1, 2, \ldots, n$. This means that X_{ji} and u_h are independent for all i and h, i.e. it implies that $\text{cov}(X_{ji}u_h) = 0$.

2. X_{ji} and u_i are contemporaneously uncorrelated, for $i = 1, 2, \ldots, n$. This means that X_{ji} and u_i are uncorrelated for all i, i.e. $\text{cov}(X_{1i}u_i) = \text{cov}(X_{2i}u_i) = \ldots \text{cov}(X_{ki}u_i) = 0$.

3. X_{ji} and u_i are neither independent nor contemporaneously uncorrelated, i.e. $\text{cov}(X_{ji}u_i) \neq 0$.

Considering the third case first we can state that in the event of (3) being true then the OLS estimators will have *none* of the desirable properties usually associated with them and this is true in both large sample as well as small sample situations.

In the event of case (2) i.e. contemporaneous uncorrelation, the OLS estimators do not retain all of their desirable properties. In particular they are not unbiased in small samples but they still retain the property of consistency and still therefore have all the desirable asymptotic properties. However, since in general there are no estimators which yield better results than the OLS estimators, these may still be used in small sample situations.

Finally, in the first case when X and u are independent, then the OLS estimators will still have all the desirable properties regardless of whether the X_js are non-stochastic and fixed in repeated samples or not. This is an important result since it means that provided we can replace assumption (A.6) with a weaker variant, i.e.

X_{ji} are u_h are statistically independent, i.e.

$$\text{cov}(X_{ji}u_h) = 0 \text{ for all } i \text{ and } h \qquad (A.6')$$

then the application of OLS will yield estimators with all the desirable properties.

Thus the breakdown of assumption (A.6) resolves itself into the question of how the explanatory variables and the disturbance variable are related. Provided assumption (A.6) can be replaced by the somewhat weaker assumption described by (A.6′) then we can proceed to apply OLS without hesitation. The detection of the breakdown of (A.6) is thus essentially a test of the independence of the Xs and the us rather than with an examination of the stochastic or non-stochastic nature of the X_js. How might we proceed to test assumption (A.6)?

One possibility which suggests itself immediately is to examine the value of the correlation coefficient between each X_j and the disturbance term values, u_i, or rather, since we do not know the u_i values, between each X_j and the $\hat{u}_i$ values which we can observe. Unfortunately this approach is not feasible since, by the construction of the model, the covariance and hence the sample correlation between the X_j and the $\hat{u}_i$ is constrained to be zero. We can demonstrate this as follows. For convenience consider a model with only one explanatory variable, X_j. Using (11.7) the correlation between X_j and the $\hat{u}_i$ is defined as

$$r_{X_j\hat{u}} = \Sigma(X_{ji} - \bar{X}_j)(\hat{u}_i - \bar{\hat{u}}_i)/\sqrt{\Sigma(X_{ji} - \bar{X}_j)^2\Sigma(\hat{u}_i - \bar{\hat{u}}_i)^2} \qquad (12.95)$$

Consider the numerator of this expression. We have previously shown, in going from equations (11.50) to (11.51), that $\bar{\hat{u}} = 0$; therefore, the numerator of (12.95) becomes

$$\Sigma(X_{ji} - \bar{X}_j)\hat{u}_i = \Sigma x_{ji}\hat{u}_i = 0$$

using (11.21b).

Thus the numerator of (12.95) is zero and the correlation between X_j and $\hat{u}$ is necessarily also zero. We cannot therefore use r as a test of assumption (A.6). In fact there are no satisfactory tests of this assumption and one can only resort to an examination of the economic theory underlying the model to see whether any of the explanatory variables is likely to be correlated with the disturbance variable.

In practice, some authors claim that some correlation (even if the association is weak) between these variables is inevitable in most economic models because of the pervasive presence of measurement error in the variables. However, we cannot in this book consider such problems. Moreover, correlation between explanatory variables and disturbance terms is a characteristic feature of simultaneous-equation models, which similarly are beyond the range of this book.

A.7: independence of Xs

Assumption (A.7) states that there should be no exact linear relationship between any combination of the explanatory variable set. If such a condition should exist, the model is said to suffer from *multicollinearity*. In practice a model is said to suffer from multicollinearity even if the relationship between the explanatory variables is not exact, and in this sense we can consider two extreme cases: the first of zero multicollinearity when there is *no* linear relationship to be found in the explanatory variable set; the second of perfect multicollinearity when the relationship is exact. In between these two limiting cases a model can suffer any degree of multicollinearity depending on how exact or inexact the linear relationship is. The two extreme cases will be rare in most economic applications but in general most economic models, particularly of the time-series type, will suffer to a greater or lesser extent from the problem. This is because, by the very nature of economic phenomena, variables frequently move up or down together, not because they are intrinsically related necessarily, but because they are often subject to the same inflationary or recessionary forces.

For example, suppose we are studying the demand for new cars over the past twenty years and two of the explanatory variables in the model are the price of new cars and average earnings. Both of these variables have increased by much the same proportion during this period and the correlation coefficient between them is likely to be quite considerably different from zero. Or suppose a model purporting to explain investment over the past twenty years contained both short-term and long-term interest rates as separate variables on the right-hand side of the equation. It is highly likely that these two interest-rate variables would have moved up and down together over the period in question and they would accordingly be correlated.

Multicollinearity is thus most likely to occur in time-series data, where most economic variables are subject to similar upward and downward pressures. In times of economic growth, for example, we are likely to find all the principal economic variables (e.g. employment, earnings, imports, prices, savings, investment, etc.) increasing together in not dissimilar amounts. Similarly, in periods of recession we may find that they all decrease together in the same way.

Although multicollinearity is more likely to occur in time-series data, it is also to be found in cross-section studies. For example, in a model relating household expenditure in some particular week to income and to wealth (as well as to other variables) for a representative cross-section of, say, 100 households, we would certainly expect income and wealth to be positively correlated with each other.

Before we examine the consequences of multicollinearity it should be noted that the problem arises out of the *sample* data rather than from any relationship in the underlying theoretical structure of the *population* of explanatory variables, since the latter are assumed to be non-stochastic by assumption (A.6′). In other words, when we specify which variables are to be included in the population equation we implicitly assume that they are all truly independent of one another (if not, this fact should be specified properly as part of the model). However, in any given *sample* it may just be the case that the explanatory variables (or at least some of them) display some form of linear relationship. However, some writers have suggested that independence in the population of variables is unlikely in practice. This being the case, the problem of multicollinearity could not be entirely a *sample* phenomenon.

Consequences

In the unlikely event of there being *perfect* collinearity in the explanatory variable set, the consequences for OLS estimation are severe; it becomes impossible to calculate the value of the estimators. In terms of the mathematics of the estimation process we end up, because of the exact relationship between the X_js, trying to divide 0 by 0, which is indeterminate (for a proof of this see, for example, Gujarati, 1978).

Intuitively, we can see that this is true if we recall that $\hat{\beta}_j$ (the partial regression parameter) represents the effect on Y of a unit change in X_i with all the other explanatory variables held constant. However, if the variables are all perfectly correlated, it will be impossible to change any of the X_j without the other variables changing at the same time. It becomes impossible to distinguish the individual and unique effects on Y of any of the explanatory variables.

We can illustrate this indeterminacy of the estimators in a somewhat different way. Assume that we are trying to explain household expenditure Y_i, using as explanatory variables household average earnings X_{2i}, household liquid assets X_{3i} and average retail prices X_{4i}. Thus we have the model

$$Y_i = \beta_1 + \beta_2 X_{2i} + \beta_3 X_{3i} + \beta_4 X_{4i} + u_i \qquad (12.96)$$

Let us suppose that over the period in question the three explanatory variables move together in such a way that

$$X_{2i} = 4X_{3i} + 0.5X_{4i} \qquad (12.97)$$

If we substitute (12.97) into (12.96) we have

$$\begin{aligned} Y_i &= \beta_1 + \beta_2(4X_{3i} + 0.5X_{4i}) + \beta_3 X_{3i} + \beta_4 X_{4i} + u_i \\ &= \beta_1 + (4\beta_2 + \beta_3)X_{3i} + (0.5\beta_2 + \beta_4)X_{4i} + u_i \qquad (12.98) \end{aligned}$$

Thus, when we come to estimate (12.98) we will be unable to obtain separate values for the estimators $\hat{\beta}_2$, $\hat{\beta}_3$ or $\hat{\beta}_4$ and will be left only with values for the composite terms $(4\beta_2 + \beta_3)$ and $(0.5\beta_2 + \beta_4)$, which in general will not provide any useful information, since it will be impossible to arrive at separate values for β_2, β_3 or β_4 from these composite terms.

Fortunately, perfect multicollinearity will be a rare event in economic data. Just as rare, however, will be a *complete* absence of any statistical relationship in the explanatory variable set. The most usual situation will thus be some degree of multicollinearity which may vary from the slight to the severe. In this event (A.7) is, strictly speaking, not violated, since it stipulates 'no *exact* relationship'. The application of OLS will in these circumstances produce estimators which have all the desirable properties, i.e. are still BLU estimators. However, this may be of very little comfort in models which suffer severe multicollinearity since although the OLS estimators may still have smaller variances than any alternative linear unbiased estimator, these variances may be quite large and provide estimators which have little precision and are as a consequence unreliable.

To see how multicollinearity inflates the variance of the estimator recall that earlier in this chapter, in our discussion of estimation in multiple regression, we established the following expressions for the OLS estimator and its standard error:

$$\hat{\beta}_j = \Sigma \hat{v}_i \hat{w}_i / \Sigma \hat{v}_i^2 \qquad (12.9)$$

$$s_{\hat{\beta}_j} = \sqrt{\hat{\sigma}^2 / \Sigma \hat{v}_i^2} \qquad (12.15)$$

where, to refresh memories, the $\hat{v}_i$ are the residuals resulting from the regression of X_j on all the other explanatory variables and the $\hat{w}_i$ are the residuals from the regression of Y on all the explanatory variables except X_j.

In this sense the $\hat{v}_i$ represent the variation in X_j which it does not share with any of the other explanatory variables; for this reason it is sometimes referred to as the *pure* variation in X_j.

Now when the degree of multicollinearity is severe, all the explanatory variables suffering from it will be very closely related, and the residuals obtained by regressing X_j on these other explanatory variables will be small, i.e. each $\hat{v}_i$ will be small, and the sum of the v_i^2s, Σv_i^2, will also be small.

This means that the denominator of equation (12.15) will be small and thus $s_{\hat{\beta}_j}$ will be large. The closer the relationship between X_j and the other explanatory variables, the larger will $s_{\hat{\beta}_j}$ tend to be.

We can show explicitly the relationship between the degree of multicollinearity and the standard errors of the estimators in the two-variable model, i.e.

$$Y_i = \beta_1 + \beta_2 X_{2i} + \beta_3 X_{3i} + u_i \tag{12.99}$$

in which the relationship between X_{2i} and X_{3i} is given by

$$X_{3i} = kX_{2i} + d_i \tag{12.100}$$

Expression (12.100) indicates that the variation in X_{3i} is not completely determined by X_{2i}: some part of it is due to the disturbance term d_i.

Now it can be shown that for this model the estimators have standard errors given by

$$s_{\hat{\beta}_2} = \frac{\hat{\sigma}}{\sqrt{\Sigma x_2^2(1 - r_{23}^2)}} \tag{12.101}$$

$$s_{\hat{\beta}_3} = \frac{\hat{\sigma}}{\sqrt{\Sigma x_3^2(1 - r_{23}^2)}} \tag{12.102}$$

In (12.101) and 12.102) r_{23} is the simple correlation coefficient between X_{2i} and X_{3i}. In the two-variable model this is a measure of the multicollinearity in the model.

By examining these equations we can see that as multicollinearity becomes more severe, i.e. as r_{23}^2 increases, the whole of the right-hand side and thus the standard error also increases. Notice that when multicollinearity is perfect, i.e. when $r_{23}^2 = 1.0$, then the $s_{\hat{\beta}_j}$ equal infinity.

The direct consequence of all this is that we are more likely to be unable to reject the null hypothesis H_0: $\beta_j = 0$, even though the parameter may well have a true value which is non-zero. This means that there will be a tendency for us to judge parameters (and hence variables) as being insignificant and drop them from the model when perhaps they should in fact be included.

The increased size of the standard errors also inevitably causes a lack of precision in the estimator value. When the $s_{\hat{\beta}_j}$s are large we can have little faith that any $\hat{\beta}_j$ will be close to the true parameter value.

There is one further important aspect of this imprecision in the estimates which we must now consider. We have seen earlier that the distribution of the OLS estimators, under the assumptions of the classical normal linear regression model, is normal. What we have not mentioned is the fact that the distribution of any particular OLS estimator is *not independent* of the distributions of the other estimators and this lack of independence is a function of and increases as the correlation between the variables increases.

We can see this by looking at equation (12.103), which, it can be shown, is equal to the covariance between the estimators in a model with two explanatory variables, such as that of equation (12.89):

$$\text{cov}(\hat{\beta}_2\hat{\beta}_3) = \frac{-\sigma^2 r_{23}}{(1 - r_{23}^2)\sqrt{\Sigma x_2^2 \Sigma x_3^2}} \qquad (12.103)$$

That is, the covariance (as was the case with the standard errors) is a function of r_{23}, the correlation between X_2 and X_3. As we have seen in the two-variable model, r_{23} can be taken as a measure of the degree of multicollinearity, so equation (12.103) means that the worse the degree of multicollinearity, the stronger is the covariance between $\hat{\beta}_2$ and $\hat{\beta}_3$, i.e. the more highly correlated they are.

As we have explained, economic variables are likely to suffer *positive* multicollinearity (moving up or down together) and so r_{23} is likely to be *positive*. This makes the right-hand side of (12.103) negative, which in turn means that $\hat{\beta}_2$ and $\hat{\beta}_3$ will be *negatively* correlated with each other. In other words, if a sample regression yields a value for $\hat{\beta}_2$ *smaller* than the true β_2, then the same sample is likely to yield a value for $\hat{\beta}_3$ *larger* than the true β_3, and vice versa. For example, suppose the true value of β_2 was 3.0 and of β_3 was 6.0. Now if a particular sample yields a value for $\hat{\beta}_2$ of 2.5 (i.e. an underestimate), then multicollinearity causes it to be likely that $\hat{\beta}_3$ will have a value of, say, 7.0 (i.e. an overestimate).

We are not talking here about *bias*: the means of the distributions of $\hat{\beta}_2$ and $\hat{\beta}_3$ are still the same; it is the values we obtain from a particular sample that are likely to be affected in this way.

This is not all; the negative correlation between the estimators, coupled with the tendency for non-zero parameters to be judged as not significantly different from zero, compounds the problem. To see this consider a model which purports to explain the savings decisions of small savers in terms of the interest rate on bank deposits i_2 and the interest rate on building society deposits i_3, i.e.:

$$S_t = \beta_1 + \beta_2 i_{2t} + \beta_3 i_{3t} + u_t \qquad (12.104)$$

where both β_2 and β_3 are truly non-zero. The explanatory variables i_2 and i_3 are likely to be highly correlated, and thus the standard errors of both estimators will be inflated. Suppose a particular sample, because of the influence of multicollinearity, yields a near-zero value for $\hat{\beta}_2$; then, because of the negative correlation between $\hat{\beta}_2$ and $\hat{\beta}_3$, this underestimation of $\hat{\beta}_2$ is likely to result in an overestimation of $\hat{\beta}_3$.

Thus this sample regression gives a result which seems to indicate that i_2, the interest rate on bank deposits, is insignificant, whereas i_3, the interest rate on building society deposits, is not only highly significant but appears to have an importance (because of the overestimate of $\hat{\beta}_3$), in terms of its effect on savings, much greater than is actually the case.

A different sample, on the other hand, might lead to $\hat{\beta}_3$ taking a near-zero value, in which case the interdependence of the estimators will result in $\hat{\beta}_2$ assuming a value larger than its true value and appearing to be significant. Whether we happen to get the first results or the second will depend entirely on which particular sample we happen to choose.

So the loss of precision in the OLS estimators takes on two dimensions: not only are the estimators prone to take on near-zero values (even when the parameters are truly non-zero), but also the value which any estimator takes depends upon the values taken in a particular sample by any other estimators it happens to be collinear with.

There is one last problem we must consider. Models with multicollinear variables have estimators which can be extremely unstable, i.e. they can be very sensitive to even small changes in the data.

A change in this regard might, for example, amount to the addition of an extra observation or two, or a correction in the value of one of the observations, or the loss of an observation because of unreliability, and so on. For models in which multicollinearity is serious such small changes in the data are likely to cause quite large changes in the estimators (sometimes even causing a change in sign). This is explained by the fact that even a small change in the data or an extra observation constitutes in effect a *different sample*, and we have seen that the inflated standard errors mean that the range of values from which $\hat{\beta}_j$ can be 'picked' is large. Furthermore, this instability is compounded by the negative correlation between the estimators.

As we have seen, the consequences of serious multicollinearity can be calamitous. We cannot tell which variables rightly belong in the model, we cannot rely on the results from any particular sample, while adding observations or changing the data slightly can cause violent changes in the estimates and their standard errors. Adding or subtracting variables can produce similar consequences. In view of the fact that multicollinearity is endemic in economic data, and serious multicollinearity is therefore a strong possibility, it becomes imperative to be able to test for multicollinearity in models that we use. This entails answering the question: 'What constitutes a "serious" degree of multicollinearity?' In the next section we shall examine these aspects of the problem.

Detection

We can state immediately that there is no generally accepted test of multicollinearity, in the sense that most regression programs will include it in their analysis. However, in the model with two explanatory variables measurement of the degree of multicollinearity is not a problem. As we have already mentioned, the zero-order (or simple) correlation between the two variables is a direct measure of the multicollinearity and will normally be available in the output of a computer regression program.

In models with more than two explanatory variables the problem is much more intractable. The zero-order correlation coefficients between all possible pairs of explanatory variables will indicate multicollinearity (if it takes a value close to 1.0) between the pair of variables in question, but if none of these correlation values is close to 1.0 it does not mean that a more complex (i.e. involving more than two variables) multicollinear structure does not exist in the model. So although the high zero-order correlation coefficients for the explanatory variables will be observable in the correlation matrix output from the computer, they will not usually prove of much help, since they are sufficient but not necessary for multicollinearity.

As a rule of thumb we might say that one indication of serious multicollinearity is when a high R^2 value coincides with no significant parameters in the estimated model, but at best this is only a rather general qualitative measure and is not foolproof since there are situations where explanatory variables are significant when R^2 is high but multicollinearity none the less exists in the model.

We could test for the seriousness of multicollinearity by re-estimating the equation with a small number of observations missing and observing the changes in the estimators and their standard errors. Alternatively, we could drop and add variables to the model and again examine the subsequent changes in the values of the estimators. These measures are also only qualitative in nature and not entirely conclusive.

One possible approach is to regress each explanatory variable in turn on all the other explanatory variables, i.e. estimate the regressions

$$X_{2i} = a + b_3X_{3i} + b_4X_{4i} + \dots \qquad R^2_{2.34\dots} = \ ; F =$$

$$X_{3i} = c + d_2X_{2i} + b_4X_{4i} + \dots \qquad R^2_{3.24\dots} = \ ; F =$$

and so on. Those regressions for which the F-value is significant are taken as displaying a multicollinear relationship (if the F-values are not printed out, they can easily be computed using the R^2-values). (For a fuller description of this approach see Farrar and Glauber, 1967.)

In practice we may find that many of these equations have significant F-statistics attached to them (because of the pervasive nature of multicollinearity in most economic data, referred to earlier). Which of these we take as being 'serious' is open to argument. Klein (1962) suggested that a variable displays troublesome multicollinearity if the R^2 for its regression on the other explanatory variables were bigger than R^2, the coefficient of determination for the regression of Y on *all* the explanatory variables. However, this test has been criticised by various writers.

In essence there is as yet no satisfactory answer as to what constitutes a serious, i.e. a harmful, degree of multicollinearity. We know what the effects of serious multicollinearity are and we know that these effects increase as the degree of multicollinearity increases. But even if we manage to identify which variables are causing the multicollinearity, should we then drop them from the model? They are after all included in the first place because economic theory indicated that they should be present for the model to be correctly specified.

Remedial measures

We have seen how potentially harmful the presence of a high degree of multicollinearity can be for the estimation process. This inevitably leads to the question: 'If we have evidence to suggest multicollinearity in a model, what can we do about it?' The immediate answer is, usually not very much. In some situations we may be able to use one of the approaches mentioned below, but for one reason or another these 'solutions' must be considered to be not usually available. We shall deal briefly with each in turn.

Dropping variables. Since multicollinearity stems from a relationship between some or all of the explanatory variables, one solution that occurs immediately is to drop the offending variables. Which variables should be dropped? One could consider dropping the variable which has the highest R^2-value, as revealed by the test described above (p. 364), and the Farrar and Glauber procedure is supposed to identify those variables which are most collinear. Alternatively, one might appeal to *a priori* knowledge of the variable set. For example, if several different rates of interest are being used to explain, say, investment or the demand for money, one would suspect these variables of being collinear, and by dropping one or more of them from the model the multicollinearity problem might be solved. In the absence of any prior information, and in circumstances in which multicollinearity is strongly suspected, Stewart (1976) has suggested as a rule of thumb the dropping of those variables having a t-value between -1 and $+1$ when OLS is applied to the original model. In normal circumstances (i.e. in the absence of multicollinearity) one would consider dropping variables from the model if their t-values lay (approximately) between -2 and $+2$, and Stewart's suggested rule is deliberately less severe in recognition of the fact that the variables in question might well otherwise be significant.

The problem is that dropping variables which rightfully belong in the model produces what is known as specification error. This inevitably leads to biased estimators for those variables left in the model, and the cure might prove worse than the disease. Besides which, the variable or variables which it would seem one might have to drop from the model might well be those in which one is most interested!

Pooling of time-series and cross-section data – the use of extraneous information. Suppose we are trying to estimate the following model, which relates the aggregate national expenditure on, say, food F_t to national disposable income Y_t, to a food price index P_t, and to an 'all other commodities' price index P_{st}, using annual time-series data, i.e.

$$F_t = \beta_1 + \beta_2 Y_t + \beta_3 P_t + \beta_4 P_{st} + u_t \tag{12.105}$$

Let us assume, to make things easier, that Y_t and P_t are highly correlated (as we might expect them to be) but that P_t and P_{st} are not (this is not a very realistic assumption).

Because of multicollinearity β_2 and β_3 are likely to prove difficult to estimate. However, suppose we also have available the results of a cross-section study of, say, 200 households (preferably with wide-ranging income levels) relating their food expenditure to their disposable income, i.e.

$$F_i = \beta_1 + \beta_2 Y_i \tag{12.106}$$

The numeric value of the estimate $\hat{\beta}_2$ should provide a reasonably reliable estimate of the income parameter since over an 'instant' in time prices will not vary much. Thus we can substitute the known value $\hat{\beta}_2$ from (12.106) into (12.105) and then proceed to estimate the model

$$(F_t - \hat{\beta}_2 Y_t) = \beta_1 + \beta_3 P_t + \beta_4 P_{st} + u'_t \tag{12.107}$$

This model contains only two explanatory variables which we have assumed are not correlated; therefore, multicollinearity is not present in (12.107).

The technique does present some difficulties, one of which is the compatibility of the β_2 from the cross-section study being an adequate substitute for the β_2 from the time-series model. Furthermore, if the price variables *are* collinear, multicollinearity will still be present in (12.107) and β_3 and β_4 will still present difficulties of estimation.

Use of a priori information. Sometimes we may have knowledge or experience which will enable us to make a good guess as to the value of one or more estimators in a model. As an example consider the following model, which relates UK household expenditure Y_i to income X_{2i} and wealth X_{3i}:

$$Y_i = \beta_1 + \beta_2 X_{2i} + \beta_3 X_{3i} \tag{12.108}$$

In this model β_2 represents the marginal propensity to consume and it is possible that we might have a good idea as to the value of this parameter in the United Kingdom. This being so, we could substitute this value for β_2 into (12.108), take the term $\hat{\beta}_2 X_{2i}$ to the left-hand side and estimate the model with wealth as the only explanatory variable, thereby removing the threat of multicollinearity. This method is equivalent to imposing a linear restriction on the model, and this reduces the variance of the remaining estimators without introducing bias. Of course, the accuracy of the value that we obtain for β_3 in the amended model depends greatly on the assumed *a priori* value of β_2.

Obtaining more data. Some authors suggest that the introduction of new data or of extra data may help reduce multicollinearity. This, however, will only be true if the additional data are not contaminated by multicollinearity in the same way as the old. What we need essentially is not more data but more information.

In practice this suggested solution is not often of much help since econometricians usually try to get hold of and use as many observations as exist in the original estimation procedure. Besides, what sort of new data (even supposing we could get some) would be most suitable? (For some ideas on this see Silvey, 1969.)

First differences. It has been suggested that taking first differences of the variables might reduce the severity of the multicollinearity. This means that in the model

$$Y_t = \beta_1 + \beta_2 X_{2t} + \beta_3 X_{3t} + u_t$$

we would estimate

$$Y_t - Y_{t-1} = \beta_2(X_{2t} - X_{2,t-1}) + \beta_3(X_{3t} - X_{3,t-1}) + u_t - u_{t-1}$$

However, this approach has the following disadvantages:

(a) It may introduce autocorrelation into the model (although this could be tested for).
(b) We lose one observation and therefore one degree of freedom. If we already have a small number of observations this may involve an unacceptably large increase in the standard errors.
(c) The method may be difficult to apply in cross-section data, where no logical ordering of the observations is obvious.

Multicollinearity and forecasting. It has been suggested by various authors that multicollinearity is not a problem if the model is to be used only for forecasting the value of the dependent variable in some future time period and not for determining the relative importance of each variable separately.

This, however, will only be true if the system of inter-correlations among the explanatory variables remains unchanged from the estimation period to the forecasting period.

Furthermore, if we want to forecast the likely value of Y_t as a result of policy changes in a certain key variable or variables, the imposition of our own specific values on the variables in question may break the pattern of interdependence characteristic of these variables. This may introduce considerable degrees of error into the forecast.

Other methods of dealing with multicollinearity have been suggested including the use of what is known as *principal components* (Bennett and Bowers, 1974). However, as well as suffering certain drawbacks, discussion of this and other approaches is beyond the scope of this course.

Summary of multicollinearity section

The problem of multicollinearity can cause serious difficulties. The standard errors of the variables involved will be inflated and there is a corresponding increase in the probability that we shall accept H_0 when we should in fact reject it, i.e. relevant variables will be seen as insignificant. Coupled with this, there is a loss in precision in the estimators; they are more likely to be seriously in error and to be seen as not different from zero. Furthermore, the value taken by each estimator is conditional upon the value taken by those estimators with which it is associated. For the usual case of positively correlated variables, if one estimate is too low, the other (in the two-variable model) will be too high, and vice versa. Finally, the estimates are very sensitive to even small changes in the data (e.g. corrections or new observations).

Detecting the presence of multicollinearity is a problem which has not been satisfactorily solved, in the sense that no general analytic procedure has been developed. One may be aware of multicollinearity by its symptoms as much as by the results of any formal test. These include a high R^2-value with no significant estimators, and instability in the estimates and their standard errors when variables are added to or dropped from the model or when small changes are made to the data.

Finally, remedial measures are not promising and most will usually not be available. We may take some comfort from the fact that forecasting may not be seriously affected by multicollinearity.

We may conclude by stating the obvious. Multicollinearity can be very troublesome for the econometrician. If its presence is suspected, great circumspection should be exercised before variables are dropped from the model on the basis of the standard errors.

12.8 Summary

This concludes our description of the multiple regression model. In this chapter we have seen how the simple regression model containing only one explanatory variable can be extended to include several such variables. In retrospect we can see that the former is merely a special case of the latter. The move from simple to multiple regression does not involve any fundamental changes in the underlying concept of the regression process. However, it does make the estimation procedure more difficult, and in practical terms models with more than one explanatory variable will be solved using an appropriate computer program.

We described one such program in this chapter (although there are several other well-established programs we might have chosen instead). Following this, we then discussed the consequences of a breakdown in each of the basic assumptions governing the classical normal linear regression model. In each case we examined the causes, the consequences for OLS estimation, detection of the breakdown and what remedial measures are available. In practice it is possible for there to be a breakdown in more than one basic assumption at a time; unfortunately we were not able to extend our discussion to include such possibilities.

There are several more areas which we did not have the space to discuss. One such is the concept of simultaneous-equation systems. However, adequate discussion of this topic can be found in many books, for example Koutsoyiannis (1977) or Kmenta (1971). Nor have we covered estimation with qualitative (or dummy) variables or the problem of lagged variables. Discussion of both these subjects can be found in either of the above two books, in Maddala (1977) and in Gujarati (1978). The choice of suitable variables for a regression model when there are a potentially large number to choose from may be aided by the use of principal components; a discussion of these (and several other interesting ideas in multivariate analysis) is to be found in Bennett and Bowers (1974).

EXERCISES

12.1 In a multiple regression model containing two explanatory variables, i.e.

$$Y_i = \beta_1 + \beta_2 X_{2i} + \beta_3 X_{3i} + u_i$$

the estimated disturbance terms are given by

$$\hat{u}_i = Y_i - \hat{Y}_i = Y_i - \hat{\beta}_1 - \hat{\beta}_2 X_{2i} - \hat{\beta}_3 X_{3i}$$

and

$$\Sigma \hat{u}_i^2 = \Sigma (Y_i - \hat{\beta}_1 - \hat{\beta}_2 X_{2i} - \hat{\beta}_3 X_{3i})^2$$

Partially differentiate this last expression with respect to $\hat{\beta}_1$, $\hat{\beta}_2$ and $\hat{\beta}_3$ and equate to zero to minimise $\Sigma \hat{u}_i^2$. Combine these three equations and solve them to provide expressions for the OLS estimators.

12.2 Table 12.4 gives the zeroorder correlation coefficients between the variables X_2, X_3 and Y in the model

$$Y_i = \beta_1 + \beta_2 X_{2i} + \beta_3 X_{3i} + \beta_4 X_{4i} + u_i$$

Table 12.4

	X_2	X_3	X_4
X_2	1.0	0.1	−0.8
X_3		1.0	0.2
X_4			1.0
Y	0.8	0.1	−0.7

Illustrate this situation using a diagram similar to that in Figure 12.1 showing the area of 'influence' of the variables.

An OLS regression equation is estimated of (a) Y on X_2, (b) Y on X_2 and X_3, (c) Y on X_2, X_3 and X_4. Discuss possible differences you might find in the signs, values and significance of $\hat{\beta}_2$, $\hat{\beta}_3$ and $\hat{\beta}_4$ in each of the three equations.

12.3 What difference, if any, would careful consideration of assumption (A.7) make to your answer to exercise 12.2?

12.4 Explain why, as variables are successively added to a model, successively increasing values of R^2 are not necessarily indicative of an increasing goodness of fit.

12.5 Table 12.5 shows the values of R^2 as variables are added to a model:

(a) Use the F-test to test the significance of R^2 in each case.
(b) Calculate $\bar{R}^2$ in each case and hence compare the relative goodness of fit of the three models.
(c) Use the F-test to examine the significance of the added variables from $k = 2$ to $k = 4$.

Table 12.5

Number of explanatory variables k	R^2
2	0.800
3	0.850
4	0.855
$n = 20$	

12.6 Use the data in exercise 12.2 to calculate the three first-order partial correlation coefficients. Compare your results with the zero-order values and comment.

12.7 Hines (1964) suggested that wage inflation could be explained by unionisation. For annual data from 1949 to 1961 Hines used OLS to obtain the following relationship:

$$\Delta W_t = -64.2058 + \underset{(0.8436)}{4.3165}\Delta T_t + \underset{(1.2869)}{1.7318}T_t \qquad R^2 = 0.675 \quad d = 3.13$$

where ΔW_t = annual rate of change in money wage rates
ΔT_t = annual rate of change in unionisation
T_t = annual level of unionisation

Get hold of a copy of Hines's paper, in particular to see how he defines his variables, and repeat the exercise using more recent data. Compare your results with those of Hines.

12.8 Draw up a table listing the consequences for the properties of the OLS estimators and for the estimation process in general of a breakdown in each of the basic assumptions governing the classical linear normal regression model (CLNRM).

12.9 The twenty values of Y and X_2 shown in Table 12.6 were generated by the non-linear model

$$Y_i = 5 + 10X_2 + 2X_2^2 + u_i$$

in which u is a well-behaved disturbance term. Draw a scatter diagram of Y against X_2 and confirm the non-linearity of the model. Estimate the parameters of the model using OLS and draw the estimated line on the scatter diagram. Plot the residual terms against X_2 and test for non-linearity.

Table 12.6

Y	13.6	20.2	33.6	35.6	28.0
X_2	1	1	2	2	2
Y	50.8	52.6	77.1	77.0	102.8
X_2	3	3	4	4	5
Y	106.0	105.2	101.8	140.2	138.0
X_2	5	5	5	6	6
Y	173.0	176.8	214.2	254.4	256.2
X_2	7	7	8	9	9

12.10 Take logarithms of the data in the previous exercise and repeat the tasks. Comment upon any differences you find.

12.11 The twenty values of Y and X_2 shown in Table 12.7 were generated by the model

$$Y_i = 5 + 2X_{2i} + u_i$$

where u is a normally distributed disturbance term with a mean of zero and a variance of $\sigma^2 = 1 + 0.1X_{2i}$. Estimate the parameters of the model and use both a graphical and analytic test of heteroskedasticity.

Table 12.7

Y	4.5	7.6	7.6	7.6	10.0
X_2	1	1	2	2	2
Y	11.3	11.3	12.6	13.5	16.0
X_2	3	3	4	4	5
Y	15.5	16.2	17.0	14.9	15.9
X_2	5	5	5	6	6
Y	20.0	18.3	19.5	27.4	25.0
X_2	7	7	8	9	9

12.12 Suggest an appropriate transformation for the model of exercise 12.11 (assuming the form of the heteroskedasticity is not known), re-estimate the transformed model and test again for heteroskedasticity.

12.13 The nineteen values of Y and X_2 shown in Table 12.8 were generated by the model

$$Y_t = 5 + 2X_{2t} + u_t$$

in which u is a normally distributed disturbance term with a mean of zero and a constant variance. The disturbance terms are generated by the first-order autoregressive scheme

$$u_t = 0.5u_{t-1} + \epsilon_t$$

in which ϵ is a well-behaved disturbance term. Estimate the parameters of the model using OLS, and test for autocorrelation using both a graphical and an analytic test.

Table 12.8

Y	7.09	7.91	9.11	7.94	7.57
X_2	1	1	2	2	2
Y	10.46	13.67	12.84	14.44	14.63
X_2	3	3	4	4	5
Y	16.40	16.31	18.88	14.80	17.81
X_2	5	5	5	6	6
Y	19.86	20.81	23.72	24.06	
X_2	7	7	8	9	

12.14 For the model in the previous exercise estimate ρ using equations (12.92), (12.93) and (12.94) and compare the values thus obtained with the true value of $\rho = 0.5$. Transform the data using a suitable method and the value of ρ from (12.94), re-estimate the model and test again for autocorrelation.

12.15 The twenty values of Y, X_2, X_3 and X_4 shown in Table 12.9 were generated by the model

$$Y_i = 5 + 2X_{2i} + X_{3i} + 3X_{4i} + u_i$$

where

$$X_{4i} = X_{2i} + X_{3i} + \epsilon_i$$

in which u and ϵ are well-behaved disturbance terms. Estimate the parameters of the model using OLS, first using all twenty observations, and second using only the first eighteen observations. Comment upon your results. Now drop X_4 from the model and repeat the exercise. Comment upon your results.

Table 12.9

Y	18.39	23.77	44.39	45.44	42.08
X_2	1	1	2	2	2
X_3	4	4	4	4	5
X_4	3.13	8.29	11.13	11.48	10.39
Y	51.64	56.36	63.58	63.35	72.60
X_2	3	3	4	4	5
X_3	5	5	5	6	6
X_4	12.88	14.12	15.86	15.75	18.20
Y	69.21	69.96	77.72	83.87	81.78
X_2	5	5	5	6	6
X_3	6	5	4	4	4
X_4	17.70	17.32	19.24	20.99	20.26
Y	90.17	90.32	98.83	110.74	112.66
X_2	7	7	8	9	9
X_3	4	4	3	3	3
X_4	22.39	22.44	24.66	27.58	28.22

12 Multiple Regression

[illegible]

Appendix: Statistical Tables

Table A.1 Areas under the normal curve

z	.00	.01	.02	.03	.04	.05	.06	.07	.08	.09
0.0	.0000	.0040	.0080	.0120	.0160	.0199	.0239	.0279	.0319	.0359
0.1	.0398	.0438	.0478	.0517	.0557	.0596	.0636	.0675	.0714	.0753
0.2	.0793	.0832	.0871	.0910	.0948	.0987	.1026	.1064	.1103	.1141
0.3	.1179	.1217	.1255	.1293	.1331	.1368	.1406	.1443	.1480	.1517
0.4	.1554	.1591	.1628	.1664	.1700	.1736	.1772	.1808	.1844	.1879
0.5	.1915	.1950	.1985	.2019	.2054	.2088	.2123	.2157	.2190	.2224
0.6	.2257	.2291	.2324	.2357	.2389	.2422	.2454	.2486	.2517	.2549
0.7	.2580	.2611	.2642	.2673	.2704	.2734	.2764	.2794	.2823	.2852
0.8	.2881	.2910	.2939	.2967	.2995	.3023	.3051	.3078	.3106	.3133
0.9	.3159	.3186	.3212	.3238	.3264	.3289	.3315	.3340	.3365	.3389
1.0	.3413	.3438	.3461	.3485	.3508	.3531	.3554	.3577	.3599	.3621
1.1	.3643	.3665	.3686	.3708	.3729	.3749	.3770	.3790	.3810	.3830
1.2	.3849	.3869	.3888	.3907	.3925	.3944	.3962	.3980	.3997	.4015
1.3	.4032	.4049	.4066	.4082	.4099	.4115	.4131	.4147	.4162	.4177
1.4	.4192	.4207	.4222	.4236	.4251	.4265	.4279	.4292	.4306	.4319
1.5	.4332	.4345	.4357	.4370	.4382	.4394	.4406	.4418	.4429	.4441
1.6	.4452	.4463	.4474	.4484	.4495	.4505	.4515	.4525	.4535	.4545
1.7	.4554	.4564	.4573	.4582	.4591	.4599	.4608	.4616	.4625	.4633
1.8	.4641	.4649	.4656	.4664	.4671	.4678	.4686	.4693	.4699	.4706
1.9	.4713	.4719	.4726	.4732	.4738	.4744	.4750	.4756	.4761	.4767
2.0	.4772	.4778	.4783	.4788	.4793	.4798	.4803	.4808	.4812	.4817
2.1	.4821	.4826	.4830	.4834	.4838	.4842	.4846	.4850	.4854	.4857
2.2	.4861	.4864	.4868	.4871	.4875	.4878	.4881	.4884	.4887	.4890
2.3	.4893	.4896	.4898	.4901	.4904	.4906	.4909	.4911	.4913	.4916
2.4	.4918	.4920	.4922	.4925	.4927	.4929	.4931	.4932	.4934	.4936
2.5	.4938	.4940	.4941	.4943	.4945	.4946	.4948	.4949	.4951	.4952
2.6	.4953	.4955	.4956	.4957	.4959	.4960	.4961	.4962	.4963	.4964
2.7	.4965	.4966	.4967	.4968	.4969	.4970	.4971	.4972	.4973	.4974
2.8	.4974	.4975	.4976	.4977	.4977	.4978	.4979	.4979	.4980	.4981
2.9	.4981	.4982	.4982	.4983	.4984	.4984	.4985	.4985	.4986	.4986
3.0	.4987	.4987	.4987	.4988	.4988	.4989	.4989	.4989	.4990	.4990

Table A.2 Critical values of the *t*-distribution

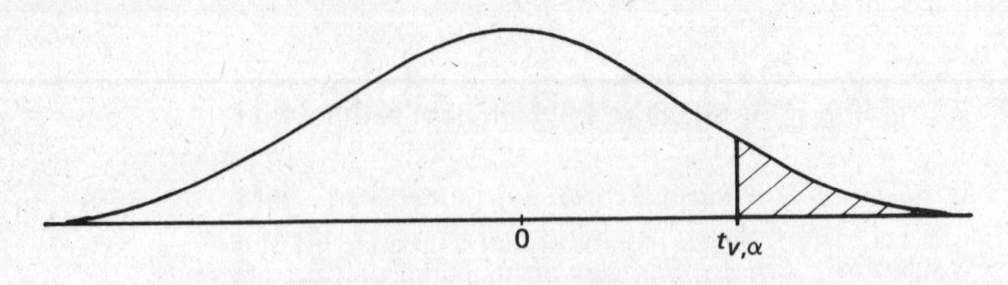

ν	$\alpha = 0.10$	$\alpha = 0.05$	$\alpha = 0.025$	$\alpha = 0.01$	$\alpha = 0.005$	ν
1	3.078	6.314	12.706	31.821	63.657	1
2	1.886	2.920	4.303	6.965	9.925	2
3	1.638	2.353	3.182	4.541	5.841	3
4	1.533	2.132	2.776	3.747	4.604	4
5	1.476	2.015	2.571	3.365	4.032	5
6	1.440	1.943	2.447	3.143	3.707	6
7	1.415	1.895	2.365	2.998	3.499	7
8	1.397	1.860	2.306	2.896	3.355	8
9	1.383	1.833	2.262	2.821	3.250	9
10	1.372	1.812	2.228	2.764	3.169	10
11	1.363	1.796	2.201	2.718	3.106	11
12	1.356	1.782	2.179	2.681	3.055	12
13	1.350	1.771	2.160	2.650	3.012	13
14	1.345	1.761	2.145	2.624	2.977	14
15	1.341	1.753	2.131	2.602	2.947	15
16	1.337	1.746	2.120	2.583	2.921	16
17	1.333	1.740	2.110	2.567	2.898	17
18	1.330	1.734	2.101	2.552	2.878	18
19	1.328	1.729	2.093	2.539	2.861	19
20	1.325	1.725	2.086	2.528	2.845	20
21	1.323	1.721	2.080	2.518	2.831	21
22	1.321	1.717	2.074	2.508	2.819	22
23	1.319	1.714	2.069	2.500	2.807	23
24	1.318	1.711	2.064	2.492	2.797	24
25	1.316	1.708	2.060	2.485	2.787	25
26	1.315	1.706	2.056	2.479	2.779	26
27	1.314	1.703	2.052	2.473	2.771	27
28	1.313	1.701	2.048	2.467	2.763	28
29	1.311	1.699	2.045	2.462	2.756	29
inf.	1.282	1.645	1.960	2.326	2.576	inf.

Table A.3 Percentage points of the *F*-distribution

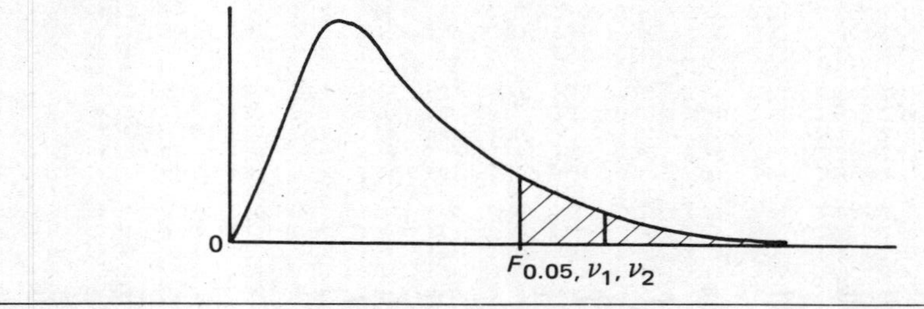

ν_1 = degrees of freedom for numerator; ν_2 = degrees of freedom for denominator

ν_2	1	2	3	4	5	6	7	8	9	10	12	15	20	24	30	40	60	120	∞
1	161	200	216	225	230	234	237	239	241	242	244	246	248	249	250	251	252	253	254
2	18.5	19.0	19.2	19.2	19.3	19.3	19.4	19.4	19.4	19.4	19.4	19.4	19.4	19.5	19.5	19.5	19.5	19.5	19.5
3	10.1	9.55	9.28	9.12	9.01	8.94	8.89	8.85	8.81	8.79	8.74	8.70	8.66	8.64	8.62	8.59	8.57	8.55	8.53
4	7.71	6.94	6.59	6.39	6.26	6.16	6.09	6.04	6.00	5.96	5.91	5.86	5.80	5.77	5.75	5.72	5.69	5.66	5.63
5	6.61	5.79	5.41	5.19	5.05	4.95	4.88	4.82	4.77	4.74	4.68	4.62	4.56	4.53	4.50	4.46	4.43	4.40	4.37
6	5.99	5.14	4.76	4.53	4.39	4.28	4.21	4.15	4.10	4.06	4.00	3.94	3.87	3.84	3.81	3.77	3.74	3.70	3.67
7	5.59	4.74	4.35	4.12	3.97	3.87	3.79	3.73	3.68	3.64	3.57	3.51	3.44	3.41	3.38	3.34	3.30	3.27	3.23
8	5.32	4.46	4.07	3.84	3.69	3.58	3.50	3.44	3.39	3.35	3.28	3.22	3.15	3.12	3.08	3.04	3.01	2.97	2.93
9	5.12	4.26	3.86	3.63	3.48	3.37	3.29	3.23	3.18	3.14	3.07	3.01	2.94	2.90	2.86	2.83	2.79	2.75	2.71
10	4.96	4.10	3.71	3.48	3.33	3.22	3.14	3.07	3.02	2.98	2.91	2.85	2.77	2.74	2.70	2.66	2.62	2.58	2.54
11	4.84	3.98	3.59	3.36	3.20	3.09	3.01	2.95	2.90	2.85	2.79	2.72	2.65	2.61	2.57	2.53	2.49	2.45	2.40
12	4.75	3.89	3.49	3.26	3.11	3.00	2.91	2.85	2.80	2.75	2.69	2.62	2.54	2.51	2.47	2.43	2.38	2.34	2.30
13	4.67	3.81	3.41	3.18	3.03	2.92	2.83	2.77	2.71	2.67	2.60	2.53	2.46	2.42	2.38	2.34	2.30	2.25	2.21
14	4.60	3.74	3.34	3.11	2.96	2.85	2.76	2.70	2.65	2.60	2.53	2.46	2.39	2.35	2.31	2.27	2.22	2.18	2.13
15	4.54	3.68	3.29	3.06	2.90	2.79	2.71	2.64	2.59	2.54	2.48	2.40	2.33	2.29	2.25	2.20	2.16	2.11	2.07
16	4.49	3.63	3.24	3.01	2.85	2.74	2.66	2.59	2.54	2.49	2.42	2.35	2.28	2.24	2.19	2.15	2.11	2.06	2.01
17	4.45	3.59	3.20	2.96	2.81	2.70	2.61	2.55	2.49	2.45	2.38	2.31	2.23	2.19	2.15	2.10	2.06	2.01	1.96
18	4.41	3.55	3.16	2.93	2.77	2.66	2.58	2.51	2.46	2.41	2.34	2.27	2.19	2.15	2.11	2.06	2.02	1.97	1.92
19	4.38	3.52	3.13	2.90	2.74	2.63	2.54	2.48	2.42	2.38	2.31	2.23	2.16	2.11	2.07	2.03	1.98	1.93	1.88
20	4.35	3.49	3.10	2.87	2.71	2.60	2.51	2.45	2.39	2.35	2.28	2.20	2.12	2.08	2.04	1.99	1.95	1.90	1.84
21	4.32	3.47	3.07	2.84	2.68	2.57	2.49	2.42	2.37	2.32	2.25	2.18	2.10	2.05	2.01	1.96	1.92	1.87	1.81
22	4.30	3.44	3.05	2.82	2.66	2.55	2.46	2.40	2.34	2.30	2.23	2.15	2.07	2.03	1.98	1.94	1.89	1.84	1.78
23	4.28	3.42	3.03	2.80	2.64	2.53	2.44	2.37	2.32	2.27	2.20	2.13	2.05	2.01	1.96	1.91	1.86	1.81	1.76
24	4.26	3.40	3.01	2.78	2.62	2.51	2.42	2.36	2.30	2.25	2.18	2.11	2.03	1.98	1.94	1.89	1.84	1.79	1.73
25	4.24	3.39	2.99	2.76	2.60	2.49	2.40	2.34	2.28	2.24	2.16	2.09	2.01	1.96	1.92	1.87	1.82	1.77	1.71
30	4.17	3.32	2.92	2.69	2.53	2.42	2.33	2.27	2.21	2.16	2.09	2.01	1.93	1.89	1.84	1.79	1.74	1.68	1.62
40	4.08	3.23	2.84	2.61	2.45	2.34	2.25	2.18	2.12	2.08	2.00	1.92	1.84	1.79	1.74	1.69	1.64	1.58	1.51
60	4.00	3.15	2.76	2.53	2.37	2.25	2.17	2.10	2.04	1.99	1.92	1.84	1.75	1.70	1.65	1.59	1.53	1.47	1.39
120	3.92	3.07	2.68	2.45	2.29	2.18	2.09	2.02	1.96	1.91	1.83	1.75	1.66	1.61	1.55	1.50	1.43	1.35	1.25
∞	3.84	3.00	2.60	2.37	2.21	2.10	2.01	1.94	1.88	1.83	1.75	1.67	1.57	1.52	1.46	1.39	1.32	1.22	1.00

Table A.3 Percentage points of the F-distribution (*contd*)

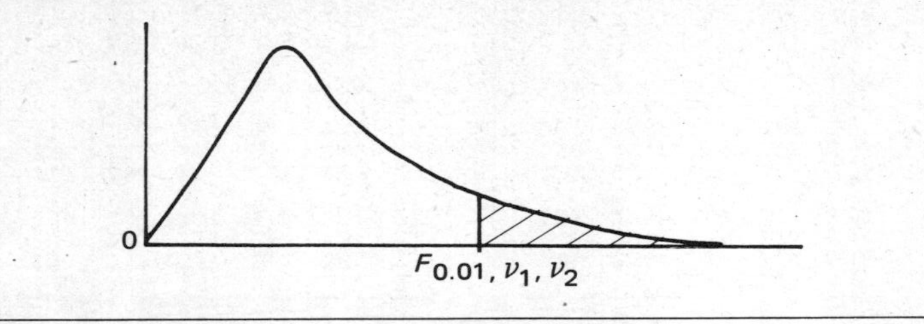

ν_1 = degrees of freedom for numerator; ν_2 = degrees of freedom for denominator

ν_2	1	2	3	4	5	6	7	8	9	10	12	15	20	24	30	40	60	120	∞
1	4052	5000	5403	5625	5764	5859	5928	5982	6023	6056	6106	6157	6209	6235	6261	6287	6313	6339	6366
2	98.5	99.0	99.2	99.2	99.3	99.3	99.4	99.4	99.4	99.4	99.4	99.4	99.4	99.5	99.5	99.5	99.5	99.5	99.5
3	34.1	30.8	29.5	28.7	28.2	27.9	27.7	27.5	27.3	27.2	27.1	26.9	26.7	26.6	26.5	26.4	26.3	26.2	26.1
4	21.2	18.0	16.7	16.0	15.5	15.2	15.0	14.8	14.7	14.5	14.4	14.2	14.0	13.9	13.8	13.7	13.7	13.6	13.5
5	16.3	13.3	12.1	11.4	11.0	10.7	10.5	10.3	10.2	10.1	9.89	9.72	9.55	9.47	9.38	9.29	9.20	9.11	9.02
6	13.7	10.9	9.78	9.15	8.75	8.47	8.26	8.10	7.98	7.87	7.72	7.56	7.40	7.31	7.23	7.14	7.06	6.97	6.88
7	12.2	9.55	8.45	7.85	7.46	7.19	6.99	6.84	6.72	6.62	6.47	6.31	6.16	6.07	5.99	5.91	5.82	5.74	5.65
8	11.3	8.65	7.59	7.01	6.63	6.37	6.18	6.03	5.91	5.81	5.67	5.52	5.36	5.28	5.20	5.12	5.03	4.95	4.86
9	10.6	8.02	6.99	6.42	6.06	5.80	5.61	5.47	5.35	5.26	5.11	4.96	4.81	4.73	4.65	4.57	4.48	4.40	4.31
10	10.0	7.56	6.55	5.99	5.64	5.39	5.20	5.06	4.94	4.85	4.71	4.56	4.41	4.33	4.25	4.17	4.08	4.00	3.91
11	9.65	7.21	6.22	5.67	5.32	5.07	4.89	4.74	4.63	4.54	4.40	4.25	4.10	4.02	3.94	3.86	3.78	3.69	3.60
12	9.33	6.93	5.95	5.41	5.06	4.82	4.64	4.50	4.39	4.30	4.16	4.01	3.86	3.78	3.70	3.62	3.54	3.45	3.36
13	9.07	6.70	5.74	5.21	4.86	4.62	4.44	4.30	4.19	4.10	3.96	3.82	3.66	3.59	3.51	3.43	3.34	3.25	3.17
14	8.86	6.51	5.56	5.04	4.70	4.46	4.28	4.14	4.03	3.94	3.80	3.66	3.51	3.43	3.35	3.27	3.18	3.09	3.00
15	8.68	6.36	5.42	4.89	4.56	4.32	4.14	4.00	3.89	3.80	3.67	3.52	3.37	3.29	3.21	3.13	3.05	2.96	2.87
16	8.53	6.23	5.29	4.77	4.44	4.20	4.03	3.89	3.78	3.69	3.55	3.41	3.26	3.18	3.10	3.02	2.93	2.84	2.75
17	8.40	6.11	5.19	4.67	4.34	4.10	3.93	3.79	3.68	3.59	3.46	3.31	3.16	3.08	3.00	2.92	2.83	2.75	2.65
18	8.29	6.01	5.09	4.58	4.25	4.01	3.84	3.71	3.60	3.51	3.37	3.23	3.08	3.00	2.92	2.84	2.75	2.66	2.57
19	8.19	5.93	5.01	4.50	4.17	3.94	3.77	3.63	3.52	3.43	3.30	3.15	3.00	2.92	2.84	2.76	2.67	2.58	2.49
20	8.10	5.85	4.94	4.43	4.10	3.87	3.70	3.56	3.46	3.37	3.23	3.09	2.94	2.86	2.78	2.69	2.61	2.52	2.42
21	8.02	5.78	4.87	4.37	4.04	3.81	3.64	3.51	3.40	3.31	3.17	3.03	2.88	2.80	2.72	2.64	2.55	2.46	2.36
22	7.95	5.72	4.82	4.31	3.99	3.76	3.59	3.45	3.35	3.26	3.12	2.98	2.83	2.75	2.67	2.58	2.50	2.40	2.31
23	7.88	5.66	4.76	4.26	3.94	3.71	3.54	3.41	3.30	3.21	3.07	2.93	2.78	2.70	2.62	2.54	2.45	2.35	2.26
24	7.82	5.61	4.72	4.22	3.90	3.67	3.50	3.36	3.26	3.17	3.03	2.89	2.74	2.66	2.58	2.49	2.40	2.31	2.21
25	7.77	5.57	4.68	4.18	3.86	3.63	3.46	3.32	3.22	3.13	2.99	2.85	2.70	2.62	2.53	2.45	2.36	2.27	2.17
30	7.56	5.39	4.51	4.02	3.70	3.47	3.30	3.17	3.07	2.98	2.84	2.70	2.55	2.47	2.39	2.30	2.21	2.11	2.01
40	7.31	5.18	4.31	3.83	3.51	3.29	3.12	2.99	2.89	2.80	2.66	2.52	2.37	2.29	2.20	2.11	2.02	1.92	1.80
60	7.08	4.98	4.13	3.65	3.34	3.12	2.95	2.82	2.72	2.63	2.50	2.35	2.20	2.12	2.03	1.94	1.84	1.73	1.60
120	6.85	4.79	3.95	3.48	3.17	2.96	2.79	2.66	2.56	2.47	2.34	2.19	2.03	1.95	1.86	1.76	1.66	1.53	1.38
∞	6.63	4.61	3.78	3.32	3.02	2.80	2.64	2.51	2.41	2.32	2.18	2.04	1.88	1.79	1.70	1.59	1.47	1.32	1.00

Table A.4 Critical values of the chi-square distribution

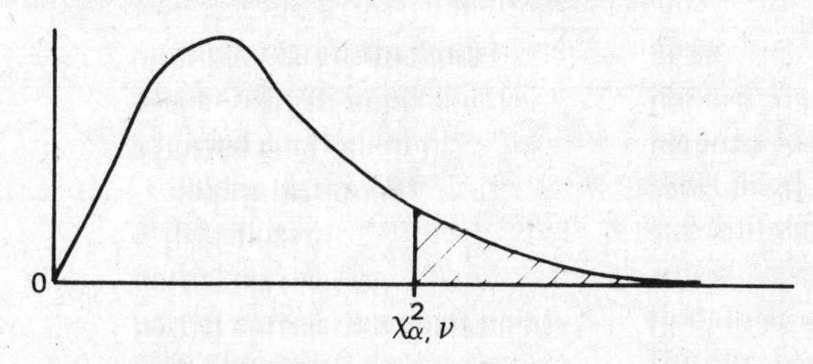

ν	$\alpha = 0.995$	$\alpha = 0.99$	$\alpha = 0.975$	$\alpha = 0.95$	$\alpha = 0.05$	$\alpha = 0.025$	$\alpha = 0.01$	$\alpha = 0.005$	ν
1	0.0000393	0.000157	0.000982	0.00393	3.841	5.024	6.635	7.879	1
2	0.0100	0.0201	0.0506	0.103	5.991	7.378	9.210	10.597	2
3	0.0717	0.115	0.216	0.352	7.815	9.348	11.345	12.838	3
4	0.207	0.297	0.484	0.711	9.488	11.143	13.277	14.860	4
5	0.412	0.554	0.831	1.145	11.070	12.832	15.086	16.750	5
6	0.676	0.872	1.237	1.635	12.592	14.449	16.812	18.548	6
7	0.989	1.239	1.690	2.167	14.067	16.013	18.475	20.278	7
8	1.344	1.646	2.180	2.733	15.507	17.535	20.090	21.955	8
9	1.735	2.088	2.700	3.325	16.919	19.023	21.666	23.589	9
10	2.156	2.558	3.247	3.940	18.307	20.483	23.209	25.188	10
11	2.603	3.053	3.816	4.575	19.675	21.920	24.725	26.757	11
12	3.074	3.571	4.404	5.226	21.026	23.337	26.217	28.300	12
13	3.565	4.107	5.009	5.892	22.362	24.736	27.688	29.819	13
14	4.075	4.660	5.629	6.571	23.685	26.119	29.141	31.319	14
15	4.601	5.229	6.262	7.261	24.996	27.488	30.578	32.801	15
16	5.142	5.812	6.908	7.962	26.296	28.845	32.000	34.267	16
17	5.697	6.408	7.564	8.672	27.587	30.191	33.409	35.718	17
18	6.265	7.015	8.231	9.390	28.869	31.526	34.805	37.156	18
19	6.844	7.633	8.907	10.117	30.144	32.852	36.191	38.582	19
20	7.434	8.260	9.591	10.851	31.410	34.170	37.566	39.997	20
21	8.034	8.897	10.283	11.591	32.671	35.479	38.932	41.401	21
22	8.643	9.542	10.982	12.338	33.924	36.781	40.289	42.796	22
23	9.260	10.196	11.689	13.091	35.172	38.076	41.638	44.181	23
24	9.886	10.856	12.401	13.848	36.415	39.364	42.980	45.558	24
25	10.520	11.524	13.120	14.611	37.652	40.646	44.314	46.928	25
26	11.160	12.198	13.844	15.379	38.885	41.923	45.642	48.290	26
27	11.808	12.879	14.573	16.151	40.113	43.194	46.963	49.645	27
28	12.461	13.565	15.308	16.928	41.337	44.461	48.278	50.993	28
29	13.121	14.256	16.047	17.708	42.557	45.722	49.588	52.336	29
30	13.787	14.953	16.791	18.493	43.773	46.979	50.892	53.672	30

Table A.5 The Durbin–Watson statistic: significance points of d_L and d_U

5 per cent

n	$k' = 1$		$k' = 2$		$k' = 3$		$k' = 4$		$k' = 5$	
	d_L	d_U	d_L	d_U	d_L	d_U	d_L	d_U	d_L	d_U
15	1.08	1.36	0.95	1.54	0.82	1.75	0.69	1.97	0.56	2.21
16	1.10	1.37	0.98	1.54	0.86	1.73	0.74	1.93	0.62	2.15
17	1.13	1.38	1.02	1.54	0.90	1.71	0.78	1.90	0.67	2.10
18	1.16	1.39	1.05	1.53	0.93	1.69	0.82	1.87	0.71	2.06
19	1.18	1.40	1.08	1.53	0.97	1.68	0.86	1.85	0.75	2.02
20	1.20	1.41	1.10	1.54	1.00	1.68	0.90	1.83	0.79	1.99
21	1.22	1.42	1.13	1.54	1.03	1.67	0.93	1.81	0.83	1.96
22	1.24	1.43	1.15	1.54	1.05	1.66	0.96	1.80	0.86	1.94
23	1.26	1.44	1.17	1.54	1.08	1.66	0.99	1.79	0.90	1.92
24	1.27	1.45	1.19	1.55	1.10	1.66	1.01	1.78	0.93	1.90
25	1.29	1.45	1.21	1.55	1.12	1.66	1.04	1.77	0.95	1.89
26	1.30	1.46	1.22	1.55	1.14	1.65	1.06	1.76	0.98	1.88
27	1.32	1.47	1.24	1.56	1.16	1.65	1.08	1.76	1.01	1.86
28	1.33	1.48	1.26	1.56	1.18	1.65	1.10	1.75	1.03	1.85
29	1.34	1.48	1.27	1.56	1.20	1.65	1.12	1.74	1.05	1.84
30	1.35	1.49	1.28	1.57	1.21	1.65	1.14	1.74	1.07	1.83
31	1.36	1.50	1.30	1.57	1.23	1.65	1.16	1.74	1.09	1.83
32	1.37	1.50	1.31	1.57	1.24	1.65	1.18	1.73	1.11	1.82
33	1.38	1.51	1.32	1.58	1.26	1.65	1.19	1.73	1.13	1.81
34	1.39	1.51	1.33	1.58	1.27	1.65	1.21	1.73	1.15	1.81
35	1.40	1.52	1.34	1.58	1.28	1.65	1.22	1.73	1.16	1.80
36	1.41	1.52	1.35	1.59	1.29	1.65	1.24	1.73	1.18	1.80
37	1.42	1.53	1.36	1.59	1.31	1.66	1.25	1.72	1.19	1.80
38	1.43	1.54	1.37	1.59	1.32	1.66	1.26	1.72	1.21	1.79
39	1.43	1.54	1.38	1.60	1.33	1.66	1.27	1.72	1.22	1.79
40	1.44	1.54	1.39	1.60	1.34	1.66	1.29	1.72	1.23	1.79
45	1.48	1.57	1.43	1.62	1.38	1.67	1.34	1.72	1.29	1.78
50	1.50	1.59	1.46	1.63	1.42	1.67	1.38	1.72	1.34	1.77
55	1.53	1.60	1.49	1.64	1.45	1.68	1.41	1.72	1.38	1.77
60	1.55	1.62	1.51	1.65	1.48	1.69	1.44	1.73	1.41	1.77
65	1.57	1.63	1.54	1.66	1.50	1.70	1.47	1.73	1.44	1.77
70	1.58	1.64	1.55	1.67	1.52	1.70	1.49	1.74	1.46	1.77
75	1.60	1.65	1.57	1.68	1.54	1.71	1.51	1.74	1.49	1.77
80	1.61	1.66	1.59	1.69	1.56	1.72	1.53	1.74	1.51	1.77
85	1.62	1.67	1.60	1.70	1.57	1.72	1.55	1.75	1.52	1.77
90	1.63	1.68	1.61	1.70	1.59	1.73	1.57	1.75	1.54	1.78
95	1.64	1.69	1.62	1.71	1.60	1.73	1.58	1.75	1.56	1.78
100	1.65	1.69	1.63	1.72	1.61	1.74	1.59	1.76	1.57	1.78

Note: k' = number of explanatory variables excluding the constant term.

Table A.5 The Durbin–Watson statistic: significance points of d_L and d_U *(contd)*

1 per cent

n	$k' = 1$		$k' = 2$		$k' = 3$		$k' = 4$		$k' = 5$	
	d_L	d_U	d_L	d_U	d_L	d_U	d_L	d_U	d_L	d_U
15	0.81	1.07	0.70	1.25	0.59	1.46	0.49	1.70	0.39	1.96
16	0.84	1.09	0.74	1.25	0.63	1.44	0.53	1.66	0.44	1.90
17	0.87	1.10	0.77	1.25	0.67	1.43	0.57	1.63	0.48	1.85
18	0.90	1.12	0.80	1.26	0.71	1.42	0.61	1.60	0.52	1.80
19	0.93	1.13	0.83	1.26	0.74	1.41	0.65	1.58	0.56	1.77
20	0.95	1.15	0.86	1.27	0.77	1.41	0.68	1.57	0.60	1.74
21	0.97	1.16	0.89	1.27	0.80	1.41	0.72	1.55	0.63	1.71
22	1.00	1.17	0.91	1.28	0.83	1.40	0.75	1.54	0.66	1.69
23	1.02	1.19	0.94	1.29	0.86	1.40	0.77	1.53	0.70	1.67
24	1.04	1.20	0.96	1.30	0.88	1.41	0.80	1.53	0.72	1.66
25	1.05	1.21	0.98	1.30	0.90	1.41	0.83	1.52	0.75	1.65
26	1.07	1.22	1.00	1.31	0.93	1.41	0.85	1.52	0.78	1.64
27	1.09	1.23	1.02	1.32	0.95	1.41	0.88	1.51	0.81	1.63
28	1.10	1.24	1.04	1.32	0.97	1.41	0.90	1.51	0.83	1.62
29	1.12	1.25	1.05	1.33	0.99	1.42	0.92	1.51	0.85	1.61
30	1.13	1.26	1.07	1.34	1.01	1.42	0.94	1.51	0.88	1.61
31	1.15	1.27	1.08	1.34	1.02	1.42	0.96	1.51	0.90	1.60
32	1.16	1.28	1.10	1.35	1.04	1.43	0.98	1.51	0.92	1.60
33	1.17	1.29	1.11	1.36	1.05	1.43	1.00	1.51	0.94	1.59
34	1.18	1.30	1.13	1.36	1.07	1.43	1.01	1.51	0.95	1.59
35	1.19	1.31	1.14	1.37	1.08	1.44	1.03	1.51	0.97	1.59
36	1.21	1.32	1.15	1.38	1.10	1.44	1.04	1.51	0.99	1.59
37	1.22	1.32	1.16	1.38	1.11	1.45	1.06	1.51	1.00	1.59
38	1.23	1.33	1.18	1.39	1.12	1.45	1.07	1.52	1.02	1.58
39	1.24	1.34	1.19	1.39	1.14	1.45	1.09	1.52	1.03	1.58
40	1.25	1.34	1.20	1.40	1.15	1.46	1.10	1.52	1.05	1.58
45	1.29	1.38	1.24	1.42	1.20	1.48	1.16	1.53	1.11	1.58
50	1.32	1.40	1.28	1.45	1.24	1.49	1.20	1.54	1.16	1.59
55	1.36	1.43	1.32	1.47	1.28	1.51	1.25	1.55	1.21	1.59
60	1.38	1.45	1.35	1.48	1.32	1.52	1.28	1.56	1.25	1.60
65	1.41	1.47	1.38	1.50	1.35	1.53	1.31	1.57	1.28	1.61
70	1.43	1.49	1.40	1.52	1.37	1.55	1.34	1.58	1.31	1.61
75	1.45	1.50	1.42	1.53	1.39	1.56	1.37	1.59	1.34	1.62
80	1.47	1.52	1.44	1.54	1.42	1.57	1.39	1.60	1.36	1.62
85	1.48	1.53	1.46	1.55	1.43	1.58	1.41	1.60	1.39	1.63
90	1.50	1.54	1.47	1.56	1.45	1.59	1.43	1.61	1.41	1.64
95	1.51	1.55	1.49	1.57	1.47	1.60	1.45	1.62	1.42	1.64
100	1.52	1.56	1.50	1.58	1.48	1.60	1.46	1.63	1.44	1.65

Note: k' = number of explanatory variables excluding the constant term.

Table A.6 Random numbers

39 65 76 45 45	19 90 69 64 61	20 26 36 31 62	58 24 97 14 97	95 06 70 99 00
73 71 23 70 90	65 97 60 12 11	31 56 34 19 19	47 83 75 51 33	30 62 38 20 46
72 20 47 33 84	51 67 47 97 19	98 40 07 17 66	23 05 09 51 80	59 78 11 52 49
75 17 25 69 17	17 95 21 78 58	24 33 45 77 48	69 81 84 09 29	93 22 70 45 80
37 48 79 88 74	63 52 06 34 30	01 31 60 10 27	35 07 79 71 53	28 99 52 01 41
02 89 08 16 94	85 53 83 29 95	56 27 09 24 43	21 78 55 09 82	72 61 88 73 61
87 18 15 70 07	37 79 49 12 38	48 13 93 55 96	41 92 45 71 51	09 18 25 58 94
98 83 71 70 15	89 09 39 59 24	00 06 41 41 20	14 36 59 25 47	54 45 17 24 89
10 08 58 07 04	76 62 16 48 68	58 76 17 14 86	59 53 11 52 21	66 04 18 72 87
47 90 56 37 31	71 82 13 50 41	27 55 10 24 92	28 04 67 53 44	95 23 00 84 47
93 05 31 03 07	34 18 04 52 35	74 13 39 35 22	68 95 23 92 35	36 63 70 35 33
21 89 11 47 99	11 20 99 45 18	76 51 94 84 86	13 79 93 37 55	98 16 04 41 67
95 18 94 06 97	27 37 83 28 71	79 57 95 13 91	09 61 87 25 21	56 20 11 32 44
97 08 31 55 73	10 65 81 92 59	77 31 61 95 46	20 44 90 32 64	26 99 76 75 63
69 26 88 86 13	59 71 74 17 32	48 38 75 93 29	73 37 32 04 05	60 82 29 20 25
41 47 10 25 03	87 63 93 95 17	81 83 83 04 49	77 45 85 50 51	79 88 01 97 30
91 94 14 63 62	08 61 74 51 69	92 79 43 89 79	29 18 94 51 23	14 85 11 47 23
80 06 54 18 47	08 52 85 08 40	48 40 35 94 22	72 65 71 08 86	50 03 42 99 36
67 72 77 63 99	89 85 84 46 06	64 71 06 21 66	89 37 20 70 01	61 65 70 22 12
59 40 24 13 75	42 29 72 23 19	06 94 76 10 08	81 30 15 39 14	81 83 17 16 33
63 62 06 34 41	79 53 36 02 95	94 61 09 43 62	20 21 14 68 86	84 95 48 46 45
78 47 23 53 90	79 93 96 38 63	34 85 52 05 09	85 43 01 72 73	14 93 87 81 40
87 68 62 15 43	97 48 72 66 48	53 16 71 13 81	59 97 50 99 52	24 62 20 42 31
47 60 92 10 77	26 97 05 73 51	88 46 38 03 58	72 68 49 29 31	75 70 16 08 24
56 88 87 59 41	06 87 37 78 48	65 88 69 58 39	88 02 84 27 83	85 81 56 39 38
22 17 68 65 84	87 02 22 57 51	68 69 80 95 44	11 29 01 95 80	49 34 35 86 47
19 36 27 59 46	39 77 32 77 09	79 57 92 36 59	89 74 39 82 15	08 58 94 34 74
16 77 23 02 77	28 06 24 25 93	22 45 44 84 11	87 80 61 65 31	09 71 91 74 25
78 43 76 71 61	97 67 63 99 61	80 45 67 93 82	59 73 19 85 23	53 33 65 97 21
03 28 28 26 08	69 30 16 09 05	53 58 47 70 93	66 56 45 65 79	45 56 20 19 47
04 31 17 21 56	33 73 99 19 87	26 72 39 27 67	53 77 57 68 93	60 61 97 22 61
61 06 98 03 91	87 14 77 43 96	43 00 65 98 50	45 60 33 01 07	98 99 46 50 47
23 68 35 26 00	99 53 93 61 28	52 70 05 48 34	56 65 05 61 86	90 92 10 70 80
15 39 25 70 99	93 86 52 77 65	15 33 59 05 28	22 87 26 07 47	86 96 98 29 06
58 71 96 30 24	18 46 23 34 27	85 13 99 24 44	49 18 09 79 49	74 16 32 23 02
93 22 53 64 39	07 10 63 76 35	87 03 04 79 88	08 13 13 85 51	55 34 57 72 69
78 76 58 54 74	92 38 70 96 92	52 06 79 79 45	82 63 18 27 44	69 66 92 19 09
61 81 31 96 82	00 57 25 60 59	46 72 60 18 77	55 66 12 62 11	08 99 55 64 57
42 88 07 10 05	24 98 65 63 21	47 21 61 88 32	27 80 30 21 60	10 92 35 36 12
77 94 30 05 39	28 10 99 00 27	12 73 73 99 12	49 99 57 94 82	96 88 57 17 91

Table A.7 Random normal numbers, $\mu = 0, \sigma = 1$

01	02	03	04	05	06	07	08	09	10
0.464	0.137	2.455	-0.323	-0.068	0.296	-0.288	1.298	0.241	-0.957
0.060	-2.526	-0.531	-0.194	-0.543	-1.558	0.187	-1.190	0.022	0.525
1.486	-0.354	-0.634	0.697	0.926	1.375	0.785	-0.963	-0.853	-1.865
1.022	-0.472	1.279	3.521	0.571	-1.851	0.194	1.192	-0.501	-0.273
1.394	-0.555	0.046	0.321	2.945	1.974	-0.258	0.412	0.439	-0.035
0.906	-0.513	-0.525	0.595	0.881	-0.934	1.579	0.161	-1.885	0.371
1.179	-1.055	0.007	0.769	0.971	0.712	1.090	-0.631	-0.255	-0.702
-1.501	-0.488	-0.162	-0.136	1.033	0.203	0.448	0.748	-0.423	-0.432
-0.690	0.756	-1.618	-0.345	-0.511	-2.051	-0.457	-0.218	0.857	-0.465
1.372	0.225	0.378	0.761	0.181	-0.736	0.960	-1.530	-0.260	0.120
-0.482	1.678	-0.057	-1.229	-0.486	0.856	-0.491	-1.983	-2.830	-0.238
-1.376	-0.150	1.356	-0.561	-0.256	-0.212	0.219	0.779	0.953	-0.869
-1.010	0.598	-0.918	1.598	0.065	0.415	-0.169	0.313	-0.973	-1.016
-0.005	-0.899	0.012	-0.725	1.147	-0.121	1.096	0.481	-1.691	0.417
1.393	-1.163	-0.911	1.231	-0.199	-0.246	1.239	-2.574	-0.558	0.056
-1.787	-0.261	1.237	1.046	-0.508	-1.630	-0.146	-0.392	-0.627	0.561
-0.105	-0.357	-1.384	0.360	-0.992	-0.116	-1.698	-2.832	-1.108	-2.357
-1.339	1.827	-0.959	0.424	0.969	-1.141	-1.041	0.362	-1.726	1.956
1.041	0.535	0.731	1.377	0.983	-1.330	1.620	-1.040	0.524	-0.281
0.279	-2.056	0.717	-0.873	-1.096	-1.396	1.047	0.089	-0.573	0.932
-1.805	-2.008	-1.633	0.542	0.250	-0.166	0.032	0.079	0.471	-1.029
-1.186	1.180	1.114	0.882	1.265	-0.202	0.151	-0.376	-0.310	0.479
0.658	-1.141	1.151	-1.210	-0.927	0.425	0.290	-0.902	0.610	2.709
-0.439	0.358	-1.939	0.891	-0.227	0.602	0.873	-0.437	-0.220	-0.057
-1.399	-0.230	0.385	-0.649	-0.577	0.237	-0.289	0.513	0.738	-0.300
0.199	0.208	-1.083	-0.219	-0.291	1.221	1.119	0.004	-2.015	-0.594
0.159	0.272	-0.313	0.084	-2.828	-0.439	-0.792	-1.275	-0.623	-1.047
2.273	0.606	0.606	-0.747	0.247	1.291	0.063	-1.793	-0.699	-1.347
0.041	-0.307	0.121	0.790	-0.584	0.541	0.484	-0.986	0.481	0.996
-1.132	-2.098	0.921	0.145	0.446	-1.661	1.045	-1.363	-0.586	-1.023
0.768	0.079	-1.473	0.034	-2.127	0.665	0.084	-0.880	-0.579	0.551
0.375	-1.658	-1.851	0.234	-0.656	0.340	-0.086	-0.158	-0.120	0.418
-0.513	-0.344	0.210	-0.736	1.041	0.008	0.427	-0.831	0.191	0.074
0.292	-0.521	1.266	-1.206	-0.899	0.110	-0.528	-0.813	0.071	0.524
1.026	2.990	-0.574	-0.491	-1.114	1.297	-1.433	-1.345	-3.001	0.479
-1.334	1.278	-0.568	-0.109	-0.515	-0.566	2.923	0.500	0.359	0.326
-0.287	-0.144	-0.254	0.574	-0.451	-1.181	-1.190	-0.318	-0.094	1.114
0.161	-0.886	-0.921	-0.509	1.410	-0.518	0.192	-0.432	1.501	1.068
-1.346	0.193	-1.202	0.394	-1.045	0.843	0.942	1.045	0.031	0.772
1.250	-0.199	-0.288	1.810	1.378	0.584	0.216	0.733	0.402	0.226
0.630	-0.537	0.782	0.060	0.499	-0.431	1.705	1.164	0.884	-0.298
0.375	-1.941	0.247	-0.491	0.665	-0.135	-0.145	-0.498	0.457	1.064
1.420	0.489	1.711	-1.186	0.754	-0.732	-0.066	1.006	-0.798	0.162
-0.151	-0.243	-0.430	-0.762	0.298	1.049	1.810	2.885	-0.768	-0.129
-0.309	0.531	0.416	-1.541	1.456	2.040	-0.124	0.196	0.023	-1.204
0.424	-0.444	0.593	0.993	-0.106	0.116	0.484	-1.272	1.066	1.097
0.593	0.658	-1.127	-1.407	-1.579	-1.616	1.458	1.262	0.736	-0.916
0.862	-0.885	-0.142	-0.504	0.532	1.381	0.022	-0.281	-0.342	1.222
0.235	-0.628	-0.023	-0.463	-0.899	-0.394	-0.538	1.707	-0.188	-1.153
-0.853	0.402	0.777	0.833	0.410	-0.349	-1.094	0.580	1.395	1.298

Bibliography

Annual Abstract of Statistics (yearly) Central Statistical Office, HMSO, London.

Bennett, S. and Bowers, D. (1974) *An Introduction to Multivariate Analysis for the Social and Behavioural Sciences*, Macmillan, London.

British Labour Statistics: Historical Abstracts, 1186–1968, Department of Employment, HMSO, London.

Common, M. S. (1976) *Basic Econometrics*, Longman, London.

Durbin, J. and Watson, G. S. (1951) 'Testing for Serial Correlation in Least Squares Regression', *Biometrika. Family Expenditure Survey* (1978) Department of Employment, HMSO, London.

Farrar, D. E. and Glauber, R. R. (1967) 'Multicollinearity in Regression Analysis: The Problem Re-visited', *Review of Economics and Statistics*, vol. 49, pp. 92–107.

Games, P. A. and Klare, G. R. (1967) *Elementary Statistics*, McGraw-Hill, New York.

General Household Survey (1973) Office of Population Censuses and Surveys, HMSO, London.

Glejser, H. (1969) 'A New Test for Heteroskedasticity', *Journal of the American Statistical Association*.

Goldfeld, S. M. and Quandt, R. F. (1965) 'Some Tests for Homoskedasticity', *Journal of the American Statistical Association*.

Gujarati, D. (1978) *Basic Econometrics*, McGraw-Hill, New York.

Hines, A. G. (1964) 'Trade Unions and Wage Inflation in the UK, 1893–1961', *Review of Economic Studies*.

Hull, C. H. and Nie, N. H. (1979) *SPSS Update*, McGraw-Hill, New York.

Johnston, J. (1972) *Econometric Methods*, McGraw-Hill, New York.

Kane, E. J. (1969) *Economic Statistics and Econometrics*, Harper & Row, New York.

Klein, L. R. (1962) *An Introduction to Econometrics*, Prentice-Hall, Englewood Cliffs, N.J.

Kmenta, J. (1971) *Elements of Econometrics*, Collier-Macmillan, London.

Koutsoyiannis, A. (1977) *Theory of Econometrics*, Macmillan, London.

Maddala, G. S. (1977) *Econometrics*, McGraw-Hill–Kogakusha, Tokyo.

Nie, N. H. *et al.* (1975) *Statistical Package for the Social Sciences*, McGraw-Hill, New York.

Park, R. E. (1966) 'Estimation with Heteroskedastic Error Terms', *Econometrica*.

Silvey, S. D. (1969) 'Multicollinearity and Imprecise Estimation', *Journal of the Royal Statistical Society*.

Stewart, J. (1976) *Understanding Econometrics*, Hutchinson, London.

Theil, H. and Nagar, A. L. (1961) 'Testing the Independence of Regression Disturbances', *Journal of the American Statistical Association*.

Tukey, J. W. (1977) *Exploratory Data Analysis*, Addison-Wesley, Reading, Mass.

Index